DATE
Chicago Public Library
C
P
L

# PRACTICAL PIPING HANDBOOK

# PRACTICAL PIPING HANDBOOK

**Otto Mendel**

**PennWell Publishing Company**
1421 South Sheridan Road/P. O. Box 1260
Tulsa, Oklahoma 74101

**Library of Congress Cataloging in Publication Data**

Mendel, Otto.
Practical piping handbook.

Includes index.
1. Pipe lines—Handbooks, manuals, etc. I. Title.
TJ930.M427 621.8'672 81-8553
ISBN 0-87814-169-3 AACR2



Printed in the United States of America

1 2 3 4 5 85 84 83 82 81

# Contents

## LIST OF TABLES

## LIST OF ILLUSTRATIONS

# Preface

Many books are available on the subject of piping, but most of them deal with a specific aspect of the subject. With todays proliferation of materials of construction for piping use, and the numerous codes, standards, and regulations that affect manufacturing processes as well as design and field erection, an overall piping handbook has been long overdue.

I have been trying to provide in simple terms an overview of what men and women are faced with, when the subject is piping. The designer who often lacks field experience and the erection foreman who has little or no theoretical knowledge both are involved in the same project, and each could furnish the other with valuable advice. This book is trying to build this bridge that is often so sorely needed, by reviewing the practical application of general information on the subject of piping. In other words, this book will hopefully impart some useful knowledge by touching on all facets of an increasingly diversified industry.

The idea of publishing this book came as the aftermath of a recent seminar on piping technology at a construction site, that was attended by engineers and erection personnel alike. During discussions in the course of the seminar it was found that know-how taken for granted by one person was not at all universally known.

For the recent graduate who is still trying to digest all the acquired theoretical wisdom, here is an introduction to the various standards and materials of construction as well as piping itself with all necessary accessories. This represents basic knowledge that is a must for any piping engineer.

The text on codes and standards is actually a condensation of material from the various issuing organizations in much more elaborate form. Likewise, manufacturers' catalogs furnish all the information on pipe, fittings, valves, and piping specialties. They are, however, mostly restricted to the vendors own specialty or product. Thus far this handbook serves as a reference work that permits the reader to reduce the number of reference works and catalogs that are normally present in a piping library. The chapters on insulation, shop fabrication, and field installation differ from the trend of condensing available source material and are intended to blend the divergent viewpoints of designers and field erectors, so that all involved will strive together to be most cost effective.

I wish to thank the many individuals and manufacturers that have been so helpful through the provision of data and illustrations, as well as the various organizations that have given permission to reprint from standards or literature. Particular thanks are due to Leo Van Amerongen for his contribution contained in the chapters on insulation shop fabrication and field installation. John Haas of Sarco Company contributed material on steam traps. Louis Harris of B,S&B Safety Systems did likewise on rupture disks, and Albert Scherm of Crane Company assisted with material supplied from Crane's various publications.

My thanks go also to Vincent Fiore for providing graphical illustrations and to Ingeborg Rupinski for her help in typing the final manuscript. Last but not least, I thank my wife Piri for her support and forbearance during the long months that it took to prepare and finalize the manuscript.

Otto Mendel
Forest Hills, N.Y.
March 1981

# 1

# Standard piping terminology

Since the early dawn of mankind, water has been one of the most precious resources, and its transportation is of the utmost importance. Early settlements were always near water so that gourds made from animal skin and then earthenware vessels could be used to transfer water to a man's dwelling. For larger quantities of water, ditches were dug; as an initial piping system, hollowed-out tree trunks were used. The Romans transported water over long distances by building long viaducts from stone and clay.

The Iron Age and modern times have first seen the development of cast iron and then steel pipe. More recently, various alloys and numerous plastic products have been introduced and are used in ever-increasing quantities. Cast-iron water lines have been in actual service since the middle of the 19th century and steel pipe followed quickly. The life of cast iron and steel pipe for certain applications has been greatly extended by the application of lining and coating materials, and the cost of a piping system can also be reduced by the use of plastics, which are known for corrosion and chemical resistance and a promise of longevity.

Today pipelines are providing convenient transportation not only for water, but also for oil, gas, chemicals, and even coal. In general, we might state that piping is used for purposes of fluid transmission. The word *pipe* normally refers to tubular products whose O.D. always meets standard sizes, although their wall thicknesses (schedule numbers) vary. While piping is mostly used for the conveyance of liquids, in certain instances tubing is preferred. Tubing sizes have not been standardized to the same extent as metallic pipe sizes, even though tubing is extensively used in such diverse applications as oil wells, heat exchangers, instrumentation, hydraulics, and the food and paper industry. Tubing, with the exception of oil-well casing, is ordinarily not threaded and is mostly specified on the basis of its outside diameter and wall thickness. Tubing is also used extensively in mechanical applications.

After referring to the conveyance of fluids, the word *fluid* needs also to be better defined. Basically, fluid means flowing as opposed to being in a solid state. While *fluid* and *liquid* are often used synonymously, the term fluid has a broader application and includes liquids, gases, and vapors. Essentially, a fluid is an element that is encountered in liquid, gaseous, or vapor form with no distinctive shape or form of its own. Common knowledge of fluid in its different states can best be used to describe what the term encompasses. Water is a fluid in the liquid state, air is the gaseous state, and steam is a vapor.

*Definition of a Liquid*—A liquid is a substance with a definite volume that has flow characteristics similar to water or slurry but whose form is subject to change as a result of changing its containing vessel, whose shape it will assume. A liquid resists compression, is not subject to appreciable volume changes when exposed to pressure variations, and remains in its subject form when pressure is removed.

*Definition of a Gas*—Gas is a substance in a gaseous state that is compressible, has no independent volume or shape, and can expand indefinitely. Any change in pressure also results in a change of volume.

*Definition of a Vapor*—Vapor is an aeriform that is thought of in conjunction with a transformed liquid or solid from which it originally derived. Steam, for instance, is derived from a liquid (water) through the application of heat and is thus transformed into a vapor. Mist is a vapor that will transform into a liquid by the abstraction of heat. In general, at the point of transformation (either from liquid to gas or from gas to liquid) the fluid substance is called a vapor.

Modern engineering technology makes use of piping in all its endeavors, be it in simple home plumbing or the sophisticated piping that is required in a nuclear-fueled power plant. With such a heavy demand for its products, the piping industry with all its satellite manufacturing facilities plays an important part in our lives. Even the experienced engineer and designer is often overcome by the diversity of the many piping products on the market and cannot always keep up with the latest developments in an industry which introduces continously new products onto the market. This brief introduction into such a diversified field can only cast a cursory glance at some of the many aspects of an industry, which is of so vital importance in any industrialized environment.

# 2

# Piping codes and standards

The first efforts of standardization of pipe sizes date back to the second quarter of the nineteenth century, when nominal sizes for iron pipe and pitches of thread were established. However, these initial standards were so broad that interchangeability of the products of different manufacturers was practically impossible. About 50 years later, Robert Briggs, who was at one time superintendent of the Pascal Iron Works in Philadelphia, prepared a paper giving detailed information about American pipe and pipe thread practice. Briggs' paper established a definite formula for external pipe threads, which became known as the Briggs Standard, and has been in use ever since. Today it is embodied in the ANSI Standard B2.1 for pipe threads and is now officially known as the American standard pipe thread.

The piping and thread tables that Briggs had initiated became the basis for interchangeable products. Modern piping tables that furnish data to the engineer or designer and which are published in reference handbooks or by the various manufacturers provide much more information but can trace their origin back to the data published by Briggs.

## PIPING CODES AND STANDARDS

Numerous piping codes and standards for practically every kind of service have been issued or are being considered by various industry associations. To facilitate their use, the standards are channeled through one central coordinating organization that adopts, classifies, rejects, or modifies the codes and standards developed by the other associations.

Many different codes and standards promulgate the basic requirements peculiar to each industry. They define materials of construction, manufacturing or fabricating methods, inspection or testing requirements, dimensional tolerances, etc., have thus become American National Standards, and have been incorporated in ANSI standards, after it has been determined that the evolvement of such a standard has been in conformance with ANSI procedures.

## AMERICAN NATIONAL STANDARDS INSTITUTE (ANSI)

The American National Standards Institute (ANSI), formerly the American Standards Association (ASA), is a federation of national associations and government departments that dates back to 1918, when it was organized by five engineering societies as the American Engineering Standards Committee. Today, more than 200 organizations representing industry, government, and consumers and 900 companies participate in its work, which is dedicated to the issuance of standards that will serve not only as voluntary guidelines to ensure uniform and reliable products but are also to be used as the basis of Federal regulations. The Occupational Safety and Health Act (OSHA), which was enacted in law by Congress in 1970, relied extensively on ANSI standards.

The Institute does not develop any standards itself; rather it provides a vehicle for standards review through the formation of sectional committees that are composed of representatives of the various organizations or groups that have a substantial interest in a particular standard. One of the important ANSI requirements for the acceptance of any standard is a complete agreement among all interested parties.

The year 1976 was a banner year for ANSI, when nearly 1,800 standards were added to its roster as compared to an average of 400 to 600 standards per year in the preceding decade. ANSI standards numbered approximately 10,000 at the end of 1980. Some of the organizations that are members of ANSI's committee on pressure piping, or are otherwise of importance in the development of standards for the piping industry, are listed below.

## AMERICAN SOCIETY OF MECHANICAL ENGINEERS (ASME)

This engineering society has been in the forefront of organizations that developed standards that have an application in the piping industry. ANSI Standard B31, the Code for Pressure Piping, was sponsored by ASME after the need for such a standard had become evident. First published in 1935 as a tentative standard, it was the culmination of a decade-long effort to produce such a document by numerous representatives of interested organizations.

ASME has also been responsible for the formulation of the various sections that comprise the ASME Boiler and Pressure Vessel Code. These were used to a large extent to assist in the development of the diverse Pressure Piping Codes for the various industries, which run parallel to some of these sections.

ASME also provides certification to manufacturers and installation contractors through the issuance of various types of certificates, which are in stamp form. These certificates provide a guaranty of some kind that the manufacturer or installer complies with the applicable ASME code in respect to design, material, fabrication, examination, testing, inspection, and installation. There are, among others, certificates for boiler manufacture and installation, nuclear facilities, and for pressure piping as part of a boiler assembly. The symbol *PP* is the designated stamp for such pressure piping.

The various ASME certifications are also accepted by governmental agencies and insurance companies as providing assurance that particular requirements have been met.

## AMERICAN SOCIETY FOR TESTING MATERIALS (ASTM)

This society has formulated a considerable number of standards, many of which deal with specifications for piping, tubing, and bolting. As such, they have been included as material specifications in the various sections of the Code for Pressure Piping B31.

### ASTM specifications

These specifications are presently divided into seven separate categories and are easily identified through the system of prefixes that are applicable to the listings within each category.

| *Prefix* | *Category* |
| --- | --- |
| A | Ferrous metals |
| B | Nonferrous metals |
| C | Ceramic, concrete, and masonry materials |
| D | Miscellaneous materials |
| E | Miscellaneous subjects |
| F | End-use materials |
| G | Corrosion, deterioration, and degradation of materials |

The specifications are divided into 48 subject parts.

## AMERICAN PETROLEUM INSTITUTE (API)

The American Petroleum Institute was established in 1919 as the first National Trade Association in the United States to encompass all branches of the petroleum industry. This organization has produced numerous standards applicable to all sectors of the petroleum industry, from the drilling of oil wells to the storage and distribution of an end product. Piping specifications published by API are the recognized standards in the oil industry.

## AMERICAN WELDING SOCIETY (AWS)

While most other organizations are concerned with specifying various basic materials and manufactured items, the American Welding Society has defined criteria that must be met when using a welding process for connecting two items of similar or dissimilar materials. With the preponderance of welded piping connections, in particular when piping is subjected to high temperature and pressure use, it is just as important to have appropriate welding requirements as it is to have material specifications.

## AMERICAN WATER WORKS ASSOCIATION (AWWA)

The specifications promulgated by this organization have long been recognized as the embodiment of today's standards for cast-iron pipe, fittings, and large valves. While earlier specifications for cast-iron pipe use a system of class designations with wall thicknesses calculated for internal pressures only, today's AWWA/ANSI specifications give wall thicknesses based on combined external loadings and internal pressure.

## MANUFACTURERS' STANDARDIZATION SOCIETY FOR THE VALVE AND FITTINGS INDUSTRY (MSS)

In order to achieve a uniformity of manufacturing practice with regard to dimensions of certain items, the MSS issued their "standard practices," some of which have meanwhile been accepted by ANSI after meeting consensus requirements. Most of these standard practices deal with dimensional standards, for which no other uniformity of design exists.

## PIPE FABRICATION INSTITUTE (PFI)

The Pipe Fabrication Institute is an organization of piping fabricators that has produced various standards otherwise not covered by standard promulgating organizations. Similar to the MSS, the PFI tries to obtain uniformity of manufacturing practices and, if possible, achieve acceptance as ANSI standards.

## NATIONAL FIRE PREVENTION AGENCY (NFPA)

This agency was organized in 1896 to promote the science and improve the methods of fire protection and assist in establishing safeguards against loss of life and property by fire. NFPA publishes the National Fire Codes (NFC), which are, to a large extent, documents that are suitable for legal adoption and enforcement and as such are often adopted by or incorporated in codes and standards of other agencies. Many of these codes have a direct bearing on pipe design and installation.

## THE CODE FOR PRESSURE PIPING, ANSI B31

Initial standards for power piping were issued in 1916 by an organization of piping fabricators. Upon this base and with the participation of representatives of fabricators, manufacturers, general industry, insurance companies, government agencies, and architect/engineers, today's code for pressure piping, ANSI B31, was built. The 1980 main code committee is comprised of more than 30 members, representing the various interest groups that have so effectively worked together in the past, with the support of the various engineering societies. The code provides the power and process industries with rules for the safe design and construction of piping systems and contains the necessary basic reference data and formulas pertaining to the following:

1. Material specifications and component standards.
2. Proper dimensional standards.
3. Design requirements for component parts.
4. Stress limitation and evaluation requirements.
5. Fabrication and construction requirements.
6. Testing and inspection requirements.

The code is concerned with the reliability and service life of piping systems within the framework of specific industries. The code is not a design handbook and does not preempt the engineer's judgment and competence.

Considering that the committee, which developed the code, comprised a multitude of members spread throughout the entire USA coming from different social backgrounds and representing a diversity of interests that runs the gamut from private industry to government and from educational to practical application, it is astonishing to review such satisfactory results. Because of the diversity of backgrounds and interest and because each committee member considers himself equally knowledgeable to his peers, it is quite apparent that opposing or conflicting views on a number of subjects require first thorough discussion and then accommodation to reach an agreement acceptable to all parties. Such a diversity of views and the resultant discussions and compromises provide the elements that ANSI nurtures and expects when looking for a consensus standard.

The Pressure Piping Code B31 is broken down into different sections for different industries, transmission distribution, and service systems. Some of these codes are presently still being prepared and are not yet operational (Table 1).

Some of these various piping codes have been evolved to parallel individual sections of the ASME Boiler and Pressure Vessel Code, as they

**TABLE 1** ANSI Code for Pressure Piping, B31

| Standard No. & Designation | Scope & Application |
|---|---|
| B31.1 Power piping | Pertains to all piping systems for electric generating stations; industrial and institutional plants; central and district heating plants and district heating systems (excluding piping covered by ASME Section I, Power Boilers) |
| B31.2 Fuel gas piping | Pertains to air and fuel gas piping for steam generating, central heating, and industrial buildings |
| B31.3 Chemical plant and petroleum refinery piping | Pertains to all fluid-carrying piping in chemical plants, petroleum refineries, and petrochemical plants. This includes steam, oils, refrigerants, chemicals, etc. |
| B31.4 Liquid petroleum transportation piping systems | Pertaining to liquid crude or refined petroleum products in cross-country pipelines |
| B31.5 Refrigeration piping | Pertaining to refrigeration piping used in package units and in commercial or public buildings. |
| B31.7 Nuclear power piping | Pertaining to all piping in a nuclear power plant that carry fluids where spillage from the system could result in radiation hazard to either plant personnel or general public (now obsolete and replaced by Section III of the ASME Boiler Code) |
| B31.8 Gas-transmission and distribution piping systems | Pertaining to gases in cross-country pipelines as well as for city distribution lines. |
| B31.9 Building service piping | In preparation |
| B31.10 Cryogenic piping | In preparation |
| B31.1–TR-116 Suggested rules for operation, maintenance, modification, replacement, and repair of ANSI B31.1 power piping systems | In draft form |
| B31 Guide for corrosion control for ANSI B31.1 power piping systems | |

can be applied to piping systems. It is logical to view the various piping codes—if not as extensions—at least to be pertinent to furtherance of the application of the applicable sections of the ASME Boiler and Pressure Vessel Code. In applying these ground rules, the Code for Power Piping B31.1 parallels ASME Section I, Power Boilers, and the allowable stress values for power piping are generally consistent with those applied for power boilers. In a similar vein, the code that is of prime interest to engineers in the chemical and petrochemical industry, namely the Chemical Plant and

Petroleum Refinery Piping Code B31.3, has been developed parallel to ASME Section III, Unfired Pressure Vessels.

Thus, the Chemical Plant and Petroleum Refinery Piping Code has become the prevailing criteria for pipe design of petrochemical plants. Most owners and insurance companies rely on the guidelines provided and accept code compliance as a guarantee of structural integrity.

All pressure piping codes are constantly being amended and updated. In addition, changes in code requirements as well as interpretations are being published in the form of code cases and as such become part of the code. The codes provide specific material specifications and standards and rely heavily on specifications provided by the various organizations that have contributed so much to general standardization practices (see Appendix 6).

## U.S. GOVERNMENT STANDARDS

While the various industry-wide standards have been fully accepted nationwide, the U.S. Government still maintains its federal specifications (FSSC), and the U.S. Navy has also its own standards (USN and BuShips). All these specifications and standards are in use mainly for federal procurement and even then have often been substituted with industry standards.

## GOVERNMENT REGULATIONS

While most construction and operation of any industrial facility are being carried out by the owner in accordance with the best industry practices and are mostly governed by adherence to those standards and codes which were discussed in this chapter, there also exist a number of governmental regulations at the local, state, and federal level. To mention just a few of these regulations, we can start at the local level, where building codes are very often specific with regard to the piping requirements for plumbing, heating, and air conditioning. State regulations often are particular as to pressure vessel and boiler construction, and often require inspection and certification of such facilities.

### Federal regulations

Federal regulations for any particular subject are published daily in the Federal Register and are compounded yearly under the various Code of Federal Regulation (CFR) title headings. Regulations pertaining to the same subject but originating in different government agencies may be published as different chapters of the same title. The U.S. Government Printing Office publishes yearly a CFR Index and Finding Aide, which is an important tool to help you find your way through the maze of governmental regulations.

### Nuclear Regulatory Commission (NRC)

This federal agency is the regulatory and inspection agency for nuclear power plants and other facilities of a similar nature. The construction and inspection requirements laid down by this agency under CFR Title 10 are a contributing factor to the good safety record of operating nuclear plants. The NRC also controls any other activity that involves a contact with radioactive materials. Inspection of pipe welds through the use of radioisotopes (X-rays) is thus governed by the regulations of this agency.

### Pipeline safety standards

The Code of Federal Regulations (CFR) has published under Title 49 regulations which have been promulgated by the Department of Transportation's Material Transport Bureau (MTB).

Chapter I, Subchapter C, deals with hazardous materials. Subchapter D refers to pipeline safety in general and to liquefied natural gas facilities in particular. Some of these regulations were adopted to comply with the Pipeline Safety Act of 1979 and replace nonregulatory industry standards that were incorporated to a large extent in standards provided by the National Fire Protection Agency (NFPA). Concurrent with the regulations issued by the MTB, the United States Coast Guard (USCG) is developing a new standard for the storage and handling of hazardous materials (including LNG) at waterfront facilities where USCG has jurisdiction.

CFR Title 49, Chapter I, Subchapter D, "Pipeline Safety," deals with the design, construction, operation, and maintenance of such facilities. Part 192 deals with pipelines for natural and other gas, Part 193 addresses liquefied natural gas facilities, and Part 195 refers to the transportation of liquids by pipeline.

While the entire subchapter is of importance, the above-listed parts discuss particular activities that must be undertaken to ensure safety for the particular application. Joining of materials, corrosion control, and test requirement each have one or more chapters devoted to the subject. Detail requirements include the checking for leakage in the areas of flanges, seals, and valves, the monitoring of sensing, warning or control devices, purging when necessary for safety, and an emergency preparedness program. Corrosion protection and control of pipelines, both internally and externally, as well as the necessary inspection are also specified. Material specifications refer mostly back to applicable ASTM, ANSI, or API standards.

# 3

# Pipe manufacturing sizes

While ANSI standard B36.10 for wrought steel pipe is by far the most widely used specification for pipe-wall thicknesses, there are a number of other standards in usage—some under the ANSI umbrella and some not included therein. The American Petroleum Institute (API) has its own standards, most of which are also referenced in B36.10. There are two other standards for piping wall thicknesses available for the design and procurement of carbon steel pipe that are seldom heard of except in the very industry which uses them. These are a standard used for gas distribution piping and the light-gauge standard that is mostly used for spiral-welded piping.

Stainless-steel piping schedule numbers, which are published as ANSI standard B36.19, are easily distinguished from carbon steel sizes through the suffix *s*, which is being used when referring to the schedule numbers for stainless-steel piping. Aluminum piping wall thicknesses are not defined in ANSI standards; however, the wall thicknesses for stainless steel piping as defined in ANSI standard B36.19 are usually applied to aluminum piping, with both NPS standards and schedule numbers used. Copper, brass, nickel, monel, and inconel piping wall thicknesses that are not defined in ANSI standards are also often referred to as having schedule 40S or schedule 80S wall thickness or to be standard and extra strong in conformance with standard NPS practice. PVC piping and certain other plastics that also have no reference point are usually referred to with schedule 40 or 80 designations. Copper tubing, however, with its many commercial applications is normally referred to by type (K, L, M, DWV, TP, or refrigeration type). Chrome-moly alloy (Martensitic steel) piping dimensions are referenced in accordance with the table of schedules applicable to wrought-steel pipe and also the NPS standards.

The above standards refer to wall thickness of pipe and fittings only. Numerous other standards, not all issued under the ANSI umbrella, deal with other dimensional data, while another group of standards addresses the material composition of a given product. References to these various standards will be made as each of the various subject matters is discussed.

A complete listing of ANSI standards is available from the American National Standards Institute, 1430 Broadway, New York, N.Y. 10018. A listing of API standards is available from American Petroleum Institute, 2101 L Street, N.W., Washington, D.C. 20037.

## NOMINAL PIPE SIZE

### NPS standard (ANSI B36.10)

NPS starts with its smallest size as 1/8 in. and then goes up the ladder until it reaches 48 in. Sizes 1/8 in. through 12 in. are still based on the old iron pipe sizes as originally catalogued by Briggs, but sizes 14 in. and larger use the pipe O.D. as the base. There are three different classifications which refer to wall thickness: standard, extra strong, and double extra strong. Standard wall thickness for sizes 1/8 in. through 10 in. denote the varying applicable dimensions, but for sizes 12 in. and larger standard wall thickness means a uniform 3/8-in. wall thickness. Extra-strong wall thickness from 1/8 in. through 6 in. varies with each size, but for sizes 8 in. and larger it denotes a uniform 1/2-in. wall thickness. Double extra-strong standards are only recognized for sizes from 1/2 in. through 12 in., with only the wall thickness for 10 in. and 12 in. pipe being the same 1 in. thickness.

### Pipe schedules (ANSI B36.10)

The piping tables in this standard list schedule numbers for piping in size from 1/8 in. to 36 in. They also list the wall thickness for NPS related sizes and furthermore list a variety of pipe-wall thicknesses for various pipe sizes that are only identifiable by delineating their wall thickness. Most piping sizes designated as NPS standard weight or extra strong are also identifiable by an appropriate overlapping schedule number.

Most noteworthy of this overlap is the designation standard weight pipe for all piping sizes 1/8 in. through 10 in., where this equals schedule 40 pipe. Similarly, piping described as extra strong is the equivalent of schedule 80 pipe in sizes 1/8 in. through 8 in. Because of this sameness in certain size ranges, it is still common practice to refer to schedule 40 piping as standard and schedule 80 piping as extra strong. Schedule numbers are not available for the entire size range indicated above but are restricted to specific sizes. The schedule numbers, which start with 10 for the lighter (thin) walls and go up to 160 for heavy (thick) wall thicknesses, are limited to the following:

| *Pipe Schedule* | *Size Range, in.* |
|---|---|
| 10 | 14–36 |
| 20 | 8–36 |
| 30 | 8–36 (not 26) |
| 40 | 1/8–36 (not 22, 26–30) |
| 60 | 8–24 |
| 80 | 1/8–24 |
| 100 | 8–24 |
| 120 | 4–24 |
| 140 | 8–24 |
| 160 | 1/2–24 |

While there is a duplicate expression in most NPS and schedule numbered pipe, there are two notable exceptions. The NPS standard weight and extra-strong wall thickness for 12-in. and 18-in. pipe cannot be found as schedule numbers. One of the most common errors occurs when 12-in. standard weight pipe is mistakenly specified as 12-in. schedule 40 pipe (Table 2).

The various wall thicknesses of API standards 5L and 5LX are also prominently featured in ANSI B36.10 and are clearly identified as API standard pipe-wall designations, which generally are divergent from those identified by schedule numbers.

## Light-gauge piping

This nomenclature for a special piping standard is mostly used for spiral-welded pipe or otherwise where minimal pressure is encountered. Its size range is from 3/4 in. to 24 in. The wall thickness of piping from 3/4 in. to 12 in. is equal to schedule 10S, which is normally used in conjunction with stainless steel piping. Piping wall thicknesses for sizes 14 in. through 24 in. is the same as schedule 10.

## API standards (API 5L)

Standard 5L is the mainstay for dimensional requirements of the API standards for metallic piping. Table 3 provides dimensions for piping from 2-3/8 in. O.D. through 36 in. Table 4 deals with piping in sizes from 1/8 in. NPS to 20 in. (Table 2). Both these standards list dimensions, weights, and test pressures for the various pipe sizes. Standard-weight pipe is comparable to ANSI NPS standard-weight pipe, with the exception that API lists several wall thicknesses for 8-in., 10-in. and 12-in. pipe. The standard regular weight also includes piping covered as standard-weight pipe, but test pressures for regular-weight pipe are higher. In addition, numerous

**TABLE 2** American Petroleum Institute Standard 5L *(Reproduced by Permission of American Petroleum Institute)*
Standard-weight threaded line pipe dimensions, weights, and test pressures* *(API Standards 5L)*

| 1 | 2 | 3 | 4 | 5 | 6 | 7 | 8 | 9 | 10 |
|---|---|---|---|---|---|---|---|---|---|
| | | | | | Calculated Weight, | | Test Pressure, psi., min. | | |
| Size: Nominal, in. | Outside Diam-eter, in. $D$ | Nominal Weight; Threads and Coupling, lb. per ft. | Wall Thick-ness, in. $t$ | Inside Diam-eter, in. $d$ | Plain End lb. per ft. $W_{pe}$ | Threads and Coupling lb. $e_w$ | Butt-Welded | Grade A | Grade B. |
| 1/8 | 0.405 | 0.25 | 0.068 | 0.269 | 0.24 | 0.20 | 700 | 700 | 700 |
| 1/4 | 0.540 | 0.43 | 0.088 | 0.364 | 0.42 | 0.20 | 700 | 700 | 700 |
| 3/8 | 0.675 | 0.57 | 0.091 | 0.493 | 0.57 | 0.00 | 700 | 700 | 700 |
| 1/2 | 0.840 | 0.86 | 0.109 | 0.622 | 0.85 | 0.20 | 700 | 700 | 700 |
| 3/4 | 1.050 | 1.14 | 0.113 | 0.824 | 1.13 | 0.20 | 700 | 700 | 700 |
| 1 | 1.315 | 1.70 | 0.133 | 1.049 | 1.68 | 0.20 | 700 | 700 | 700 |
| 1 1/4 | 1.660 | 2.30 | 0.140 | 1.380 | 2.27 | 0.60 | 1000 | 1000 | 1100 |
| 1 1/2 | 1.900 | 2.75 | 0.145 | 1.610 | 2.72 | 0.40 | 1000 | 1000 | 1100 |
| 2 | 2.375 | 3.75 | 0.154 | 2.067 | 3.65 | 1.20 | 1000 | 1000 | 1100 |
| 2 1/2 | 2.875 | 5.90 | 0.203 | 2.469 | 5.79 | 1.80 | 1000 | 1000 | 1100 |

| | | | | | | | | | |
|---|---|---|---|---|---|---|---|---|---|
| 3 | 3.500 | 7.70 | 0.216 | 3.068 | 7.58 | 1.80 | 1000 | 1000 | 1100 |
| 3½ | 4.000 | 9.25 | 0.226 | 3.548 | 9.11 | 3.20 | 1200 | 1200 | 1300 |
| 4 | 4.500 | 11.00 | 0.237 | 4.026 | 10.79 | 4.40 | 1200 | 1200 | 1300 |
| 5 | 5.563 | 15.00 | 0.258 | 5.047 | 14.62 | 5.60 | .... | 1200 | 1300 |
| 6 | 6.625 | 19.45 | 0.280 | 6.065 | 18.97 | 7.20 | .... | 1200 | 1300 |
| 8 | 8.625 | 25.55 | 0.277 | 8.071 | 24.70 | 14.80 | .... | 1160 | 1350 |
| 8 | 8.625 | 29.35 | 0.322 | 7.981 | 28.55 | 14.00 | .... | 1340 | 1570 |
| 10 | 10.750 | 32.75 | 0.279 | 10.192 | 31.20 | 20.00 | .... | 930 | 1090 |
| 10 | 10.750 | 35.75 | 0.307 | 10.136 | 34.24 | 19.20 | .... | 1030 | 1200 |
| 10 | 10.750 | 41.85 | 0.365 | 10.020 | 40.48 | 17.40 | .... | 1220 | 1430 |
| 12 | 12.750 | 45.45 | 0.330 | 12.090 | 43.77 | 32.60 | .... | 930 | 1090 |
| 12 | 12.750 | 51.15 | 0.375 | 12.000 | 49.56 | 30.80 | .... | 1060 | 1240 |
| 14D | 14.000 | 57.00 | 0.375 | 13.250 | 54.57 | 24.60 | .... | 960 | 1120 |
| 16D | 16.000 | 65.30 | 0.375 | 15.250 | 62.58 | 30.00 | .... | 840 | 980 |
| 18D | 18.000 | 73.00 | 0.375 | 17.250 | 70.59 | 35.60 | .... | 750 | 880 |
| 20D | 20.000 | 81.00 | 0.375 | 19.250 | 78.60 | 42.00 | .... | 680 | 790 |

Calculated weights shall be determined in accordance with the following formula:

$$W_L = (w_{pe} \times L) + e_w$$

*where*

$W_L$ = *calculated weight of a piece of pipe of length L, lb (kg).*
$w_{pe}$ = *plain-end weight, lb/ft (kg/m).*
$e_w$ = *weight gain or loss due to end finishing, lb (kg). For plain-end pipe,* $e_w$ *equals zero.*

*Reprinted courtesy API.

**TABLE 3** Regular weight and special* plain-end line pipe dimensions, weights, and test pressures *(API Standard 5L)***

| 1 | 2 | 3 | 4 | 5 | 6 | 7 | 8 | 9 |
|---|---|---|---|---|---|---|---|---|
| | | | | Test Pressure, psi, min. | | | | |
| | | | | Grade A | | Grade B | | |
| Size: Outside Diameter, in. D | Plain-End Weight, lb. per ft. | Wall Thickness in. t | Inside Diameter, in. d | Std. | Alt. | Std. | Alt. | Butt-Welded |
| *2 3/8 | 2.03 | 0.083 | 2.209 | 1260 | .... | 1470 | .... | .... |
| *2 7/8 | 2.47 | 0.083 | 2.709 | 1040 | .... | 1210 | .... | .... |
| *3 1/2 | 3.03 | 0.083 | 3.334 | 850 | .... | 1000 | .... | .... |
| *3 1/2 | 4.51 | 0.125 | 3.250 | 1290 | .... | 1500 | .... | 1000 |
| *3 1/2 | 5.57 | 0.156 | 3.188 | 1600 | .... | 1870 | .... | 1000 |
| 3 1/2 | 6.65 | 0.188 | 3.124 | 1930 | .... | 2260 | .... | 1000 |
| 3 1/2 | 7.58 | 0.216 | 3.068 | 2220 | .... | 2500 | .... | .... |
| 3 1/2 | 8.68 | 0.250 | 3.000 | 2500 | .... | 2500 | .... | .... |
| 3 1/2 | 9.66 | 0.281 | 2.938 | 2500 | .... | 2500 | .... | .... |
| *4 | 3.47 | 0.083 | 3.834 | 750 | .... | 870 | .... | .... |
| *4 | 5.17 | 0.125 | 3.750 | 1120 | .... | 1310 | .... | .... |
| *4 | 6.40 | 0.156 | 3.688 | 1400 | .... | 1640 | .... | .... |
| *4 | 7.65 | 0.188 | 3.624 | 1690 | .... | 1970 | .... | 1200 |
| *4 | 9.11 | 0.226 | 3.548 | 2030 | .... | 2370 | .... | .... |
| *4 | 10.01 | 0.250 | 3.500 | 2250 | .... | 2500 | .... | .... |
| *4 | 11.16 | 0.281 | 3.438 | 2500 | .... | 2500 | .... | .... |
| *4 1/2 | 3.92 | 0.083 | 4.334 | 660 | .... | 770 | .... | .... |
| *4 1/2 | 5.84 | 0.125 | 4.250 | 1000 | .... | 1170 | .... | 800 |
| *4 1/2 | 6.56 | 0.141 | 4.218 | 1130 | .... | 1320 | .... | .... |
| *4 1/2 | 7.24 | 0.156 | 4.188 | 1250 | .... | 1460 | .... | 1000 |
| 4 1/2 | 7.95 | 0.172 | 4.156 | 1380 | .... | 1610 | .... | .... |
| 4 1/2 | 8.66 | 0.188 | 4.124 | 1500 | .... | 1750 | .... | 1200 |
| 4 1/2 | 9.32 | 0.203 | 4.094 | 1620 | .... | 1890 | .... | .... |
| 4 1/2 | 10.01 | 0.219 | 4.062 | 1750 | .... | 2040 | .... | 1200 |
| 4 1/2 | 10.79 | 0.237 | 4.026 | 1900 | .... | 2210 | .... | .... |
| 4 1/2 | 11.35 | 0.250 | 4.000 | 2000 | .... | 2330 | .... | .... |
| 4 1/2 | 12.66 | 0.281 | 3.938 | 2250 | .... | 2500 | .... | .... |
| 4 1/2 | 13.96 | 0.312 | 3.876 | 2500 | .... | 2500 | .... | .... |
| 4 1/2 | 19.00 | 0.438 | 3.624 | 2500 | .... | 2500 | .... | .... |
| 4 1/2 | 22.51 | 0.531 | 3.438 | 2500 | .... | 2500 | .... | .... |
| *5 9/16 | 4.86 | 0.083 | 5.397 | 540 | .... | 630 | .... | .... |
| *5 9/16 | 9.01 | 0.156 | 5.251 | 1010 | .... | 1180 | .... | .... |
| *5 9/16 | 10.79 | 0.188 | 5.187 | 1220 | .... | 1420 | .... | .... |
| *5 9/16 | 12.50 | 0.219 | 5.125 | 1420 | .... | 1650 | .... | .... |
| *5 9/16 | 14.62 | 0.258 | 5.047 | 1670 | .... | 1950 | .... | .... |
| *5 9/16 | 15.85 | 0.281 | 5.001 | 1820 | .... | 2120 | .... | .... |
| *5 9/16 | 17.50 | 0.312 | 4.939 | 2020 | .... | 2360 | .... | .... |

*Special plain-end pipe. All others are regular-weight plain-end pipe.
**Reproduced courtesy API.

| 1 | 2 | 3 | 4 | 5 | 6 | 7 | 8 | 9 |
|---|---|---|---|---|---|---|---|---|
| | | | | Test Pressure, psi, min. | | | | |
| | | | | Grade A | | Grade B | | |
| Size: Outside Diameter, in. D | Plain-End Weight, lb. per ft. | Wall Thickness in. t | Inside Diameter, in. d | Std. | Alt. | Std. | Alt. | Butt-Welded |
| *5⁹⁄₁₆ | 19.17 | 0.344 | 4.875 | 2230 | .... | 2500 | .... | .... |
| *5⁹⁄₁₆ | 27.04 | 0.500 | 4.563 | 2500 | .... | 2500 | .... | .... |
| *5⁹⁄₁₆ | 32.96 | 0.625 | 4.313 | 2500 | .... | 2500 | .... | .... |
| *6⁵⁄₈ | 5.80 | 0.083 | 6.459 | 450 | 560 | 530 | 660 | .... |
| *6⁵⁄₈ | 8.68 | 0.125 | 6.375 | 680 | 850 | 790 | 990 | .... |
| *6⁵⁄₈ | 9.76 | 0.141 | 6.343 | 770 | 960 | 890 | 1120 | .... |
| *6⁵⁄₈ | 10.78 | 0.156 | 6.313 | 850 | 1060 | 990 | 1240 | .... |
| 6⁵⁄₈ | 11.85 | 0.172 | 6.281 | 930 | 1170 | 1090 | 1360 | .... |
| 6⁵⁄₈ | 12.92 | 0.188 | 6.249 | 1020 | 1280 | 1190 | 1490 | .... |
| 6⁵⁄₈ | 13.92 | 0.203 | 6.219 | 1100 | 1380 | 1290 | 1610 | .... |
| 6⁵⁄₈ | 14.98 | 0.219 | 6.187 | 1190 | 1490 | 1390 | 1740 | .... |
| 6⁵⁄₈ | 17.02 | 0.250 | 6.125 | 1360 | 1700 | 1580 | 1980 | .... |
| 6⁵⁄₈ | 18.97 | 0.280 | 6.065 | 1520 | 1900 | 1780 | 2220 | .... |
| 6⁵⁄₈ | 21.04 | 0.312 | 6.001 | 1700 | 2120 | 1980 | 2470 | .... |
| 6⁵⁄₈ | 23.08 | 0.344 | 5.937 | 1870 | 2340 | 2180 | 2500 | .... |
| 6⁵⁄₈ | 25.03 | 0.375 | 5.875 | 2040 | 2500 | 2380 | 2500 | .... |
| 6⁵⁄₈ | 32.71 | 0.500 | 5.625 | 2500 | 2500 | 2500 | 2500 | .... |
| 6⁵⁄₈ | 36.39 | 0.562 | 5.501 | 2500 | 2500 | 2500 | 2500 | .... |
| 6⁵⁄₈ | 40.05 | 0.625 | 5.375 | 2500 | 2500 | 2500 | 2500 | .... |
| 6⁵⁄₈ | 45.35 | 0.719 | 5.187 | 2500 | 2500 | 2500 | 2500 | .... |

| 1 | 2 | 3 | 4 | 5 | 6 | 7 | 8 |
|---|---|---|---|---|---|---|---|
| | | | | Test Pressure, psi, min. | | | |
| | | | | Grade A | | Grade B | |
| Size: Outside Diameter, in. D | Plain-End Weight, lb. per ft. | Wall Thickness, in. t | Inside Diameter, in. d | Std. | Alt. | Std. | Alt. |
| 8⁵⁄₈ | 16.94 | 0.188 | 8.249 | 780 | 980 | 920 | 1140 |
| 8⁵⁄₈ | 19.66 | 0.219 | 8.187 | 910 | 1140 | 1070 | 1330 |
| 8⁵⁄₈ | 22.36 | 0.250 | 8.125 | 1040 | 1300 | 1220 | 1520 |
| 8⁵⁄₈ | 24.70 | 0.277 | 8.071 | 1160 | 1450 | 1350 | 1690 |
| 8⁵⁄₈ | 27.70 | 0.312 | 8.001 | 1300 | 1630 | 1520 | 1900 |
| 8⁵⁄₈ | 28.55 | 0.322 | 7.981 | 1340 | 1680 | 1570 | 1960 |
| 8⁵⁄₈ | 30.42 | 0.344 | 7.937 | 1440 | 1790 | 1680 | 2090 |
| 8⁵⁄₈ | 33.04 | 0.375 | 7.875 | 1570 | 1960 | 1830 | 2280 |
| 8⁵⁄₈ | 38.30 | 0.438 | 7.749 | 1830 | 2290 | 2130 | 2500 |
| 8⁵⁄₈ | 48.40 | 0.562 | 7.501 | 2350 | 2500 | 2500 | 2500 |
| 8⁵⁄₈ | 53.40 | 0.625 | 7.375 | 2500 | 2500 | 2500 | 2500 |
| 8⁵⁄₈ | 60.71 | 0.719 | 7.187 | 2500 | 2500 | 2500 | 2500 |

**TABLE 3** *continued*

| 1 | 2 | 3 | 4 | 5 | 6 | 7 | 8 |
|---|---|---|---|---|---|---|---|
| | | | | Test Pressure, psi, min. | | | |
| | | | | Grade A | | Grade B | |
| Size: Outside Diameter, in. *D* | Plain-End Weight, lb. per ft. | Wall Thickness, in. *t* | Inside Diameter, in. *d* | Std. | Alt. | Std. | Alt. |
| *10¾ | 21.21 | 0.188 | 10.374 | 630 | 790 | 730 | 920 |
| 10¾ | 24.63 | 0.219 | 10.312 | 730 | 920 | 860 | 1070 |
| 10¾ | 28.04 | 0.250 | 10.250 | 840 | 1050 | 980 | 1220 |
| 10¾ | 31.20 | 0.279 | 10.192 | 930 | 1170 | 1090 | 1360 |
| 10¾ | 34.24 | 0.307 | 10.136 | 1030 | 1290 | 1200 | 1500 |
| 10¾ | 38.23 | 0.344 | 10.062 | 1150 | 1440 | 1340 | 1680 |
| 10¾ | 40.48 | 0.365 | 10.020 | 1220 | 1530 | 1430 | 1780 |
| 10¾ | 48.24 | 0.438 | 9.874 | 1470 | 1830 | 1710 | 2140 |
| 10¾ | 61.15 | 0.562 | 9.626 | 1880 | 2350 | 2200 | 2500 |
| 10¾ | 67.58 | 0.625 | 9.500 | 2090 | 2500 | 2440 | 2500 |
| 10¾ | 77.03 | 0.719 | 9.312 | 2410 | 2500 | 2500 | 2500 |
| 10¾ | 86.18 | 0.812 | 9.126 | 2500 | 2500 | 2500 | 2500 |
| *12¾ | 25.22 | 0.188 | 12.374 | 530 | 660 | 620 | 770 |
| *12¾ | 29.31 | 0.219 | 12.312 | 620 | 770 | 720 | 900 |
| 12¾ | 33.38 | 0.250 | 12.250 | 710 | 880 | 820 | 1030 |
| 12¾ | 37.42 | 0.281 | 12.188 | 790 | 990 | 930 | 1160 |
| 12¾ | 41.45 | 0.312 | 12.126 | 880 | 1100 | 1030 | 1280 |
| 12¾ | 43.77 | 0.330 | 12.090 | 930 | 1160 | 1090 | 1360 |
| 12¾ | 45.58 | 0.344 | 12.062 | 970 | 1210 | 1130 | 1420 |
| 12¾ | 49.56 | 0.375 | 12.000 | 1060 | 1320 | 1240 | 1540 |
| 12¾ | 57.59 | 0.438 | 11.874 | 1240 | 1550 | 1440 | 1880 |
| 12¾ | 73.15 | 0.562 | 11.626 | 1590 | 1980 | 1850 | 2310 |
| 12¾ | 80.93 | 0.625 | 11.500 | 1760 | 2210 | 2060 | 2500 |
| 12¾ | 88.63 | 0.688 | 11.374 | 1940 | 2430 | 2270 | 2500 |
| 12¾ | 96.12 | 0.750 | 11.250 | 2120 | 2500 | 2470 | 2500 |
| *14 | 36.71 | 0.250 | 13.500 | 640 | 800 | 750 | 940 |
| *14 | 41.17 | 0.281 | 13.438 | 720 | 900 | 840 | 1050 |
| 14 | 45.61 | 0.312 | 13.376 | 800 | 1000 | 940 | 1170 |
| 14 | 50.17 | 0.344 | 13.312 | 880 | 1110 | 1030 | 1290 |
| 14 | 54.57 | 0.375 | 13.250 | 960 | 1210 | 1120 | 1410 |
| 14 | 63.44 | 0.438 | 13.124 | 1130 | 1410 | 1310 | 1640 |
| 14 | 72.09 | 0.500 | 13.000 | 1290 | 1610 | 1500 | 1880 |
| 14 | 80.66 | 0.562 | 12.876 | 1450 | 1810 | 1690 | 2110 |
| 14 | 89.28 | 0.625 | 12.750 | 1610 | 2010 | 1880 | 2340 |
| 14 | 97.81 | 0.688 | 12.624 | 1770 | 2210 | 2060 | 2500 |
| 14 | 106.13 | 0.750 | 12.500 | 1930 | 2410 | 2250 | 2500 |
| 14 | 114.37 | 0.812 | 12.376 | 2090 | 2500 | 2440 | 2500 |
| *16 | 42.05 | 0.250 | 15.500 | 560 | 700 | 660 | 820 |
| *16 | 47.17 | 0.281 | 15.438 | 630 | 790 | 740 | 920 |
| 16 | 52.27 | 0.312 | 15.376 | 700 | 880 | 820 | 1020 |
| 16 | 57.52 | 0.344 | 15.312 | 770 | 970 | 900 | 1130 |
| 16 | 62.58 | 0.375 | 15.250 | 840 | 1050 | 980 | 1230 |

| 1 | 2 | 3 | 4 | 5 | 6 | 7 | 8 |
|---|---|---|---|---|---|---|---|
| | | | | Test Pressure, psi, min. | | | |
| | | | | Grade A | | Grade B | |
| Size: Outside Diameter, in. *D* | Plain-End Weight, lb. per ft. | Wall Thickness, in. *t* | Inside Diameter, in. *d* | Std. | Alt. | Std. | Alt. |
| 16 | 72.80 | 0.438 | 15.124 | 990 | 1230 | 1150 | 1440 |
| 16 | 82.77 | 0.500 | 15.000 | 1120 | 1410 | 1310 | 1640 |
| 16 | 92.66 | 0.562 | 14.876 | 1260 | 1580 | 1480 | 1840 |
| 16 | 102.63 | 0.625 | 14.750 | 1410 | 1760 | 1640 | 2050 |
| 16 | 112.51 | 0.688 | 14.624 | 1550 | 1940 | 1810 | 2260 |
| 16 | 122.15 | 0.750 | 14.500 | 1690 | 2110 | 1970 | 2460 |
| 16 | 131.71 | 0.812 | 14.376 | 1830 | 2280 | 2130 | 2500 |
| *18 | 47.39 | 0.250 | 17.500 | 500 | 620 | 580 | 730 |
| *18 | 53.18 | 0.281 | 17.438 | 560 | 700 | 660 | 820 |
| 18 | 58.94 | 0.312 | 17.376 | 620 | 780 | 730 | 910 |
| 18 | 64.87 | 0.344 | 17.312 | 690 | 860 | 800 | 1000 |
| 18 | 70.59 | 0.375 | 17.250 | 750 | 940 | 880 | 1090 |
| 18 | 82.15 | 0.438 | 17.124 | 880 | 1100 | 1020 | 1280 |
| 18 | 93.45 | 0.500 | 17.000 | 1000 | 1250 | 1170 | 1460 |
| 18 | 104.67 | 0.562 | 16.876 | 1120 | 1400 | 1310 | 1640 |
| 18 | 115.98 | 0.625 | 16.750 | 1250 | 1560 | 1460 | 1820 |
| 18 | 127.21 | 0.688 | 16.624 | 1380 | 1720 | 1610 | 2010 |
| 18 | 138.17 | 0.750 | 16.500 | 1500 | 1880 | 1750 | 2190 |
| 18 | 149.06 | 0.812 | 16.376 | 1620 | 2030 | 1890 | 2370 |
| *20 | 52.73 | 0.250 | 19.500 | 450 | 560 | 520 | 660 |
| *20 | 59.18 | 0.281 | 19.438 | 510 | 630 | 590 | 740 |
| 20 | 65.60 | 0.312 | 19.376 | 560 | 700 | 660 | 820 |
| 20 | 72.21 | 0.344 | 19.312 | 620 | 770 | 720 | 900 |
| 20 | 78.60 | 0.375 | 19.250 | 680 | 840 | 790 | 980 |
| 20 | 91.51 | 0.438 | 19.124 | 790 | 990 | 920 | 1150 |
| 20 | 104.13 | 0.500 | 19.000 | 900 | 1120 | 1050 | 1310 |
| 20 | 116.67 | 0.562 | 18.876 | 1010 | 1260 | 1180 | 1480 |
| 20 | 129.33 | 0.625 | 18.750 | 1120 | 1410 | 1310 | 1640 |
| 20 | 141.90 | 0.688 | 18.624 | 1240 | 1550 | 1440 | 1810 |
| 20 | 154.19 | 0.750 | 18.500 | 1350 | 1690 | 1580 | 1970 |
| 20 | 166.40 | 0.812 | 18.376 | 1460 | 1830 | 1710 | 2130 |
| *22 | 58.07 | 0.250 | 21.500 | 410 | 510 | 480 | 600 |
| *22 | 65.18 | 0.281 | 21.438 | 460 | 570 | 540 | 670 |
| 22 | 72.27 | 0.312 | 21.376 | 510 | 640 | 600 | 740 |
| 22 | 79.56 | 0.344 | 21.312 | 550 | 700 | 660 | 820 |
| 22 | 86.61 | 0.375 | 21.250 | 610 | 770 | 720 | 890 |
| 22 | 100.86 | 0.438 | 21.124 | 720 | 900 | 840 | 1050 |
| 22 | 114.81 | 0.500 | 21.000 | 820 | 1020 | 950 | 1190 |
| 22 | 128.67 | 0.562 | 20.876 | 920 | 1150 | 1070 | 1340 |
| 22 | 142.68 | 0.625 | 20.750 | 1020 | 1280 | 1190 | 1490 |
| 22 | 156.60 | 0.688 | 20.624 | 1130 | 1410 | 1310 | 1640 |
| 22 | 170.21 | 0.750 | 20.500 | 1230 | 1530 | 1430 | 1790 |
| 22 | 183.75 | 0.812 | 20.376 | 1330 | 1660 | 1550 | 1940 |

**TABLE 3** *continued*

| 1 | 2 | 3 | 4 | 5 | 6 | 7 | 8 |
|---|---|---|---|---|---|---|---|
| | | | | Test Pressure, psi, min. | | | |
| Size: Outside Diameter, in. *D* | Plain-End Weight, lb. per ft. | Wall Thickness, in. *t* | Inside Diameter, in. *d* | Grade A | | Grade B | |
| | | | | Std. | Alt. | Std. | Alt. |
| *24 | 63.41 | 0.250 | 23.500 | 380 | 470 | 440 | 550 |
| *24 | 71.18 | 0.281 | 23.438 | 420 | 530 | 490 | 610 |
| 24 | 78.93 | 0.312 | 23.376 | 470 | 580 | 550 | 680 |
| 24 | 86.91 | 0.344 | 23.312 | 520 | 640 | 600 | 750 |
| 24 | 94.62 | 0.375 | 23.250 | 560 | 700 | 660 | 820 |
| 24 | 110.22 | 0.438 | 23.124 | 660 | 820 | 770 | 960 |
| 24 | 125.49 | 0.500 | 23.000 | 750 | 940 | 880 | 1090 |
| 24 | 140.68 | 0.562 | 22.876 | 840 | 1050 | 980 | 1230 |
| 24 | 156.03 | 0.625 | 22.750 | 940 | 1170 | 1090 | 1370 |
| 24 | 171.29 | 0.688 | 22.624 | 1030 | 1290 | 1200 | 1500 |
| 24 | 186.23 | 0.750 | 22.500 | 1120 | 1410 | 1310 | 1640 |
| 24 | 201.09 | 0.812 | 23.376 | 1220 | 1520 | 1420 | 1780 |
| *26 | 68.75 | 0.250 | 25.500 | 350 | 430 | 400 | 500 |
| *26 | 77.18 | 0.281 | 24.438 | 390 | 490 | 450 | 570 |
| 26 | 85.60 | 0.312 | 25.376 | 430 | 540 | 500 | 630 |
| 26 | 94.26 | 0.344 | 25.312 | 480 | 600 | 560 | 690 |
| 26 | 102.63 | 0.375 | 25.250 | 520 | 650 | 610 | 760 |
| 26 | 119.57 | 0.438 | 25.124 | 610 | 760 | 710 | 880 |
| 26 | 136.17 | 0.500 | 25.000 | 690 | 870 | 810 | 1010 |
| 26 | 152.68 | 0.562 | 24.876 | 780 | 970 | 910 | 1130 |
| 26 | 169.38 | 0.625 | 24.750 | 870 | 1080 | 1010 | 1260 |
| 26 | 185.99 | 0.688 | 24.624 | 950 | 1190 | 1110 | 1390 |
| 26 | 202.25 | 0.750 | 24.500 | 1040 | 1300 | 1210 | 1510 |
| 28 | 110.64 | 0.375 | 27.250 | 480 | 600 | 560 | 700 |
| 28 | 128.93 | 0.438 | 27.124 | 560 | 700 | 660 | 820 |
| 28 | 146.85 | 0.500 | 27.000 | 640 | 800 | 750 | 940 |
| 28 | 164.69 | 0.562 | 26.876 | 720 | 900 | 840 | 1050 |
| 28 | 182.73 | 0.625 | 26.750 | 800 | 1000 | 940 | 1170 |
| 30 | 118.65 | 0.375 | 29.250 | 450 | 560 | 520 | 660 |
| 30 | 138.29 | 0.438 | 29.124 | 530 | 660 | 610 | 770 |
| 30 | 157.53 | 0.500 | 29.000 | 600 | 750 | 700 | 880 |
| 30 | 176.69 | 0.562 | 28.876 | 670 | 840 | 790 | 980 |
| 30 | 196.08 | 0.625 | 28.750 | 750 | 940 | 880 | 1090 |
| 32 | 126.66 | 0.375 | 31.250 | 420 | 530 | 490 | 620 |
| 32 | 147.64 | 0.438 | 31.124 | 490 | 620 | 570 | 720 |
| 32 | 168.21 | 0.500 | 31.000 | 560 | 700 | 660 | 820 |
| 32 | 188.70 | 0.562 | 30.876 | 630 | 790 | 740 | 920 |
| 32 | 209.43 | 0.625 | 30.750 | 700 | 880 | 820 | 1030 |
| 34 | 134.67 | 0.375 | 33.250 | 400 | 500 | 460 | 580 |
| 34 | 154.00 | 0.438 | 33.124 | 460 | 580 | 540 | 680 |

| 1 | 2 | 3 | 4 | 5 | 6 | 7 | 8 |
|---|---|---|---|---|---|---|---|
| | | | | Test Pressure, psi, min. | | | |
| | | | | Grade A | | Grade B | |
| Size: Outside Diameter, in. *D* | Plain-End Weight, lb. per ft. | Wall Thickness, in. *t* | Inside Diameter, in. *d* | Std. | Alt. | Std. | Alt. |
| 34 | 178.89 | 0.500 | 33.000 | 530 | 660 | 620 | 770 |
| 34 | 200.70 | 0.562 | 32.876 | 600 | 740 | 690 | 870 |
| 34 | 222.78 | 0.625 | 32.750 | 660 | 830 | 770 | 970 |
| 36 | 142.68 | 0.375 | 32.250 | 380 | 470 | 440 | 550 |
| 36 | 166.35 | 0.438 | 35.124 | 440 | 550 | 510 | 640 |
| 36 | 189.57 | 0.500 | 35.000 | 500 | 620 | 580 | 730 |
| 36 | 212.70 | 0.562 | 34.876 | 560 | 700 | 660 | 820 |
| 36 | 236.13 | 0.625 | 34.750 | 620 | 780 | 730 | 910 |

additional wall thicknesses for the same outside diameter pipe are covered in the regular-weight standard.

## Stainless-steel pipe schedules (ANSI B36.19)

Stainless-steel piping schedules sometimes overlap with piping schedules for carbon steel piping but are always identified with the suffix *S*. For instance, we might refer to some 4-in. stainless-steel pipe as schedule 40S. There is also a common use of schedules 5S and 10S, which do not appear in carbon steel pipe schedules. Schedule 5S pipe wall thicknesses are covered in sizes 1/2 in. through 30 in., and schedule 10S ranges from 1/8 in. through 30 in. Only the sizes 24 in. and 30 in. of schedule 10S overlap with the regular schedule 10 wall thickness. Schedule 40S and schedule 80S both range from 1/8 in. to 12 in. While schedule 40S wall thickness equals schedule 40 in all sizes except 12 in., 10-in. and 12-in. sizes in schedule 80S do match those of schedule 80.

Pipe and fittings with heavier wall thicknesses than those indicated in B36.19 are also manufactured on a regular basis and often relate to dimensions listed in B36.10. Stainless steel tubing is also very much in demand for specialized applications in the process industries where its light wall thicknesses are preferred for economic, corrosive, or extra clean service with minimal pressure or temperature requirements. Manufacturing standards for such applications are available from the various fabricators, who specialize in this type of product (Table 4).

**TABLE 4** Dimensions and Weights of Stainless Steel Pipe, ANSI B36.19
*(Reproduced by permission of American Society of Mechanical Engineers)*

| Nominal | | Pipe Schedule | | | | | | | |
|---|---|---|---|---|---|---|---|---|---|
| | | 5S | | 10S | | 40S | | 80S | |
| Pipe Size | Outside Diameter | Wall | Wt | Wall | Wt | Wall | Wt | Wall | Wt |
| 1/8 | .405 | | | .049 | .19 | .068 | .24 | .095 | .31 |
| 1/4 | .540 | | | .065 | .33 | .088 | .42 | .119 | .54 |
| 3/8 | .675 | | | .065 | .42 | .091 | .57 | .126 | .74 |
| 1/2 | .840 | .065 | .54 | .083 | .67 | .109 | .85 | .147 | 1.09 |
| 3/4 | 1.050 | .065 | .69 | .083 | .86 | .113 | 1.13 | .154 | 1.47 |
| 1 | 1.315 | .065 | .87 | .109 | 1.40 | .133 | 1.68 | .179 | 2.17 |
| 1 1/4 | 1.660 | .065 | 1.11 | .109 | 1.81 | .140 | 2.27 | .191 | 3.00 |
| 1 1/2 | 1.900 | .065 | 1.28 | .109 | 2.09 | .145 | 2.72 | .200 | 3.63 |
| 2 | 2.375 | .065 | 1.61 | .109 | 2.64 | .154 | 3.65 | .218 | 5.02 |
| 2 1/2 | 2.875 | .083 | 2.48 | .120 | 3.53 | .203 | 5.79 | .276 | 7.66 |
| 3 | 3.500 | .083 | 3.03 | .120 | 4.33 | .216 | 7.58 | .300 | 10.25 |
| 3 1/2 | 4.000 | .083 | 3.48 | .120 | 4.97 | .226 | 9.11 | .318 | 12.51 |
| 4 | 4.500 | .083 | 3.92 | .120 | 5.61 | .237 | 10.79 | .337 | 14.98 |
| 5 | 5.563 | .109 | 6.36 | .134 | 7.77 | .258 | 14.62 | .375 | 20.78 |
| 6 | 6.625 | .109 | 7.60 | .134 | 9.29 | .280 | 18.97 | .432 | 28.57 |
| 8 | 8.625 | .109 | 9.93 | .148 | 13.40 | .322 | 28.55 | .500 | 43.39 |
| 10 | 10.750 | .134 | 15.23 | .165 | 18.70 | .365 | 40.48 | .500 | 54.74 |
| 12 | 12.750 | .156 | 22.22 | .180 | 24.20 | .375 | 49.56 | .500 | 65.42 |
| 14 | 14.000 | .156 | 23.04 | .188 | 27.71 | | | | |
| 16 | 16.000 | .165 | 27.88 | .188 | 31.72 | | | | |
| 18 | 18.000 | .165 | 31.40 | .188 | 35.73 | | | | |
| 20 | 20.000 | .188 | 39.74 | .218 | 46.02 | | | | |
| 24 | 24.000 | .218 | 55.32 | .250 | 63.35 | | | | |
| 30 | 30.000 | .250 | 79.36 | .312 | 98.83 | | | | |

All dimensions are given in inches
All pipe weights are given in pounds and are given for a linear foot of carbon steel pipe with plain ends.
The different grades of stainless steel pipe vary in weight from 5% less to 3% more, depending on material composltlon.

## Aluminum piping

Aluminum piping is being produced with wall thicknesses to meet stainless steel schedule numbers as defined in ANSI standard B36.19. In addition, aluminum piping manufacturing standards often meet NPS standards or have wall thicknesses that do not fit any of the above categories. Pipe is manufactured in standard size range from 1/2 in. through 24 in., meeting ANSI specifications 5S, 10S, 40S, and 80S. Common manufacturing range meeting other requirements is standardized between 1/2 in. and 12 in.

## Cast-iron piping (ANSI A.21.1)

Among the various specifications for cast-iron piping, ANSI standard A.21.1, often referred to as ANSI/AWWA C101 and "Thickness Design of Cast Iron Pipe", stands out as a major reference work and is also referred to in some of the other ANSI standards for cast-iron piping. Unlike steel piping standards, the standards for cast-iron piping not only refer to wall thickness but also deal with manufacturing methods and pipe use (gas or water). Considerable attention is devoted to installation conditions and the depth of the proposed pipe burial.

Pipe wall thickness is designated as a thickness class number, which is not to be confused with another designation known as a pressure class. With so many variables available in an ANSI standard, it is clear that any design or purchase recommendation must be very precise in order to ensure that the desired material will be furnished for its specified use. Cast-iron thickness classes range from class 20 to class 30 and are similar to pipe schedules. Each of the various thickness classes is applicable to each pipe size. While Table 5, which illustrates the standard thickness classes for cast iron pipe, lists only pipe sizes up to 48 in., other standards are available for pipe sizes up to 60 in. and beyond that size. In addition to other ANSI standards for cast-iron pipe, standards exist for pipe fittings and various connecting methods, such as bell and spigot, mechanical joint, push joint, river crossing joint, or flanged joint.

**TABLE 5** Standard thickness classes of cast-iron pipe

| Pipe Size | Thickness for Standard Class Number—*in.* | | | | | | | | | | |
|---|---|---|---|---|---|---|---|---|---|---|---|
| *in.* | 20 | 21 | 22 | 23 | 24 | 25 | 26 | 27 | 28 | 29 | 30 |
| 3 | | | 0.32* | 0.35 | 0.38 | 0.41 | 0.44 | 0.48 | 0.52 | 0.56 | 0.60 |
| 4 | | | 0.35* | 0.38 | 0.41 | 0.44 | 0.48 | 0.52 | 0.56 | 0.60 | 0.65 |
| 6 | | 0.35* | 0.38 | 0.41 | 0.44 | 0.48 | 0.52 | 0.56 | 0.60 | 0.65 | 0.70 |
| 8 | 0.35* | 0.38 | 0.41 | 0.44 | 0.48 | 0.52 | 0.56 | 0.60 | 0.65 | 0.70 | 0.76 |
| 10 | 0.38* | 0.41 | 0.44 | 0.48 | 0.52 | 0.56 | 0.60 | 0.65 | 0.70 | 0.76 | 0.82 |
| 12 | 0.41* | 0.44 | 0.48 | 0.52 | 0.56 | 0.60 | 0.65 | 0.70 | 0.76 | 0.82 | 0.89 |
| 14 | 0.43 | 0.48* | 0.51 | 0.55 | 0.59 | 0.64 | 0.69 | 0.75 | 0.81 | 0.87 | 0.94 |
| 16 | 0.46 | 0.50* | 0.54 | 0.58 | 0.63 | 0.68 | 0.73 | 0.79 | 0.85 | 0.92 | 0.99 |
| 18 | 0.50 | 0.54* | 0.58 | 0.63 | 0.68 | 0.73 | 0.79 | 0.85 | 0.92 | 0.99 | 1.07 |
| 20 | 0.53 | 0.57* | 0.62 | 0.67 | 0.72 | 0.78 | 0.84 | 0.91 | 0.98 | 1.06 | 1.14 |
| 24 | 0.58 | 0.63* | 0.68 | 0.73 | 0.79 | 0.85 | 0.92 | 0.99 | 1.07 | 1.16 | 1.25 |
| 30 | 0.68* | 0.73 | 0.79 | 0.85 | 0.92 | 0.99 | 1.07 | 1.16 | 1.25 | 1.35 | 1.46 |
| 36 | 0.75* | 0.81 | 0.87 | 0.94 | 1.02 | 1.10 | 1.19 | 1.29 | 1.39 | 1.50 | 1.62 |
| 42 | 0.83* | 0.90 | 0.97 | 1.05 | 1.13 | 1.22 | 1.32 | 1.43 | 1.54 | 1.66 | 1.79 |
| 48 | 0.91* | 0.98 | 1.06 | 1.14 | 1.23 | 1.33 | 1.44 | 1.56 | 1.68 | 1.81 | 1.95 |

Reprinted from ANSI/AWWA Standard C101-67, *American National Standard for Thickness Design of Cast-Iron Pipe,* by permission. Copyright 1967, the American Water Works Association.

## Ductile iron piping (ANSI A21.50)

Wall thickness requirements for ductile iron piping follow a similar approach as that indicated for cast-iron piping, with thickness class numbers ranging from 50 to 56 because of the higher properties of this type of piping (see Table 6).

Specification ANSI A21.50 denotes the varying pipe-wall dimensions and manufacturing methods. Ductile iron pipe is available for similar connecting methods as cast iron pipe and is normally furnished in sizes from 3 in. through 24 in.

**TABLE 6** Standard Thickness Classes of Ductile-Iron Pipe

| Size *in.* | Outside Diameter—*in.* | Thickness Class 50 | 51 | 52 | 53 | 54 | 55 | 56 |
|---|---|---|---|---|---|---|---|---|
| | | Thickness—*in.* | | | | | | |
| 3 | 3.96 | — | 0.25 | 0.28 | 0.31 | 0.34 | 0.37 | 0.40 |
| 4 | 4.80 | — | 0.26 | 0.29 | 0.32 | 0.35 | 0.38 | 0.41 |
| 6 | 6.90 | 0.25 | 0.28 | 0.31 | 0.34 | 0.37 | 0.40 | 0.43 |
| 8 | 9.05 | 0.27 | 0.30 | 0.33 | 0.36 | 0.39 | 0.42 | 0.45 |
| 10 | 11.10 | 0.29 | 0.32 | 0.35 | 0.38 | 0.41 | 0.44 | 0.47 |
| 12 | 13.20 | 0.31 | 0.34 | 0.37 | 0.40 | 0.43 | 0.46 | 0.49 |
| 14 | 15.30 | 0.33 | 0.36 | 0.39 | 0.42 | 0.45 | 0.48 | 0.51 |
| 16 | 17.40 | 0.34 | 0.37 | 0.40 | 0.43 | 0.46 | 0.49 | 0.52 |
| 18 | 19.50 | 0.35 | 0.38 | 0.41 | 0.44 | 0.47 | 0.50 | 0.53 |
| 20 | 21.60 | 0.36 | 0.39 | 0.42 | 0.45 | 0.48 | 0.51 | 0.54 |
| 24 | 25.80 | 0.38 | 0.41 | 0.44 | 0.47 | 0.50 | 0.53 | 0.56 |
| 30 | 32.00 | 0.39 | 0.43 | 0.47 | 0.51 | 0.55 | 0.59 | 0.63 |
| 36 | 38.30 | 0.43 | 0.48 | 0.53 | 0.58 | 0.63 | 0.68 | 0.73 |
| 42 | 44.50 | 0.47 | 0.53 | 0.59 | 0.65 | 0.71 | 0.77 | 0.83 |
| 48 | 50.80 | 0.51 | 0.58 | 0.65 | 0.72 | 0.79 | 0.86 | 0.93 |
| 54 | 57.10 | 0.57 | 0.65 | 0.73 | 0.81 | 0.89 | 0.97 | 1.05 |

Reprinted from ANSI 21.50–1976 (AWWA C150–76), *American National Standard for the Thickness Design of Ductile-Iron Pipe,* by permission. Copyright 1976, the American Water Works Association.

## Nonferrous metals

Piping in the various metals, such as brass, copper, nickel, monel, inconel, etc., for which there is no ANSI standard, are mostly manufactured in accordance with the wall thickness designations listed in ANSI standard B36.19 for stainless-steel piping. However, the use of NPS wall thicknesses is also not uncommon. Pipe is commonly available in sizes 12 in. and below.

## PIPING DESCRIPTION

Piping on purchase orders is often described in various forms, and it is not always easy to identify what pipe is being called for. A standard piping description should start out by listing the piping material, followed by pipe size and schedule number of wall thickness, then a description of piping manufactured such as seamless or lap welded. Then the required pipe-end preparation should be shown, and as the last item the piping specification should be listed.

One sample of such a piping description is galvanized carbon steel pipe, 2-in. NPS extra strong, butt welded, threaded and coupled, ANSI A-120 (Galv. C.S. Pipe 2 in. XH, BW, T&C, ANSI A-120). Or stainless steel pipe, Type 304 ELC, fully annealed 4-in. Schedule 5 welded, plain ends ANSI A-312 (S.S. Pipe 304 ELC fully annealed, 4 in. Sch. 5 welded, P.E. ANSI A-312).

## TUBING STANDARDS

Tubing is generally specified by first indicating O.D., wall thickness, and gauge (if applicable), followed by material designation and manufacturing process, and finally the applicable ANSI or other standard. In many instances, the exact tubing length is also specified—for example, when ordering heat exchanger tubing. It is also important to list the wall thickness in thousandths of an inch (three digits after the decimal point).

Tubing is generally designated by its actual outside diameter, a practice that does not apply to piping 12 in. and smaller. Very often, the wall thickness of steel tubing is also stipulated in accordance with the Birmingham Wire Gauge, or BWG for short. This British standard has gained wide acceptance in the US and has actually been recognized as an acceptable standard for thicknesses by an act of Congress.

Tubing is in wide use with a diversity of applications. Small-diameter tubing is very much in demand for instrumentation, hydraulic, plumbing, and heating service, while large-diameter tubing is used to a large extent in oil-well applications. Other tubing usage includes heat exchanger, condenser, and boiler tubes. In the petrochemical industry, heavy-wall tubing is often preferred to piping. Tubing materials run the whole gamut of metals available, and include but are not limited to steel and stainless steel, copper and copper alloys, bronze, aluminum, nickel, monel and inconel, incolloy, and hastelloy.

The use of tubing is not restricted to pressure applications. Tubing is also very much in demand for mechanical use, where its shape is not restricted to the round form.

## Pipe dimensions*

The following dimensions are usually provided in properties-of-pipe tables to aid pipe users and designers in various calculations:

*External Pipe Diameter (OD)*—These are used in making piping layouts and for determining space requirements for them in pipe troughs, chases, pipe racks, etc., as well as to determine the sizes of openings or holes through which the pipe must pass. Required in piping flexibility formulas, O.D. dimensions are necessary in determining pipe stress, which is independent of pipe schedule number.

*Internal Pipe Diameter (ID)*—Knowledge of this dimension is required in maximum-stress formulas. The I.D. raised to a power is used in equations for flow calculations.

*Pipe-Wall Thickness*—This dimension furnishes the reference needed to determine the nearest commercial pipe schedule number after the thickness has been calculated from the piping code formulas.

*Weight of Pipe per Lineal Foot*—Used in determining the total weight of a fabricated pipe spool, the unit value must include the additional weights of the other piping components, such as flanges and fittings. This weight usually includes the internal weight of water to determine requirements of pipe hangers and other means of support.

*Circumference of Pipe, External and Internal*—These are used to calculate pipe surface areas for such applications as boiler tubes exposed to the fire or the water side.

*Cross-Sectional Area*—Formulas for calculating structural-stress problems include this dimension.

*Pipe Length/Sq Ft of Surface*—This is used to calculate the heating and cooling surfaces of pipe coils and their radiation effects. Painting requirements can also be determined.

*Length of Pipe/Cu Ft or per U.S. Gal. Capacity*—Such data come in handy when the total holding capacity of a pipeline or the time required to fill it must be known. Sometimes the holding capacity of a line that must be dumped in an emergency must be known.

*Weight of Water/Ft of Pipe*—This value is added to the weight of bare pipe plus the weight of the insulation to determine hanger load requirements. When pipelines are installed in open trenches, the soil bearing pressure must be checked to see if the soil will support the pipe without additional supports.

*Moment of Inertia and Section Modulus*—These are terms included in formulas for structural and piping flexibility problems. The values are used in stress formulas determining the maximum spacing for supports so that they will be within the allowable bending stress for the piping.

*Courtesy *Chemical Engineering,* June 17, 1968. © McGraw-Hill.

# 4

# Metallic piping

With the advent of the iron age, it became inevitable that iron piping would rank as one of the many achievements that took place in modern history. The history of the use of cast iron pipe goes back more than 300 years, and the progress since then has brought metallic products into a prominent position for the transportation of fluids. In today's industrialized world, the need for piping products has become enormous. With ever-increasing operating pressure and temperature ratings, the need for new products has increased. So it is not surprising that old metallic piping products are being improved and that new ones are being developed constantly.

The variety of metallic piping materials, be they standard shelf items or manufactured for a special purpose, make it a necessity to provide a general oversight over these materials only. They do not allow space to give the detailed information that some materials merit and which is required when making an engineering decision regarding their use. When a selection must be made and a decision is required as to the better suitability of one material, today's technical libraries and the information provided by manufacturers assist greatly in providing the background data that are a prerequisite for the choice of the most appropriate material. The American Society for Testing Materials (ASTM) has done a yeoman's job in providing specifications and listings for a large variety of metallic piping materials for a spectrum of services. Most of these specifications have been incorporated in ANSI's roster of standards. Together with standards issued by other organizations, most of them are now used as reference standards in various sections of the ANSI Code for Pressure Piping. However, not all standards are incorporated in ANSI. In particular, standards for the petroleum industry as developed by the American Petroleum Institute (API) remain as a guide for that industry. On the other hand, the various manufacturers and particularly those that furnish corrosion or temperature resisting piping materials are providing on a routine basis information regarding their various products and indicate corrosion resistance in specified applications with ratings indicative of the products' performance. Appendix 7 has been compiled from data issued by various manufacturers and can serve as general guidance in selecting a material, a choice that should be confirmed through competent technical advice.

The various grades of steel represent the lion's share of all metallic piping materials in use today. The continuous construction and operation of piping systems, based on a proven service record, is indicative of the resistance to innovation until convinced that a new or improved product will perform better. This better performance can be either operational or costwise, but it always takes a considerable length of time to prove the better suitability of the new product.

When a new piping product is being considered, it is only natural that one will want to know the full details of its manufacture, size range, application, and so forth. That same knowledge is, in fact, also required when the usage of a well-known material with previous application under similar conditions is contemplated. In a similar fashion as a calculator prevents us to use our mental capabilities to solve arithmetical problems, so does the use of a previously prepared specification permit an engineer to specify piping materials without having a good knowledge of their actual composition and without knowing some subtleties that might restrict or enhance their application.

## PIPING MATERIALS AND THEIR MANUFACTURING METHODS

Different manufacturing processes are used for pipe and in the production of fittings and flanges. Therefore, it may be necessary to purchase pipe from one manufacturer, fittings and flanges from another, and valves yet again from one or more different sources. The pipe supply houses fill a very important role in this respect in that they can furnish all piping materials from a single distribution point. The chapter on metallic piping deals only with the methods of pipe fabrication and the standards applicable to such piping. Fittings, flanges, and valves are the subject of separate chapters.

### Cast-iron pressure piping

Two basic methods are being used in the production of cast-iron pressure piping, namely casting by pouring the molten iron into stationary molds, which may be either vertical or horizontal, or the use of centrifugal molds, which rotate while the molten iron is being poured, the centrifugal force thus exerted produces a perfect casting product.

Centrifugal casting, which is a modern adaption of a century-old art, is further identified through the differentiation of the type of mold that is being employed in the process. The mold is so called because it determines the shape of finished product. In centrifugal casting, it can be either sand lined or it may be a water-cooled metal mold, the latter a process that has been in commercial use in the US for over 50 years. The iron for pressure pipe casting may have a varying carbon content from 3% to 3.75%. Low carbon content results in hard iron, and high carbon produces soft iron. Minimal quantities or sulphur, phosphorus, and manganese are also present in cast iron. Foundry records are required to determine the exact chemical composition of the cast iron, and some standards set a limit of permissible sulphur and phosphorous content. However, the main criteria for conformance to the former AWWA and now ANSI specifications is the meeting of various acceptance tests as prescribed in these standards, which are written for the different manufacturing methods.

Piping for water is normally coated inside and outside with a very thin layer of tar or cement lined at the option of the purchaser. Centrifugally cast cast-iron piping is furnished in sizes up to 48 in. and in length up to 20 feet, depending on the method of manufacture. There is practically no limit on pipe sizes for pit-cast piping; however, ANSI specifications for this process are no longer applicable. The following specifications are applicable to centrifugally cast piping:

ANSI/AWWA C106-75 for pipe cast in metal molds
ANSI/AWWA C108-75 for pipe cast in sand-lined molds

### Cast-iron soil pipe

This type of piping normally produced through the pit-casting process is used only for plumbing and other nonpressure drainage application. Soil piping is normally designated as standard or extra-heavy weight and is manufactured in lengths up to 5 feet. Material specifications refer to ANSI/ASTM A74.

### Ductile cast-iron piping

Ductile cast-iron pipe combines the long life of cast iron with the physical strength of mild steel and has the ability to bend under stress. Its corrosion resistance is about the same as cast iron, and it is virtually unbreakable in day-to-day service. Ductile iron results from the addition of magnesium alloy to molten iron with low sulphur and phosphorous content. It is manufactured by the centrifugal casting process, mostly using metal molds.

The physical and chemical requirements for ductile cast-iron pipe for general service are spelled out in ANSI/AWWA C-151, while ANSI A21.52 deals specifically with ductile cast-iron pipe for gas service, where it is used to a large extent.

## Wrought steel piping

While the use of the terminology "carbon steel pipe" is accepted practice, the correct term "wrought steel pipe", which is indicative of the manufacturing process as opposed to cast-iron pipe, is also referred to in many specifications. Unfortunately, there often arises confusion with wrought-iron pipe, which is a specialized product and which is being dealt with separately.

Steel is classified as carbon steel when its content of manganese, silicon, and copper is within prescribed guidelines, and where there is no minimum content specified for aluminum, chromium, columbium, molybdenum, nickel, titanium, tungsten, vanadium, or zirconium.

Some of the various manufacturing methods for carbon steel pipe are described in more detail below. However, unlike cast-iron piping, the ANSI and API specifications for carbon steel piping in some instances permit more than one method of manufacture for its product. ANSI specifications for carbon steel pipe were mostly promulgated under the auspices of the American Society of Testing Materials (ASTM) and are referred to mostly as ANSI/ASTM standard designations. API specifications were developed especially for the petroleum industry. These specifications are very detailed, and a sample index of some of the topics addressed in promulgating such a specification is given below:

1. Scope
2. Process
3. Chemical composition
4. Ladle analysis
5. Check analysis
6. Tensile properties
7. Bending properties
8. Flattening tests
9. Hydrostatic tests
10. Test specimens and method of testing
11. Number of tests
12. Retests
13. Dimensions
14. Permissible variations in weight and dimensions
15. Lengths
16. Ends
17. Finish
18. Marking
19. Inspection
20. Rejection

There are six basically different methods of pipe manufacturing in use that produce most of today's steel piping. Similar manufacturing methods are used to produce other metallic piping.

## Butt-welded pipe (furnace welded)

This pipe is manufactured from flat strips of steel called *skelp,* with square or slightly bevelled edges. The skelp is mostly produced from steel with high phosphorous content, which is best suitable for furnace welding. It is furnace heated full length to welding temperature and then drawn through a funnel welding die, which forms and welds the pipe both in one step. As an alternate measure, welding rolls can also be employed (Fig. 1).

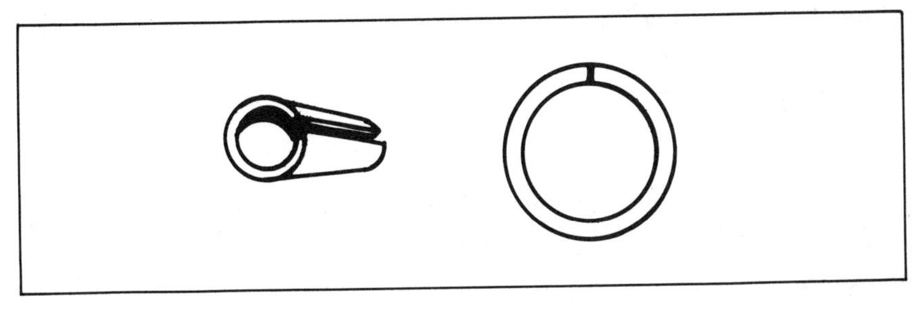

**FIG. 1. Butt-welded pipe.**

Additional rolling then straightens and finishes the pipe. This pipe is normally manufactured in sizes from $^1/_8$ in. through 4 in., and is lowest in cost among the various types of steel piping available for use in pressure systems. ANSI/ASTM specifications A-53 and A-120 relate to this type of piping.

## Lap-welded pipe (furnace welded)

Lap-welded pipe is also manufactured from skelp, but the ends, which have been scarfed, overlap in this process instead of being butted together.

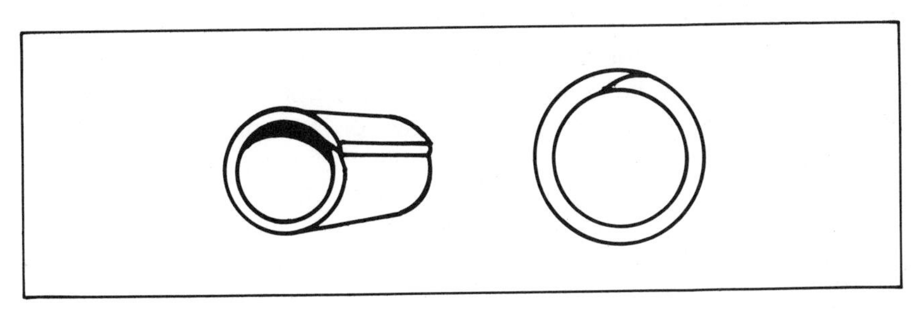

**FIG. 2. Lap-welded pipe.**

The skelp is first heated and shaped into a tubular form then reheated to welding temperature, slid over a mandrel, and welded through the compression of two grooved welding rolls that compress the pipe and achieve a furnace weld. Additional rolling completes the manufacturing process (Fig. 2). Pipe sizes normally range from 4 in. to 16 in., and most manufacturing is done to meet ANSI/ASTM specifications A53 and A120.

## Electric fusion welded pipe (EFW)

In this process, plate with suitably prepared edges is first hot or cold rolled into a tubular shape. The resulting longitudinal opening is then welded together, with or without additional filler material being deposited at the same time (Fig. 3). Electric arc welding can be manually or auto-

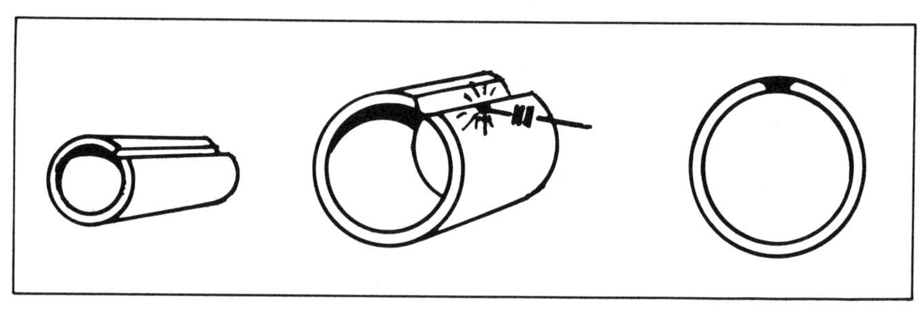

**FIG. 3. Electric arc-welded pipe (single-welded joint).**

matically performed and may be of the single or double-weld joint type, depending on plate thickness. Minimum size for this type of pipe is normally 4 in., but there is practically no upper size limit for this type of pipe. ANSI/ASTM specifications A-134, A-139, and A-672 are applicable for this manufacturing process.

## Electric resistance welded pipe (ERW)

Similar to the electric fusion-welding process, plate is first rolled into a tubular form. The welding operation is then performed at the same time while the tube is being compressed by two or more pressure rollers. The whole operation can be performed without preheating the plate or pipe since the welding process employed does not require such a prerequisite. Pipe in sizes from 1/2 in. to 30 in. is normally available and manufactured in accordance with this procedure and meeting ANSI/ASTM A-53 or A-135 or API 5L.

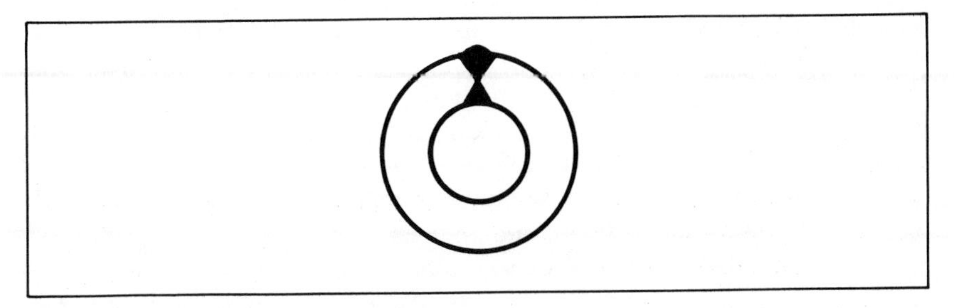

**FIG. 4. Electric arc-welded pipe (double-welded joint).**

## Seamless pipe

Two different processes are in use for the production of seamless piping and tubing products. They are identified as the hot-piercing process (Fig. 5) and the cupping process (Fig. 6).

The hot piercing process starts with a round bar, billet, or bloom (all different names for a similarly unfinished steel product), which is heated to a temperature of over 2,000°F. It is then pierced and forced over a short mandrel by revolving rolls. The initial product is a short, thick-walled pipe

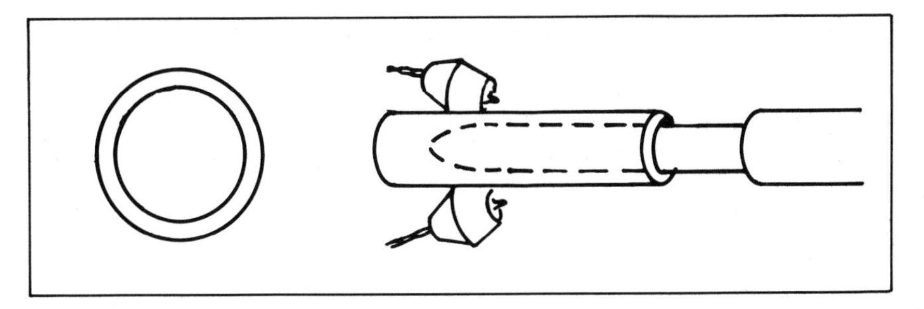

**FIG. 5. Seamless pipe (hot pierced).**

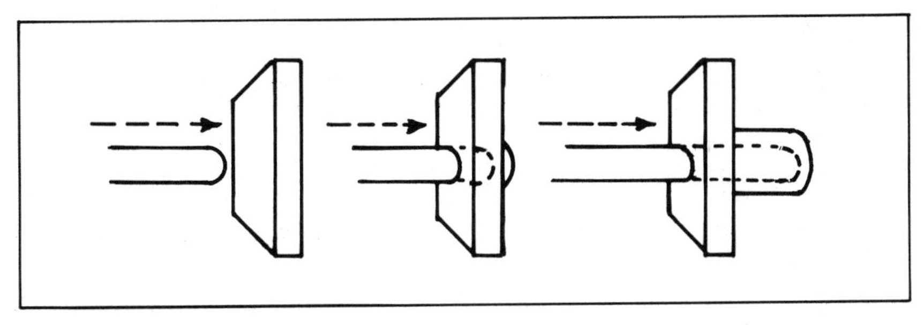

**FIG. 6. Seamless pipe (hot cupped).**

that, through a continuing process of either hot rolling or hot drawing, is finally brought to the desired size.

In the alternate hot-cupping method, a steel plate heated to forging temperature is placed against a bottom die and a round nosed plunger is pushed through. The emerging cup is repeatedly heated and forced through smaller dies, while a closed end remains. This closed end is finally cut off and the resultant pipe is straightened.

While seamless steel pipe can be manufactured to meet various other specifications, ANSI/ASTM A-106 and API 5L are widely preferred standards and are regarded as the mainstay of the power and petroleum industry.

## Spiral-welded pipe

As the name implies, steel strips are spirally wound to form long cylinders (Fig. 7). The edges of the steel, which may abut or overlap, are then butt welded or fillet welded together by the electric arc method. This pipe,

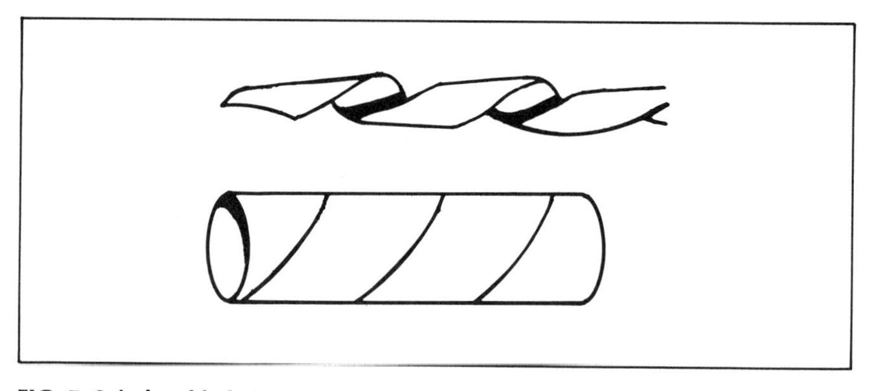

**FIG. 7. Spiral-welded pipe.**

which is mostly manufactured as a thin-walled product, is available in sizes up to 48 in. and in lengths up to 60 feet long. ANSI/ASTM specification A-211 and API specification 5LS were specially incepted for the production of this type of pipe.

## Carbon steel tubing

The very extensive use that is being made of large-sized steel tubing in the drilling and maintenance of oil and gas wells requires incorporation in this chapter. In particular, the American Petroleum Institute (API) has promulgated standards that are applicable to the use of casing and tubing in the drilling industry. Since, however, different but previously discussed manufacturing methods are permitted under these standards, only a listing of the various applicable specifications follows:

API 5A Casing, Tubing and Drill Pipe
API 5AC Restricted Yield Strength Casing and Tubing
API 5AX High Strength Casing, Tubing and Drill Pipe

## Wrought-iron piping

As previously pointed out, genuine wrought-iron pipe is a derivation of carbon steel pipe, which has better corrosion resisting properties. The word "genuine" is often used in order to ensure that the correct material is being referred to. The manufacturing process is either through butt or lap welding. A low carbon content and the inclusion of about 3% of iron silicate (slag) tend to give the finished pipe a tough, fibrous structure and lend substance to the claim for better corrosion resistance. ANSI standards for the manufacture of this type of piping (A-72) are now obsolete since alternate products have replaced the use of genuine wrought-iron pipe.

## Intermediate alloy steel pipe for low-temperature service

While stainless steel and aluminum are often used for low-temperature service, carbon steel piping with nickel and/or various other alloy content is also often used for this purpose. The manufacturing methods may be either seamless or welded; however, special testing is required during and after manufacture. This is known as the notched-bar impact test. Rigid specifications must be complied with in order to achieve satisfactory results. ANSI/ASTM specifications A-333 deals with varying grades for this service (Table 7).

## Intermediate ferritic alloy steels

Alloy piping is the nomenclature normally used for chrome-moly alloy steel piping that is the workhorse for high pressure, high-temperature piping in power plants, and oil refineries. The various types of this pipe are usually identified by their P-number, such as P-11 or P-22, or by the percentage of chrome and molybdenum content, such as 1¼ CR–½ moly or 2¼ CR–½ moly. The various grades all contain some molybdenum, while most but not all contain chrome. ANSI/ASTM specification A-335 covers the material requirements for this type of pipe, which is also known as the P-group (Table 8).

**TABLE 7** ANSI/ASTM A 333 chemical requirements for steel pipe for low temperature service

| Chemical Requirements | | | | | | | |
|---|---|---|---|---|---|---|---|
| Element | Composition, % | | | | | | |
| | Grade 1[A] | Grade 3 | Grade 4 | Grade 6[A] | Grade 7 | Grade 8 | Grade 9 |
| Carbon, max | 0.30 | 0.19 | 0.12 | 0.30 | 0.19 | 0.13 | 0.20 |
| Manganese | 0.40–1.06 | 0.31–0.64 | 0.50–1.05 | 0.29–1.06 | 0.90 max | 0.90 max | 0.40–1.06 |
| Phosphorus, max | 0.05 | 0.05 | 0.04 | 0.048 | 0.04 | 0.045 | 0.045 |
| Sulfur, max | 0.06 | 0.05 | 0.04 | 0.058 | 0.05 | 0.045 | 0.050 |
| Silicon | ... | 0.18–0.37 | 0.08–0.37 | 0.10 min | 0.13–0.32 | 0.13–0.32 | ... |
| Nickel | ... | 3.18–3.82 | 0.47–0.98 | ... | 2.03–2.57 | 8.40–9.60 | 1.60–2.24 |
| Chromium | ... | ... | 0.44–1.01 | ... | ... | ... | ... |
| Copper | ... | ... | 0.40–0.75 | ... | ... | ... | 0.75–1.25 |
| Aluminum | ... | ... | 0.04–0.30 | ... | ... | ... | ... |

[A]For each reduction of 0.01% carbon below 0.30%, an increase of 0.05% maganese above 1.06% would be permitted to a maximum of 1.35% manganese.

**TABLE 8** ANSI/ASTM A 335 chemical requirements for low and intermediate alloy piping

Chemical Requirements

| Grade | UNS Designation[A] | Composition, % | | | | | | | |
|---|---|---|---|---|---|---|---|---|---|
| | | Carbon | Manganese | Phosphorus, max | Sulfur, max | Silicon | Chromium | Molybdenum | Titanium or Columbium |
| P1 | K11522 | 0.10–0.20 | 0.30–0.80 | 0.045 | 0.045 | 0.10–0.50 | ... | 0.44–0.65 | ... |
| P2 | K11547 | 0.10–0.20 | 0.30–0.61 | 0.045 | 0.045 | 0.10–0.30 | 0.50–0.81 | 0.44–0.65 | ... |
| P5 | K41545 | 0.15 max | 0.30–0.60 | 0.030 | 0.030 | 0.50 max | 4.00–6.00 | 0.45–0.65 | ... |
| P5b | K51545 | 0.15 max | 0.30–0.60 | 0.030 | 0.030 | 1.00–2.00 | 4.00–6.00 | 0.45–0.65 | ... |
| P5c | K41245 | 0.12 max | 0.30–0.60 | 0.030 | 0.030 | 0.50 max | 4.00–6.00 | 0.45–0.65 | ...[B] |
| P7 | K61595 | 0.15 max | 0.30–0.60 | 0.030 | 0.030 | 0.50–1.00 | 6.00–8.00 | 0.44–0.65 | ... |
| P9 | K81590 | 0.15 max | 0.30–0.60 | 0.030 | 0.030 | 0.25–1.00 | 8.00–10.00 | 0.90–1.10 | ... |
| P11 | K11597 | 0.15 max | 0.30–0.60 | 0.030 | 0.030 | 0.50–1.00 | 1.00–1.50 | 0.44–0.65 | ... |
| P12 | K11562 | 0.15 max | 0.30–0.61 | 0.045 | 0.045 | 0.50 max | 0.80–1.25 | 0.44–0.65 | ... |
| P15 | K11578 | 0.15 max | 0.30–0.60 | 0.030 | 0.030 | 1.15–1.65 | ... | 0.44–0.65 | ... |
| P21 | K31545 | 0.15 max | 0.30–0.60 | 0.030 | 0.030 | 0.50 max | 2.65–3.35 | 0.80–1.06 | ... |
| P22 | K21590 | 0.15 max | 0.30–0.60 | 0.030 | 0.030 | 0.50 max | 1.90–2.60 | 0.87–1.13 | ... |

[A]New designation established in accordance with ASTM E 527 and SAE J1086. Recommended Practice for Numbering Metals and Alloys (UNS).
[B]Grade P5c shall have a titanium content of not less than 4 times the carbon content and not more than 0.70%; or a columbium content of 8 to 10 times the carbon content.

## Austenitic stainless steel pipe

This is the group of chrome-nickel alloys also being referred to as Series 300 stainless steels. Piping manufactured from this alloy is produced in both welded and seamless forms. Various manufacturers have adopted special welding and finishing methods, which include grinding down weld beads in welded pipe. Most piping requires annealing and pickling after fabrication, and these requirements are incorporated in the piping specifications.

The most common grades of austenitic stainless steel, such as 304, 316, 321, or 347 contain various amounts of chrome, nickel and other metals (Table 9). There are also different material compositions in a certain type itself. For instance, by reducing the carbon content in the metal, the material is qualified to be graded ELC (extra-low carbon). ELC material allows better weldable quality.

The size range for stainless steel pipe today is no more limited than carbon steel pipe use. Stainless steel is being used in chemical plants, paper mills, and numerous other applications because of its resistance to corrosive attacks. In the dairy industry, it is preferred because it can also be an exceedingly clean material. Some of the most often used specifications are ANSI/ASTM A-312, A-358, A-376, and A-409.

## Ferritic stainless-steel pipe

This refers to the piping that is also known as Class 400 stainless steel. It is composed of hardenable and nonhardenable alloys. The varying grades of piping in this classification have either no nickel or a very minimal nickel content and a minimum of 12% chrome content. The weldability of these materials is not the best and leaves much room for improvement. Use of piping of this class is, therefore, held to a minimum, where welding is a fabrication requirement. ANSI/ASTM specification A-405 is an applicable standard.

## Abrasion-resistant ferritic alloys

This type of piping, which is produced in a similar process as cast iron, is mostly known by the various trade names that manufacturers have bestowed on it, such as Ni-hard, Ashcolite, Xtec, or Nuvalloy. It is best recognized by its high nickel content. This alloy achieves an exceptional brinnell hardness, which may go as high as 750. This particular hardness withstands the abrasive actions of slurry-like fluids, and thus makes it a favorite material for power-plant bottom ash disposal or mine tailings.

Installation of pipelines made of this material is mostly connected with mechanical connectors, which permit the pipe to be rotated as much as

**TABLE 9** Chemical requirements for austenitic stainless steel pipe, ANSI/ASTM A 312

| | Chemical Requirements | | | | | | | | | | | | |
|---|---|---|---|---|---|---|---|---|---|---|---|---|---|
| | Composition, % | | | | | | | | | | | | |
| Grade | Carbon, max[G] | Manganese, max[G] | Phosphorus, max | Sulfur, max | Silicon, max | Nickel | Chromium | Molybdenum | Titanium | Columbium plus Tantalum | Tantalum, max | Nitrogen[F] | Vanadium |
| TP 304 | 0.08 | 2.00 | 0.040 | 0.030 | 0.75 | 8.00–11.0 | 18.0–20.0 | ... | ... | ... | ... | ... | ... |
| TP 304H | 0.04–0.10 | 2.00 | 0.040 | 0.030 | 0.75 | 8.00–11.0 | 18.0–20.0 | ... | ... | ... | ... | ... | ... |
| TP 304L | 0.035[A] | 2.00 | 0.040 | 0.030 | 0.75 | 8.00–13.0 | 18.0–20.0 | ... | ... | ... | ... | ... | ... |
| TP 304N | 0.08 | 2.00 | 0.040 | 0.030 | 0.75 | 8.00–11.0 | 18.0–20.0 | ... | ... | ... | ... | 0.10–0.16 | ... |
| TP 304LN | 0.035 | 2.00 | 0.040 | 0.030 | 0.75 | 8.00–11.0 | 18.0–20.0 | ... | ... | ... | ... | 0.10–0.16 | ... |
| TP 309 | 0.15 | 2.00 | 0.040 | 0.030 | 0.75 | 12.0–15.0 | 22.0–24.0 | ... | ... | ... | ... | ... | ... |
| TP 310 | 0.15 | 2.00 | 0.040 | 0.030 | 0.75 | 19.0–22.0 | 24.0–26.0 | ... | ... | ... | ... | ... | ... |
| TP 316 | 0.08 | 2.00 | 0.040 | 0.030 | 0.75 | 11.0–14.0 | 16.0–18.0 | 2.00–2.00 | ... | ... | ... | ... | ... |
| TP 316H | 0.04–0.10 | 2.00 | 0.040 | 0.030 | 0.75 | 11.0–14.0 | 16.0–18.0 | 2.00–3.00 | ... | ... | ... | ... | ... |
| TP 316L | 0.035[A] | 2.00 | 0.040 | 0.030 | 0.75 | 10.0–15.0 | 16.0–18.0 | 2.00–3.00 | ... | ... | ... | ... | ... |
| TP 316N | 0.08 | 2.00 | 0.040 | 0.030 | 0.75 | 11.0–14.0 | 16.0–18.0 | 2.00–3.00 | | | | 0.10–0.16 | ... |
| TP 316LN | 0.035 | 2.00 | 0.040 | 0.030 | 0.75 | 11.0–14.0 | 16.0–18.0 | 2.00–3.00 | ... | ... | ... | 0.10–0.16 | ... |
| TP 317 | 0.08 | 2.00 | 0.040 | 0.030 | 0.75 | 11.0–14.0 | 18.0–20.0 | 3.00–4.00 | ... | ... | ... | ... | ... |
| TP 321 | 0.08 | 2.00 | 0.040 | 0.030 | 0.75 | 9.00–13.0 | 17.0–20.0 | ... | [B] | ... | ... | ... | ... |
| TP 321H | 0.04–0.10 | 2.00 | 0.040 | 0.030 | 0.75 | 9.00–13.0 | 17.0–20.0 | ... | [C] | ... | ... | ... | ... |
| TP 347 | 0.08 | 2.00 | 0.040 | 0.030 | 0.75 | 9.00–13.0 | 17.0–20.0 | ... | ... | [D] | ... | ... | ... |
| TP 347H | 0.04–0.10 | 2.00 | 0.040 | 0.030 | 0.75 | 9.00–13.0 | 17.0–20.0 | ... | ... | [E] | ... | ... | ... |
| TP 348 | 0.08 | 2.00 | 0.040 | 0.030 | 0.75 | 9.00–13.0 | 17.0–20.0 | ... | ... | [D] | 0.10 | ... | ... |
| TP 348H | 0.04–0.10 | 2.00 | 0.040 | 0.030 | 0.75 | 9.00–13.0 | 17.0–20.0 | ... | ... | [E] | 0.10 | ... | ... |
| TP XM-10 | 0.08 | 8.00–10.00 | 0.040 | 0.030 | 1.00 | 5.50–7.50 | 19.00–21.50 | ... | ... | ... | ... | 0.15–0.40 | ... |
| TP XM-11 | 0.04 | 8.00–10.00 | 0.040 | 0.030 | 1.00 | 5.50–7.50 | 19.00–21.50 | ... | ... | ... | ... | 0.15–0.40 | ... |
| TP XM-15 | 0.08 | 2.00 | 0.030 | 0.030 | 1.50–2.50 | 17.50–18.50 | 17.0–19.0 | ... | ... | ... | ... | ... | ... |
| TP XM-19 | 0.060 | 4.00–6.00 | 0.040 | 0.030 | 1.00 | 11.50–13.50 | 20.50–23.50 | 1.50–3.00 | ... | 0.10–0.30 | ... | 0.20–0.40 | 0.10–0.30 |
| TP XM-29 | 0.060 | 11.50–14.50 | 0.060 | 0.030 | 1.00 | 2.25–3.75 | 17.0–19.0 | ... | ... | ... | ... | 0.20–0.40 | ... |

[A]For small diameter or thin walls or both, where many drawing passes are required, a carbon maximum of 0.040% is necessary in grades TP 304L and TP 316L. Small outside diameter tubes are defined as those less than 0.500 in. (12.7 mm) in outside diameter and light wall tubes as those less than 0.049 in. (1.24 mm) in average wall thickness (0.044 in. (1.12 mm) in minimum wall thickness).
[B]The titanium content shall be not less than five times the carbon content and not more than 0.70%.
[C]The titanium content shall be not less than four times the carbon and not more than 0.60%.
[D]The columbium plus tantalum content shall be not less than ten times the carbon content and not more than 1.00%.
[E]The columbium plus tantalum content shall be not less than eight times the carbon content and not more than 1.0%.
[F]The method of analysis for nitrogen shall be a matter of agreement between the purchaser and manufacturer.
[G]Maximum, unless otherwise indicated.

180° to give the system a longer life after the bottom of the pipe has suffered from some erosion. No ANSI specification for size or material composition exists for this type of piping, and manufacturers' standards have to be relied upon for specifying materials.

## Nickel and nickel alloys

The various grades of nickel and monel pipe are manufactured as either seamless or welded pipe, as are inconel and hastelloy, all of which contain more than 60% nickel. Welding of these materials no longer presents any problems. They are normally furnished in a low-temperature stress relieved temper but can also be furnished in the drawn or soft-annealed state.

Other alloys with an approximate 25% nickel and 25%–30% chrome content again are mostly recognizable by their trade names, such as carpenter 20, durimet 20, or worthite. All are also of a weldable quality.

These types of pipe are mostly referred to under the same terminology for pipe wall thickness that is applicable to stainless steel, although the terminology of standard or extra strong is also in usage, particular in smaller sizes where both standards refer to the same thickness. Specifications for nickel pipe are ANSI/ASTM B-161 and for some of the alloys ANSI/ASTM B-167, B-407, and B-513, as well as others.

## Aluminum

Several grades of aluminum pipe are on the market today. They range from almost pure aluminum through various alloys, which may contain magnesium, silicon, and even chrome. Most aluminum piping is fabricated as seamless drawn or extruded piping, but welded piping is not uncommon, although it is not covered by an ANSI specification. The various grades of aluminum piping, which were previously known by a number with an S suffix, have been renamed:

| *Old Designation* | *New Number* |
|---|---|
| 3S | 3003 |
| 61S | 6061 |
| 63S | 6063 |
| A54S | 5154 |
| 2S | 1100 |

Welding of aluminum piping, which used to present some problems years ago, is now an every-day practice. Specification ANSI/ASTM B-241, B-210 (for tubing) and B-345 are the most widely used, as well as specifications B-234, which is applicable to tubing. Pipe is normally available in sizes up to 20 in. For wall-thickness designations, the schedule numbers of the ANSI standard for wrought steel pipe B36.10 as well as the ANSI standard for stainless steel pipe ANSI B36.19 (for Schedules 5S and 10S only) are being used.

### Copper and its alloys

Copper and its various alloys are often used in tubing as well as in pipe sizes for pressure and nonpressure applications. While some seamless copper pipe and tubing is manufactured in a manner similar to steel piping, another method involves cold drawing and later annealing. Piping is usually referred to as standard and extra strong, while the various tubing designations are also indicative of wall thickness. Pipe and most tubing can all be furnished in 20-foot straight length, but types K and L tubing can also be furnished in coils, and ACR or refrigeration tubing is furnished only in coiled form.

## SERVICE DESIGNATIONS FOR TUBING

Types K & L tubing are for water service.
Type M tubing is for interior heating and nonpressure applications.
Type DMV tubing is for drainage, waste, and vent lines above ground.
Type ACR tubing is for refrigeration and air conditioning service.
Type TP threadless pipe is for plumbing.
ANSI/ASTM specification B-42 covers copper pipe.
ANSI/ASTM specification B-466 covers cupronickel pipe.
ANSI/ASTM specification B-315 covers copper-silicon (Everdur pipe).

A number of other specifications are also applicable to copper and its various alloys.

## METALLIC PIPING APPLICATION

After reviewing a variety of piping materials, their methods of manufacture, and their commercial availability, it now behooves us to take a closer look at the practical uses of some of the materials.

Most of the referenced materials have already been proven in a variety of applications and have a good performance track record when compared to another material in a specific service where temperature, pressure and corrosion resistance or a combination thereof are part and parcel of a material and cost evaluation process.

**Listing of various metallic materials of construction for piping systems and some of their applications***

| Material | Application |
|---|---|
| Admiralty metal | — Used in exchanger tubes in contact with fresh and salt water. Resistant to corrosive waters. |
| Aluminum 1100 | — Low in strength, most readily welded aluminum type. Resistant to formaldehyde, ammonia, dyes, phenol, hydrogen sulfide. Used in food plants. |
| Aluminum 3003 | — Better mechanical properties because of manganese content, used for chemical plants. |
| Aluminum 6061 or 6063 | — Containing silicon and magnesium additives, highest resistance to corrosion. |
| Brass, red | — Resistant to corrosive waters. |
| Bronze, silicon | — Resistant to brines, sulfite solutions, sugar solutions and organic acids. |
| Copper | — Resistant to corrosive waters. |
| Cupro-Nickel, 70-30 | — Similar to metal used in U.S. nickel coin. Resistant to salt water corrosion, even when flowing at high velocity. |
| Duriron, Corrosiron | — Available only as cast piping. Resists nitric, sulfuric and acetic acids. Used for corrosive waste and pressure applications. |
| Durimet 20, Worthite, Carpenter 20 | — Extremely resistant to sulfuric acid. |
| Hastelloy B | — Resistant to corrosive effects of boiling hydrochloric acid and wet hydrochloric-acid gas. |
| Hastelloy C | — Resistant to phosphoric, acetic, formic, nitric and sulfurous acids, as well as free chlorine. Resists thermal shock at high temperatures to 1800F. |
| Incoloy | — Resistant to oxidation and carburation at elevated temperatures. |
| Inconel | — Resistant to oxidizing and reducing solutions at elevated temperatures. |
| Iron, cast | — Good resistance to corrosion. Extensively used for water and gas distribution, and for sewage systems. |

*Adapted from J. A. Masek's "Metallic Piping." *Chemical Engineering,* 17 June 1968.

Iron, ductile — Similar anti-corrosive properties as cast iron, but has stronger mechanical properties.

Lead — High resistance to corrosion when in soluble coating forms such as lead sulfate, carbonate or phosphate.

Monel 400 — High strength nickel based alloy generally recommended under reducing rather than oxidizing conditions. Can handle alkaline solutions and airfree acids, whenever copper contamination is no problem.

Monel 500 — Retains strength also at elevated temperatures.

Nickel — Highly resistant to most corrosive environments, also available in a version containing low carbon.

NI-Hard — Castings used in slurry service, because of abrasion resistant properties.

Stainless Steel, Austenitic, Type 304 — Most common of the stainless steels used in process piping. Resists corrosion and provides sanitary conditions for food and drug industries. Extra low carbon (ELC) grade provides better weldability and resists intergranular corrosion.

Stainless Steel, Austenitic, Types 316, 321 or 347 — Also available in ELC grades. These materials provide better corrosion resistance at elevated temperatures and pressures.

Tantalum — Resistant to nitric and other acids.

Titanium — High corrosion resistance in oxidizing media. Resists hypochlorites, 30% sulfuric acid and perchlorites, also resists abrasion and erosion by cavitation.

## PIPING IDENTIFICATION

It is fairly easy to differentiate between carbon steel and stainless steel or copper pipe. However, when it is necessary to identify one carbon steel grade from another carbon steel or alloy piping grade, difficulties become apparent. Many of the various piping specifications require marking pipe; however, this is often done only at the end of a length of pipe.

Stainless steel piping is mostly marked continuously along the entire length of pipe so that even after cutting the remaining pipe is easily identified. In power plants and oil refineries, where often various grades of carbon steel and alloy steel piping are used, it is often essential that such piping be color coded for easier determination of piping grade. Such color coding during the construction process is often achieved by painting rings around the pipe or painting stripes lengthwise. Color coding during construction is not applied on a uniform basis, although the Pipe Fabrication Institute (PFI) has issued a standard especially for that purpose.

Installed piping systems are also often color coded by the owner, and this is mostly achieved by painting the entire piping. Unless a metallic protection is covering insulation, insulated piping is often painted just as well as otherwise bare piping. Pipelines with metallic insulation protection as well as otherwise not identifiable piping are often provided with identification bands at lineal intervals that identify the fluid stream.

## PIPING SYSTEM SPECIFICATIONS

The various specifications that have been referred to all apply to pipe or tubing and its respective wall thickness. Bear in mind that specifications for a complete piping system involve more than just these specifications. Normally, fittings, flanges, valves, bolts, and gaskets all differ in their material composition for the pipe that form part of the same piping system in which these items are incorporated. There may also be differences in specifications between various pipe sizes. As an example, as to what such a specification for a single piping system might look like, refer to Table 10.

One of the important reference points in a piping-system specification is the designation of the pipe material itself. A comprehensive listing of various specifications for metallic pipe which have been published by ANSI and API is provided in Appendix 8.

**TABLE 10** Standard piping system specifications

| Item | Size Range In. | Sch/Wall Rating | ID Tag | Material | Type Ends | Catalog References or Remarks | Note |
|---|---|---|---|---|---|---|---|
| Pipe | 1/2 to 3/4 | Sch. 80 | P1A-2 | A53, B | PE | Seamless C.S. | a |
| | 1 to 6 | Sch. 40 | P1A-2 | A53, B | BE | Seamless C.S. | a |
| | 8 to 20 | 3/8″ W | P1A-2 | A53, B | BE | Seamless C.S. | a |
| | 22 to 24 | 3/8″ W | P1D-6 | A155, C55 | BE | EFW Carbon Steel | a |
| Nipple | 1/2 to 2 | XS | N1A-1 | A106, B | Thr'd | Seamless C.S. | a |
| Valves Globe | 1/2 to 2 | 600 LB | 600VGL3 | Frgd CS | SW | XYZ-SW-10 | m |
| | 2 1/2 to 8 | 150 LB | 150VGL5 | Cast CS | RF | XYZ-FE-3 | |
| Gate | 1/2 to 2 | 600 LB | 600VG7 | Frgd CS | SW | YZX-SW-8 | m |
| | 2 1/2 | 150 LB | 150VG5 | Cast CS | RF | YZX-FE-6 | |
| | 3 to 10 | 150 LB | 150VG8 | Cast CS | RF | YZX-FE-4 | |
| | 12 to 24 | 150 LB | 150VG9 | Cast CS | RF | YZX-FE-2 | |
| Angle | 1/2 to 2 | 600 LB | 600VA3 | Frgd CS | SW | ZXY-SW-10 | |
| | 2 1/2 to 8 | 150 LB | 150VA5 | Cast CS | RF | ZXY-FE-34 | |
| Plug | 1/2 to 2 | 300 LB | 300VP3 | Cast CS | SW | XYZ-SW-10 | d, e |
| | 2 1/2 to 4 | 150 LB | 150VP7 | Cast CS | RF | XYZ-FE-7 | e |
| | 6 to 8 | 150 LB | 150VP8 | Cast CS | RF | XYZ-FE-8 | e |
| | 10 to 12 | 150 LB | 150VP9 | Cast CS | RF | XYZ-FE-12 | e |
| | 14 to 24 | 150 LB | 150VP4 | Cast CS | RF | XYZ-FE-15 | e |
| Check | 1/2 to 2 | 600 LB | 600VC3 | Frgd CS | SW | YZX-SW-8 | Lift |
| | 2 1/2 to 14 | 150 LB | 150VC6 | Cast CS | RF | YZX-FE-12 | Swing |
| Fittings | 1/2 to 2 | 3000 LB | FSW-5 | A105 | SW | BPH-15 | f |
| | 2 1/2 to 6 | Sch. 40 | BWS-7 | A234 | BW | CPH | f |
| | 8 to 24 | 3/8″ W | BWS-7 | A234 | BW | CPH | f |
| | 1/2 to 2 | 3000 LB | FSS-5 | A105 | Scr'd | BPH-10 | g |

**TABLE 10** *continued*

| Item | Size Range In. | Sch/Wall Rating | ID Tag: | Material | Type Ends | Catalog References or Remarks | Note |
|---|---|---|---|---|---|---|---|
| Unions | 1/2 to 2 | 3000 LB | USS-5 | A105 | Scr'd | BPH-12 | |
| Flanges | 1/2 to 2 | 150 LB | 150FG2 | A181 | SW-RF | ABC | l |
| | 2 1/2 to 24 | 150 LB | 150FG4 | A181 | SO-RF | ABC | l |
| | 2 1/2 to 24 | 300 LB | 300FG6 | A181 | SO-RF | ABC | h, 1 |
| Gaskets | 1/2 to 24 | 1/16″ | GAS-5 | Fiber | Ring | DEF-15 | i |
| | 1/2 to 24 | 0.175″ | GAS-7 | SS 304 | Spiral | EFG-8 | i, j |
| Bolts | — | — | Bolt-6 | A193, B7 Studs—A194, 2H Nuts | | | k |
| Joints | 1/2 to 2 | Coupling or Unions | | | | | |
| | 2 1/2 to 24 | Flanged or Welded | | | | | |

a. Specify schedule, pipe size and actual length.
b. Gear operated.
c. Maximum Temperature 350° F.
d. Furnished with 6″ long nipples on each end (A106 Grade B); screwed in and sealwelded to valve, exposed end of nipple Plain End for Socket-Welding.
e. Maximum Temperature 500° F.
f. Specify size and type of fitting.
g. For Instrumentation connections of Thermometers, Thermocouples and Thermowells only. Specify size of couplings.
h. For equipment connections.
i. Specify nominal pipe size and Flange rating and facing.
j. Must be used at temperatures above 750° F.
k. Specify size and length.
l. For blind flanges use Tag: 150FG23
m. For vents, drains and instrument connections, "Special" Screwed end valves may be used.

| Item | Size Range | Wall, Schedule or Rating | Material | Type Ends | Catalog No. | Notes |
|---|---|---|---|---|---|---|
| Pipe | 1/2"–1 1/2" | Sch. 80 | A-106 B | P.E. | | Smls |
| Pipe | 2"–10" | Sch. 40 | A-53 B | B.E. | | Smls |
| Pipe | 12"–24" | 3/8" (Std) | A-53 B | B.E. | | Smls |
| Fittings | 1/2"–1 1/2" | 3000 lbs | A-105 | S.W. | | Sch 80 |
| Fittings | 2"–10" | Sch. 40 | A-234 | B.W. | | |
| Fittings | 12"–24" | 3/8" (Std) | A-234 | B.W. | | |
| Flanges | 1/2"–1 1/2" | 150 lbs | A-181 | S.W. | | R.F. |
| Flanges | 2"–24" | 150 lbs | A-181 | S.O. | | R.F. |
| Globe valves | 1/2"–2" | 600 lbs | F.C.S. | S.W. | XYZ-17 | |
| Globe valves | 2 1/2"–8" | 150 lbs | Cast Stl | W.E. | XYZ-19 | |
| Gate valves | 1/2"–2" | 600 lbs | F.C.S. | S.W. | XYZ-25 | |
| Gate valves | 2 1/2"–24" | 150 lbs | Cast Stl | W.E. | XYZ-27 | |
| Check valves | 1/2"–2" | 600 lbs | F.S. | S.W. | XYZ-31 | Swing |
| Check valves | 2 1/2"–12" | 150 lbs | Cast Stl | W.E. | XYZ-43 | Swing |
| Plug valves | 1/2"–1 1/2" | 300 lbs | Cast Stl | Scrd | XYZ-52 | |
| Plug valves | 2"–24" | 150 lbs | Cast Stl | Flgd | XYZ-54 | R.F. |
| Gaskets | 1/2"–24" | 0.175" thk | SS Tp 304 | | TP-05 | |
| Bolting studs | | | A 193 B7 | | | |
| Bolting nuts | | | A 194 2H | | | |

Note: Gasket ID & OD to meet flange raised face dimensions

# 5

# Nonmetallic piping

Nonmetallic piping systems are often the least costly solution for many piping problems in the initial installation. But such materials should be used with caution, particularly in untried applications, and are recommended primarily for economic considerations.

In many instances, nonmetallic piping is obviously preferable because corrosion-resistant qualities and an inherent resistance to fluid contamination through absorption of metallic substances from the fluid-carrying pipe can be considered as very beneficial. These qualities can easily be translated into cost factors when performing a value analysis based on life-cycle costing.

## VALUE ANALYSIS

Cost or value analysis is of particular importance when reviewing the application of nonmetallic piping because a variety of factors that do not have to be considered in comparable metallic systems can and do play a significant role. Some of these factors, which are also quite difficult to evaluate from a cost point, are listed:

1. *Ignitibility*—Refers to the possibility of ignition or burning of the piping material itself through instantaneous combustion or by being affected from a fire in the vicinity of the location of the piping system.
2. *Breakage Susceptibility*—Considers the possibility of the pipe being broken accidentally prior to or during the installation or breakage occurring when the system is in usage.
3. *Support Requirements*—Are a necessity but to a lesser degree for metallic piping systems too and are indicative of the additional requirements that a fully operating nonmetallic piping system may demand in order to resist deflection outside of permissible limits.
4. *Thermal Shock Resistance*—Implies the ability of the piping to withstand considerable temperature differentials within a relative short time span.

5. *External Pressure Resistance*—Is of particular importance when reviewing underground piping systems and alludes to the capability to withstand external pressures without rupture or deformation, both when in service or prior to the introduction of fluid into a system.

Besides the above factors, other considerations that are applicable to metallic piping as well should also be included in a value analysis, and here the pros and cons of various materials should permit an easier evaluation.

1. *Corrosion Resistance*—The term so often mentioned that refers to the ability of a piping material to transmit a given fluid and resist any impairment resulting from the contact between piping and fluid. Corrosion resistance is a requirement for handling a multitude of chemicals.
2. *High- or Low-Temperature Resistance*—Refers to the ability of the piping material to withstand the temperature of any fluid that may be conveyed in the piping or the temperature of the surrounding elements.
3. *Construction Techniques*—Vary with each type of piping material and even within a material group, depending on what type of connection is being used. Lightweight plastic piping and an assumption of faster field installation may be offset by a lack of experience in handling a given pipe material.

## MATERIAL SELECTION

As of late, the use of nonmetallic piping and in particular plastic piping has been increasing steadily. The use of wood and clay products for water transportation dates back to early civilizations. Concrete, glass, stoneware, graphite, carbon, rubber, fabric, and asbestos-cement are some of the other materials available today. And then there are the plastic materials which had their beginning only in the 1930s and which abound at this time. The Society of the Plastic Industry Inc. lists more than 100 piping manufacturers and basic materials suppliers in its piping sections which comprise the Plastics Pipe Institute and the Machine-made Reinforced Plastic Pipe Producers.

To find the right piping material for any given process is a herculean task, and no engineer can be blamed when he falls back on materials that have a proven track record. The old adage "Be neither the last nor the first to try a new thing" describes perfectly the situation in which many a designer of piping systems finds himself at this time.

With new plastic piping products being constantly introduced, nobody wants to miss a good thing. However, one also needs some reassurance that the new product is not lacking in some of its requirements for a given piping system. And even after discovering a new piping material used heretofore, there comes the problem of value analysis. For something new to be introduced, it must show some cost effectiveness. Of course, an added impetus to look for a new material is when there is dissatisfaction with the material in use. Maintenance cost for a piping system may be high, corrosion or erosion may require an early replacement, installation practices may be unsatisfactory, and delays in material delivery can be very costly. All of these and other factors contribute to looking for a substitute, which may be just waiting around the corner to be discovered and put into use.

## PIPING MATERIALS

*Concrete Pipe*—There are two major classes of concrete pipe, sewer pipe and pressure pipe. The sewer pipe may be fabricated from plain concrete or be reinforced with wire mesh, while the pressure-type pipe, which is mostly used for water supply, is always reinforced with steel caging.

Joints for sewer pipe are mostly of the bell-joint type, which can be finished by applying cement mix, while pressure pipe joints preponderantly use rubber or plastic O-rings as an effective method of sealing grooved joints. There also exist methods today to manufacture concrete pipe at the point of installation as cast-in-place concrete pipe.

There are practically no size limitations on concrete pipe and fittings and transition pieces for connections to other materials can easily be furnished. While sewer pipe can be cut to a required size at the point of installation if necessary, pressure pipe has to be fully designed and furnished to exact length requirements. ANSI/ASTM specification C361 applies to pressure piping, while C14, C76, and C506 are some of the standards applicable to sewer pipe.

*Clay Pipe*—This pipe is mostly known as vitrified clay pipe or VCP and is manufactured in sizes up to 42 in. and is preponderantly used for sewage disposal. There are two types of this pipe being manufactured, namely standard and extra-strength pipe. The joints are mostly bell and spigot type, but compression couplings can also be used.

Besides the regular bell joints, which are filled with cement mix for sealing, the pipe is also manufactured with a plastic sealing ring already installed in the bell which acts as a compression joint when the end of another pipe is shoved in it. A common specification for clay pipe is ANSI/ASTM-C700.

*Wood Pipe*—Wood pipe is mostly manufactured from wood staves, but laminated plywood is also used. It is not very much in use today except for certain specialty applications (i.e., salt water handling for the extraction of salt or bromide). However, small water-storage tanks fabricated from wood staves are still very much in use.

*Graphite and Carbon Pipe*—This piping is being used in specialty applications where its resistance to high temperatures is of prime importance. The material is brittle, and great care must be used during its installation, particularly when joining two flanges together. The material is machineable, and piping is mostly threaded or flanged to make the necessary connections. There are only the manufacturers' standards available as a guide to size and wall thickness selection.

*Asbestos Cement*—This piping is much better known by one of its trade names, Transite. It has been in use for a considerable time now and is easy to install. Asbestos-cement pipe is being made for pressure-type applications and for sewer pipe. The standard length for this pipe material is 13 feet, and connections are being made through the use of couplings with rings inserted that act as a pressure seal.

While installation of this piping is relatively cheap because of its simplicity, testing or rather repairing leaks in pressure-type joints can be very costly. Cast-iron fittings are mostly used for directional changes in otherwise straight-pressure pipelines. The pressure piping is largely used in underground applications since the couplings cannot hold the pressure in above-ground installations unless all piping is solidly anchored.

ANSI/ASTM specifications C-428 and C-644 apply to nonpressure sewer pipe, while specification C-668 is a standard for transmission pipe. Separate ANSI/ASTM specifications are available for connecting and testing methods.

*Rubber Piping*—We are much more familiar with the term rubber hose than rubber pipe. The term pipe itself refers mostly to rigid materials, while hose implies flexibility. Rubber piping, which was previously used in a great many applications, has been replaced in most instances by plastic piping materials.

*Glass Pipe*—The major applications of glass piping are in laboratory use. But it is also used in process plants, where its cleanliness and high permissible working temperature (up to 200°C) are definite assets. Piping is also furnished as an armored or reinforced glass, which promises better protection against breakage.

Piping must be fully factory prefabricated for flanged field assembly (Fig. 8). Glass is one of the materials where possible breakage and support requirements have to be considered before selecting its use. While no ANSI standards are available for this product, manufacturers' catalogs with ample information regarding their products are readily attainable.

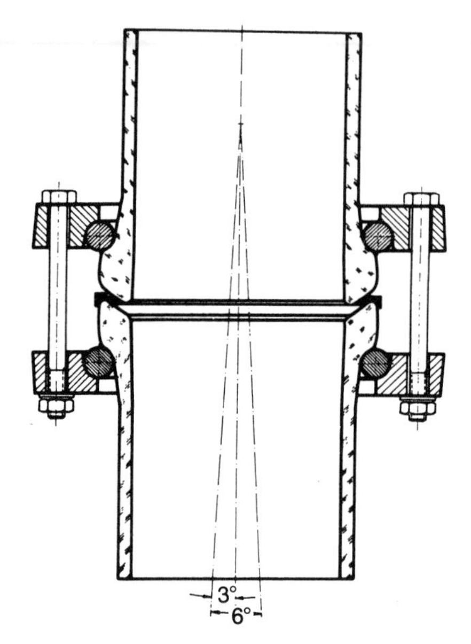


**FIG. 8. Glass pipe, flanged joint** *(courtesy OI/Schott Process Systems).*

## PLASTIC PIPING

Because of its low cost, good corrosion resistance, light weight, and ease of handling, plastic piping has succeeded, against great odds, to make considerable inroads into the American piping market. Long life expectancy of plastic piping, while achieved under laboratory conditions, has still to be proven in actual field installations. This spectacular leap into prominence and its acceptance by skeptical users of the wide variety of plastic materials within a relatively short period of time is an indication as to how well these materials must perform. Among the most widely used plastic piping materials available today are the following:

1. Polyvinyl Chloride (PVC)*
2. Polypropelene (PP)
3. Chlorinated Polyvinyl Chloride (CPVC)*
4. Polyethylene (PE)*
5. Acrylonitrile–Butadiene–Styrene (ABS)*
6. Polybutylene (PB)*
7. Glass fiber reinforced epoxy vinyl ester or polyester (FRP) or reinforced thermosetting line pipe (RTRP)*
8. Chlorinated Polyester*
9. Polyvinyledene Chloride*

10. Polyvinyledene Fluoride (PVDF)
11. Fluorinated Ethylene-Propylene (FEP)
12. Polytetra Fluoro Ethylene (TFE)
13. Reinforced or Armored Phenol Formaldehyde
14. Reinforced or Armored Furfuryl Alcohol
15. Polyester Lined Epoxy
16. Cellulose Acetate Butyrate (CAB)*
17. Poly Sulfone

(Piping materials marked with an asterisk (*) are those for which one or more ANSI/ASTM specifications have been developed and which are generally encompassed in D-2513, while ANSI/ASTM specification F-412 deals with plastic piping terminology.)

API specifications 5A for casing and tubing and 5LE, 5CP, and 5LR for piping have been promulgated for the use of plastics in the petroleum industry (Table 11). Those plastic piping materials for which no ANSI or API specifications as yet exist are mostly available under their various trade names.

As with most nonmetallic piping materials, manufacturers publish their own recommendations for use, size specifications, and installation techniques for their products. Pipe sizes and wall thicknesses are sometimes referred to by NPS sizes and schedule numbers, like for PVC piping.

Size limitations for some types of plastic piping have all but disappeared. Cooling-water lines for power plants in sizes of 96 in. (8 feet) diameter and larger made from plastic materials are already in service. Because of the specialized nature of the materials, pipe, fittings, flanges, and sometimes valves are mostly fabricated at the same location and from the same base material and by the same casting or molding process. Other manufacturing methods include the extrusion process or lamination.

In general, the various plastics follow standard piping connecting practices, with specialized adaptions for the various products. There are variations of threaded, flanged, and mechanical-joint connections, as well as four types of joining, namely adhesive bonding, butt fusing, solvent cementing, and heat fusing (welding).

The cementing or fusing can be performed as a socket or bell-type operation (Fig. 9), while the adhesive bonding and heat or butt fusing is applicable to the butting together of two pipes. Cementing of socket or bell-type connections is mostly done through the application of special resins and requires well-prepared and cleaned surfaces. Cleaning and preparation of piping exterior and bell interior is mostly accomplished through the application of a special solvent, which acts as a cleaner and primer and prepares plastic surfaces for the final cementing operation. Methods of

**TABLE 11** ANSI/ASTM, AWWA, and API specifications for nonmetallic piping

| | | |
|---|---|---|
| ANSI/ASTM | F-412 | Plastic piping systems, definition of terms relating to |
| ANSI/ASTM | D-2513 | Thermoplastic gas pressure pipe, tubing, and fittings |
| ANSI/ASTM | D-2662 | Polybutylene (PB) plastic pipe (SDR-PR) |
| ANSI/ASTM | D-1503 | Cellulose acetate butyrate (CAB) plastic pipe (SDR-PR) and tubing |
| ANSI/ASTM | D-2646 | Cellulose acetate butyrate (CAB) plastic pipe (SDR-PR) and tubing |
| ANSI/ASTM | D-2680 | Acrylonitrile-butadiene-styrene (ABS) composite sewer piping |
| ANSI/ASTM | D-2661 | Acrylonitrile-butadiene-styrene (ABS) plastic drain, waste, and vent pipe and fittings |
| ANSI/ASTM | D-2282 | Acrylonitrile-butadiene-styrene (ABS) plastic pipe (SDR-PR) |
| ANSI/ASTM | D-1527 | Acrylonitrile-butadiene-styrene (ABS) plastic pipe schedules 40 and 80 |
| ANSI/ASTM | F-441 | Chlorinated poly (vinylchloride) (CPVC) plastic pipe schedules 40, 80 and 120 |
| ANSI/ASTM | F-442 | Chlorinated poly (vinylchloride) (CPVC) plastic pipe (SDR-PR) |
| ANSI/ASTM | D-2672 | Bell-end poly (vinylchloride) (PVC) pipe |
| ANSI/ASTM | D-2241 | Poly (vinylchloride) (PVC) and chlorinated poly (vinylchloride) (CPVC) plastic pipe (SDR-PR) |
| ANSI/ASTM | D-1785 | Poly (vinylchloride) (PVC) and chlorinated poly (vinylchloride) (CPVC) plastic pipe schedules 40, 80 and 120 |
| ANSI/ASTM | D-2665 | Poly (vinylchloride) (PVC) plastic drain, waste, and vent pipe fittings |
| ANSI/ASTM | F-443 | Bell-end chlorinated poly (vinylchloride) (CPVC) pipe |
| ANSI/ASTM | D-2729 | PVC sewer and drain pipe |
| ANSI/ASTM | D-3034 | PVC sewer pipe |
| ANSI/ASTM | D-2104 | Polyethylene (PE) plastic pipe, schedule 40 |
| ANSI/ASTM | D-2239 | Polyethylene (PE) plastic pipe (SDR-PR) |
| AWWA | C 300 | Reinforced concrete pressure pipe, steel cylinder type, for water and liquids |
| AWWA | C 301 | Prestressed concrete pressure pipe, steel cylinder type, for water and liquids |
| AWWA | C 400 | Asbestos-cement distribution pipe 4″–16″ for water and liquids |
| AWWA | C 402 | Asbestos-cement transmission pipe 18″–42″ for water and liquids |
| AWWA | C 900 | Polyvinylchloride (PVC) pressure pipe 4″–12″ for water |
| AWWA | C 901 | Polyethylene (PE) pressure pipe, tubing and fittings, ½″–3″ for water |
| AWWA | C 902 | Polybutylene (PB) pressure pipe, tubing and fittings, ½″–3″ for water |
| AWWA | C 950 | Glass fiber-reinforced thermofitting-resin pressure pipe |
| API | 5LE | Polyethylene pipe |
| API | 5LP | Thermoplastic pipe (PVC and CPVC) |
| API | 5LR | Reinforced thermosetting pipe (RTRP) |

(Suffixes indicating latest year of issue have not been included in specification number).

joining two straight ends together (butt joint) differ greatly, depending on the plastic material. Polypropylene can be fused together in an operation similar to welding, where heat is applied and even filler rod is added. Adhesive binding for various plastics calls for the wrapping of the butt joint with one or more layers of a specially prepared resin-soaked envelope that

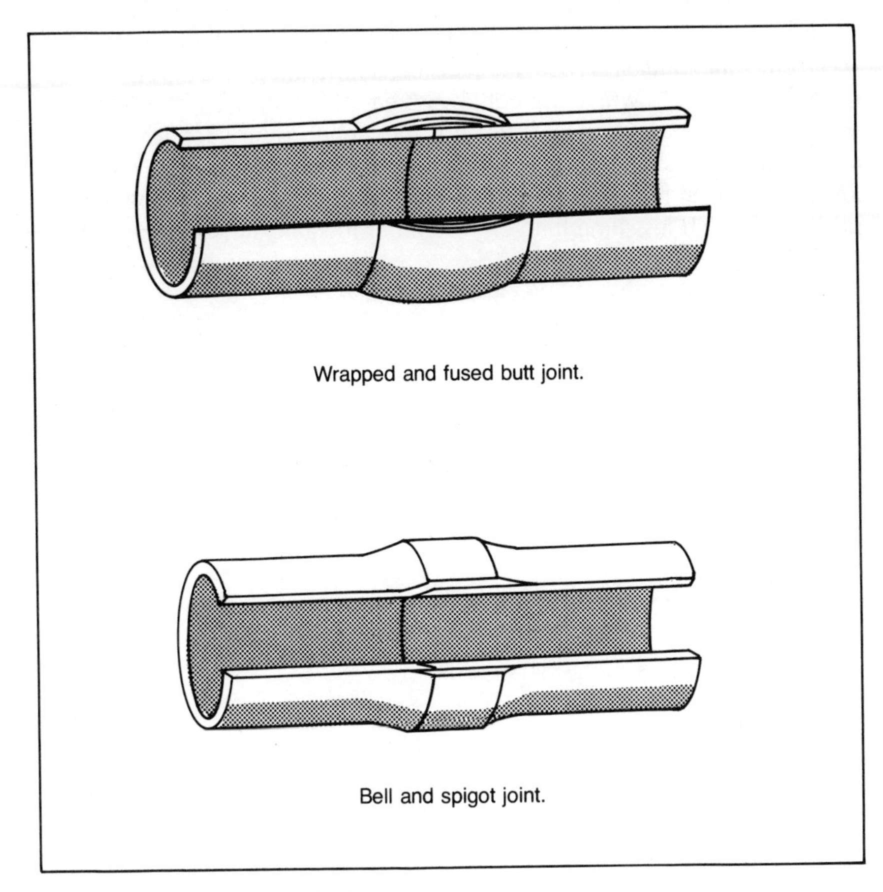


**FIG. 9. Plastic pipe, bonded joints.**

may or may not be heated and then develops into a strong connection after having been cured for a specified time.

Joining of plastic materials can be a simple and speedy operation if carried out correctly by someone with experience in this particular work. However, many of the fitters who perform this task lack the necessary experience or training so that a relatively simple job becomes a time-consuming chore. It is therefore important to get the proper assistance and training for personnel involved in the installation of a piping system so that they can master the required joining technique and have initial supervision by someone who is well versed in the particular requirements of that operation.

Application of some of the bonding resins must be performed within strict time limitations. When the outside temperature is above a certain mark, requirements which are not always easy to comply with and requirements that are often conveniently not mentioned by a supplier until the time of actual installation is at hand. On the other hand, once the art of joining has been mastered, installing plastic piping systems can be accomplished in much less time than can a similar metallic system.

# 6

# Lined and coated pipe systems

Like the many piping materials that are available, there is also a bewildering array of linings available. Most piping systems that employ lined pipe do so because such a system is the most economical way to handle a particular corrosive fluid. However, certain linings (in particular, thermosetting linings) are often used to increase flow efficiency or decrease flow resistance since the flow rates in piping systems lined for that purpose tend to remain constant for much longer periods of time and have economic advantages over alternative materials of construction.

Nearly all of the lined-pipe systems have metallic piping, mostly steel piping as the main structural component, but there is also some use of aluminum, cast iron, and some nonmetallic materials. When an external corrosive environment is encountered, the economic advantages of lined pipe no longer prevail except under specific conditions, e.g., when a coating or wrapping is being used. For most lined piping systems, no standard specifications exist. Instead, the manufacturers' guidelines are mostly accepted.

Most of the lined piping systems are based on standards that have been developed by a manufacturer that produces an assortment of lined pipe, fittings, and valves that may be available from warehouse stock. Some of this standardized lined piping may be cut and fabricated in the field. However, many lined systems require prefabricated metallic piping components that are then lined and do not permit any alteration after application of the liner material.

## BITUMASTIC LINING AND COATING

One of the first uses of a liner was the application of coal tar, asphalt, or bitumen. These materials are still in use today and are being applied to steel, cast iron, concrete, and wood. Besides lining, these materials are also used for coating steel and cast-iron piping. These lined and coated piping systems are most often used for underground water distribution.

Steel pipe lined with bitumen is often welded and then hand patched internally where the weld seam is located. The only alternative in high-pressure lines would be to use flanged connections. Mechanical connectors (Dresser or Victaulic couplings) can be used only within the limitations of their respective pressure ratings.

*Cement Lining*—These linings are normally applied to steel or cast-iron piping. Such a lining is often preferred when the piping system is to be used for the distribution of brackish or salt water. Cement lining in fabricating shops is mostly limited to 25–30 feet of pipe length. Special machinery has been developed to do such lining on a production basis, but the lining in fittings still has to be applied by hand. Cement-lined piping is normally welded in the field, and the weld seams are touched up by hand after completion. There also exists today the capability to do field-linings of steel piping systems larger than 28 in. after installing the steel shell. Field modification of fabricated cement-lined piping can be accomplished when absolutely necessary but normally is not recommended.

*Lead Lining* —There are two types of lead-lined pipe and fittings available: bonded lining and expanded lead linings. Their names imply the difference between the two. The piping with bonded lining is normally used when the temperature of the conveyed fluid is above 200°F., or when temperature or pressure fluctuates. It is also used for vacuum service since there is a possible danger of collapse of the expanded lining. Lead-lined or solid lead fittings and valves are also available.

Piping up to 12 in. diameter is normally available as a standard item, while larger piping requires custom fabrication. Pipe is normally flanged, with flanges either threaded or welded to the steel pipe. Flanged lead-lined fittings are either based on cast iron or cast-steel material as the main structural component.

*Glass Lining*—Glass-lined pipe, fittings, and valves have been available for a long time and are normally furnished in flanged form. Small-diameter piping can also be field cut and fabricated. When the flanged piping is glass lined, the lining normally extends over the flange facing, which may not always be smooth as would be required for a good mating of the corresponding surfaces. Glass-lined piping provides good resistance against many different corrosive substances and at the same time is able to withstand relatively high temperatures. However, these good qualities are more than offset by low impact and thermal-shock resistance.

*Rubber Lining*—There are two types of rubber linings: soft rubber-lined pipe, well known for its abrasion resistance, and hard rubber-lined pipe, mostly used to convey strong acid and alkali solutions. The working pressures of these types of pipe are only limited by the pressure ratings of the encasing steel or cast-iron pipe, while temperatures should not exceed 180° F.

Synthetic rubbers are also used as linings for special applications for which they are deemed superior to other materials. Rubber-lined pipe, fittings, and valves are readily available as standard manufactured items in sizes up to 12 in. with different liner thicknesses. On pipe flanges, the rubber also covers the flanges, similar to lead or glass-lined piping. While such a rubber facing could be used directly as a gasket, this is not recommended since rubber tends to stick together and one of the facings might be damaged during repair operations.

Besides rubber lining, some piping is also manufactured using rubber as pipe covering to protect the piping from internal as well as external corrosive elements. Such a use is very often found in the chlorinating or prechlorination tanks of water-treatments systems.

*Brick-lined Piping*—One of the oldest methods of protecting metal from the fluids being transferred is brick lining. Most of the liners previously mentioned are used as a shield against the corrosive properties of any given fluid. Brick lining, when using acid-resisting brick, serves the same purpose, but brick lining as protection against heat is a slightly different matter.

Mostly, piping is insulated on the outside in order to preserve the temperature of the conducted fluid, but brick lining with fire brick achieves the same goal while protecting the steel shell at the same time. Brick lining in contrast to other pipe linings by its very nature is seldom installed prior to shipment to its final destination but is mostly constructed at the plant site. By the same token, there are minimum dimensions below which installing brick lining is a near impossibility.

Acid-resisting bricks are most often installed in large ducts that transmit corrosive gases at slightly elevated temperatures. The steel duct surface is first covered with lead sheeting that resists any corrosion should a leak develop in the brick lining. The bricks are then installed over the lead lining, and acid-proof mortar is used between the bricks.

A variety of fire bricks are available to serve as brick lining and temperature insulators. These bricks are installed in similar fashion as the acid-resisting bricks—only instead of the corrosion-resisting lead sheeting, an insulating lining is sometimes installed beneath the fire brick.

Until the early 1950s, the six linings described above were about everything that was used as a lining material. However, since then numerous other nonmetallic as well as metallic linings or claddings have gained wide acceptance.

*Fluorocarbon Liners*—These materials, which are better known under their tradenames of Teflon and Kynar, can handle a wide range of temperature and corrosive conditions and are suitable to a far wider range of applications than any other plastic lining. While Teflon and Kynar are the two most commonly known brand names, there also exist several variations of Teflon such as TFE, FEP, PFA, or KEL-F, each with its own preferred use for certain applications. For instance, TFE, PFA, and FEP are mostly used as lining for piping systems, while KEL-F is sought after as a diaphragm in lined valves because of its high flexibility.

Manufacturing of lined pipe sections can be accomplished by inserting an extruded tube into a prefabricated pipe section and then flaring the tube out over the flange faces or by swaging the steel tubing and locking in the liner. The linings in fittings are usually molded in place. These materials are normally available in sizes through 12 in., with larger sizes handled as special orders

*Thermoplastic Linings*—PVC, polyethylene, polypropylene, and polyvinylidene chloride are some of the materials covered under this heading. (Fluorocarbons are also thermoplastics but have been described separately because of their special physical properties.) These thermoplastic-lined piping systems have been adapted for field fabrication according to various manufacturers' methods. One manufacturer in particular has devised special appurtenances in order to make field fabrication as easy as can possibly be achieved for a lined steel pipe.

Since these are moldable materials, manufacture of molded valves and fittings, besides loose or bonded liners, presents no problem. In general, the corrosion resistance of pipe lined with thermoplastics is the same as solid thermoplastic pipe, but the lined pipe can handle higher pressure ranges, depending on the pressure rating of the metallic shell. For pipe envelopes, steel or aluminum are being mostly used as a supportive framework. The special field-fabricating techniques were first introduced for polyvinylidene chloride-lined pipe. Besides special gaskets, special cutting tools are required for this purpose.

*Thermosetting Linings*—Epoxies, polyesters, and phenolics are the most widely used products in this group, and there exist a multitude of trade names for the various products which abound on the market. These linings are either sprayed on prefabricated flanged piping sections and fittings and are then baked to produce a hard glass-like surface or are force-cured.

Piping sections must be carefully prepared prior to spraying, as must be done for most other lining applications as well; a particular requirement is that weld beads have to be ground perfectly smooth. Since these linings are normally not very thick, they are most vulnerable at their connecting points and particular care has to be exercised during bolt-up operations.

More recently, various plastic piping materials have been introduced that serve as an outer shell and are internally lined with different plastics. One such newly developed piping material that is supposed to demonstrate an exceptional abrasion resistance in slurry handling consists of an outer shell of fiberglass-reinforced epoxy resin, lined with high alumina ceramic spheres in a matrix of epoxy. This should be well suited for handling bottom ash slurry in power plants or mine tailings. Another combination that was introduced mainly to replace cast iron for water piping service is composed of an inner core of PVC wound with continuous roving fiberglass and bonded with epoxy resin.

*Galvanizing*—This method of protecting both the inside as well as the piping exterior has been around for a long time. Originally initiated to provide a rust-proof interior of carbon steel pipe for use in water-distribution piping, the manufacturing method of submerging or hot dipping the entire pipe length at the same time provides a protective outer surface. Thus, galvanized piping became some time ago a preferred type of piping for sanitary services. Since the galvanized piping does not lend itself to welding but rather must rely on threaded or flanged connections, this type of piping has of late lost some of its popularity and has been substituted by different piping materials in many of its applications.

*Cladded Piping and Plating*—Because of the tremendous price difference between carbon steel and stainless steel or nickel—as well as some of the more exotic materials such as titanium, tantalum, zirconium, and silver—steel piping clad with the more expensive material has been introduced as a money saver. Mostly, only plate is cladded originally and then formed and welded to produce pipe.

Cladding itself can be performed by various methods, such as homogeneous bonding or spot welding. Loose, thin wall liners of some of the expensive materials are also sometimes inserted into steel piping, and in this way the anticorrosive properties of the exotic materials are combined with the pressure ratings of the exterior steel pipe. A variety of these materials can be manufactured according to individual specifications: however, welding in the field of these specially developed systems still presents problems. Plating of nickel and silver for special applications has also been performed, with plating silver against a background of cupro nickel more successful than against steel. Electroless-plated nickel will also result in a better corrosion-resistant product than electroplated nickel, which may develop porosity when exposed to certain chemicals.

*Pipe Wrapping*—Besides the various pipe coatings that have been referred to, miscellaneous other methods are being used to protect the piping exterior from a possible corrosive environment.

The best-known application is called coating and wrapping and is mostly used for the protection of steel pipe. Application can be done in the shop, or in the instance of pipelines can be a field operation. Initially, the pipe is coated with a coal tar or bitumastic primer, then wrapped with kraft paper, followed with another coat of bitumastic and another layer of kraft paper. Specifications may vary from a single layer to as many as three layers of bitumastic and kraft paper.

AWWA specifications 7A.5 and 7A.6 are in wide use as industry standards.

Other external protection may be provided by using a plastic tape, which is spirally wrapped around the pipe and is mostly applied at the site of pipe installation. Another type of corrosion protection is offered by manufacturers of steel pipe, who furnish their product encased in plastic or accomplish the same result through spray coating or similar processes.

## GENERAL

The large variety of lined piping systems that have been discussed all have in common the special consideration that has to be given to instrument connections and other small-side connections. While some of the piping can be cut to size at a late stage at the plant site, most of the lined piping systems must consist of prefabricated spools. Tees must normally be installed for any small-side outlet, or a thick solid spacer of similar material as the liner material with threaded hole on the side may be used. Most often, lined piping is complemented with smaller-diameter piping of compatible full-strength material without a metallic envelope such as glass, rubber, lead, polyvinylidene, or PVC whenever piping runs from side outlets are required.

Whatever lined piping system may be used, it is of great importance that all details of pipe wall and liner thickness, fittings, and valves as well as the type of pipe connections are fully detailed in the drawings and specifications in order to avoid installation difficulties at the plant site. This can, of course, very often be accomplished by using any given manufacturers standards as guidelines for field construction and providing the applicable literature.

# 7

# Pipe tracing and jacketing

Pipe tracing is essentially a heat-transfer method used to assist in the transfer of fluids that have high freezing points and that become viscous or solid at normal temperatures. Pipe tracing may also prevent the freezing of water lines during periods when outside temperatures would affect such a piping system. The same methodology is also used in various other heat-transfer applications. There are four most commonly used methods to accomplish this:

1. Conventional steam tracing
2. Jacketed piping
3. Internal tracing
4. Electric-resistance heated systems.

## STEAM TRACING

This is the most widely used method of heating piping systems. It is mostly accomplished by attaching some small pipe or tubing to the pipe which conveys the fluid that is to be heated. Copper or steel tubing or pipe is mostly used to convey the steam, which is supplied under pressure at one end of the pipe or tubing with condensate removed at the other end. Depending on the size of the pipeline conveying the fluid that has to be heated, and on the heating-temperature requirements, one or more tracing lines are banded to the process pipeline and then covered with insulation in order to achieve a better heat transfer.

The tracing lines are often covered with plastic cement which has a high thermal conductivity and good bonding characteristics and can be supplied in paste form ready for application. Such a plastic cement forms an efficient thermal connection that fills the voids that normally are present between the heating tubing and the process pipe. Since the plastic cement is a good heating conductor, the entire surface of the heating tubing is put to use. By using such a plastic cement, the number of tracing lines required to heat a larger-size pipeline can also be reduced and savings can be accomplished.

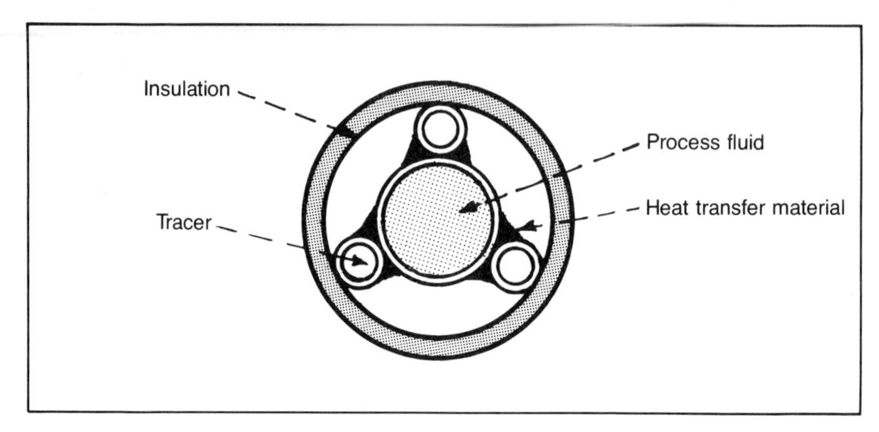


**FIG. 10. Steam-traced pipe.**

When large systems or long pipelines require steam tracing, it may become necessary to install an additional steam and condensate system to serve as a reservoir of steam supply which is required at intermediate points and also to accommodate the condensate, if it is being accumulated and returned to a single point for future use (Fig. 10).

## JACKETED PIPING

This consists of an inner and outer piping system. Normally, the inner pipe is used to transfer the process fluid that requires heating or cooling and the outer piping or jacket is used to handle the heating or cooling fluid. While we mostly refer to such systems as steam jacketing because steam is the heating media most commonly used, the jacketing pipe may handle other heating or cooling fluids as well.

Jacketed pipe is in such wide use that there are various commercial establishments that deal principally in the manufacture of prefabricated jacketed piping systems. While the most common piping material for both inner and outer piping is carbon steel, there is no restriction as to the materials used as long as there is a welding compatibility of both the inner and outer piping material and provisions are made for any different coefficient of expansion. It is an accepted practice to use an inner piping of stainless steel because of its corrosion resistance to the process fluid being handled and to use a jacket of carbon steel for the steam transfer. While there are very few lined piping systems fabricated in jacketed form except on special order, glass-lined piping is available as a commercial standard.

While the heat transfer characteristics of jacketed pipe are extremely good, repair costs can be high, especially when a leak develops in an inner pipe. This may be nearly impossible to pinpoint accurately. Jacketed pipe is mostly fabricated in flanged sections with special flanges welded both against the inner and outer pipe. This flanging arrangement permits a continuous system of the inner piping, while the flange acts as a blind flange for the jacket piping. Another method of fabrication makes use of regular flanges for the process fluid, and the jacketing pipe terminates with a pipe reducer just ahead of the flange (Fig. 11).

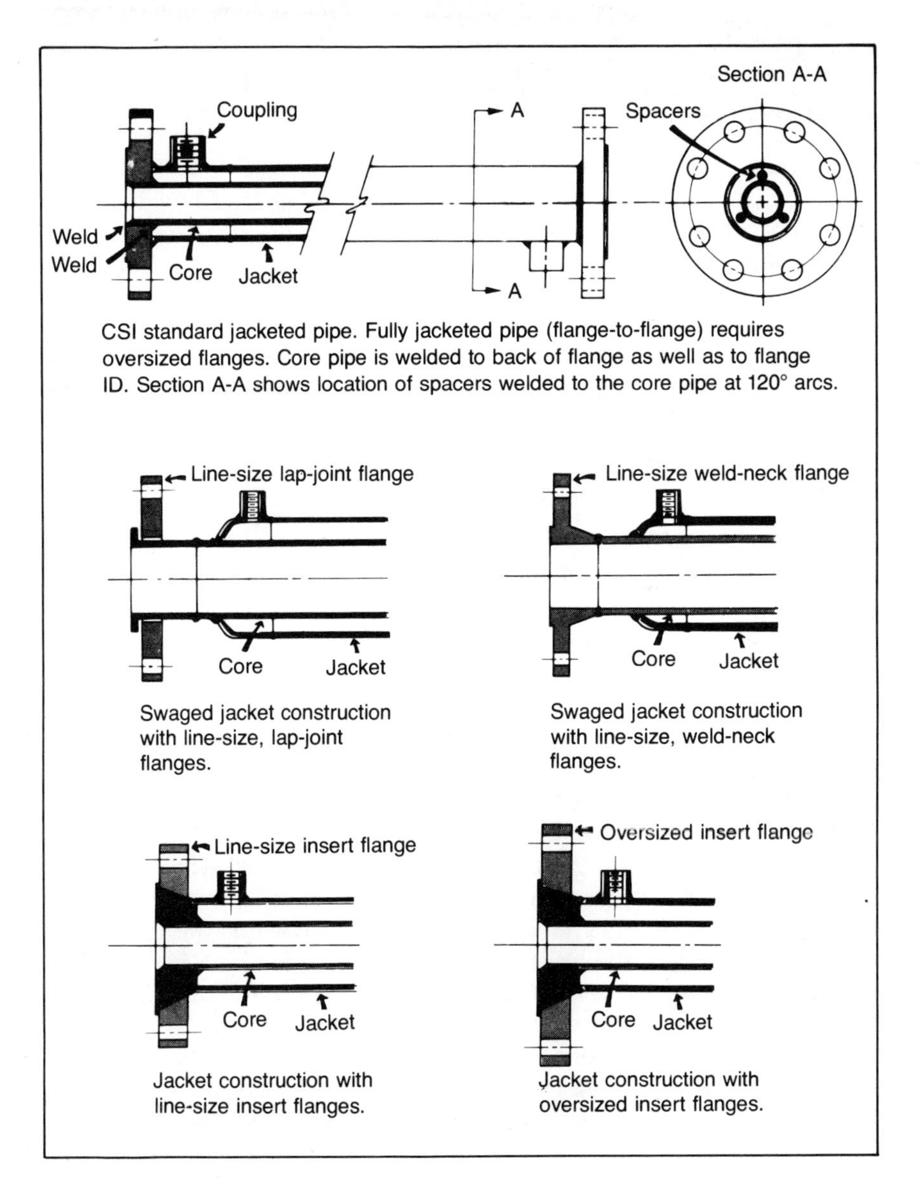


**FIG. 11. Jacketed pipe** *(courtesy Controls Southeast Inc.)*

Wherever two piping sections are flanged together, a special *jumper* must be installed to provide continuous steam or liquid flow. Instead of a jumper, the connection near the flange may also be used to remove condensate and introduce fresh steam. A deviation of the jacketed pipe concept has been introduced by one of the aluminum pipe manufacturers. Traceline aluminum pipe has two separate piping systems built into one single pipe. This system, which is supplied with special appurtenances for this use, permits welding without the necessity of flanges.

## INTERNAL TRACING

Internal tracing is the cheapest method of heat transfer but also presents the possibility of contamination of the process fluid in case the tracing line is ruptured or becomes corroded. The tracing line consists of steam tubing running through the center of the process pipe. For certain applications, it is better if the tracing-line material is manufactured from stainless steel or other corrosion-resistant material that might be dissimilar from the process pipe. In such case, ample provisions must be made for the different expansion rates of these materials. The expansion and contraction of an internal tracer, depending on how often the heating cycle is interrupted, also results in constant flexing of the metal that very often culminates in failure through metal fatigue.

## ELECTRIC-RESISTANCE HEATING

While this is not something new, it has very often been found commercially unattractive for process plants because of the high cost of power involved. More often, it has been put to use in power plants, where the cost of steam might be higher than the energy that is generated. Like a steam distribution system that must accompany a steam-traced piping system, a special power-distribution system must also be installed for most electric heat-tracing systems.

Various electric heating cables have been developed and are commercially available. For lower temperatures not exceeding 300°F, a strip of ribbon consisting of several resistance wires molded from synthetic elastic heat-resistant material into continuous form has proven very effective. For higher temperatures, other electric heating cables have been attractive. In most of these cables, the resistance wire is surrounded by a mineral-filled insulating material and is then enclosed in a metal tube. That metal tube enclosing the resistance wire and insulation may be made of copper or various nickel alloys. In general, the installation methods of electric tracing are similar to steam tracing in that the electric resistance cable is fastened to the fluid carrying process pipe and then covered with insulation. The good thermal conductivity and bonding characteristics of special plastic cement are just as well applicable to electric heating cables as they are to steam tracing.

Electric heating can be much better controlled than heating with steam through the use of temperature-sensing elements that are bonded to the process pipe and act as surface-contact thermostatic controls. Through the use of controlling devices, the amount of electrical power consumed can be limited to the actual requirements to maintain the process fluid at a desired temperature. In such a way, overheating and burn-out of the electric tracer can also be avoided.

One manufacturer makes use of the so-called skin effect and says that his method only requires power input every 20 miles. The skin-effect phenomenon is utilized in the skin-effect current tracing system of pipeline heat tracing. An ordinary steel pipe of 3/4 or 1 in. size becomes a heating element when alternating current of commercial frequency flows through it and is concentrated near the inner surface of this tube by the presence

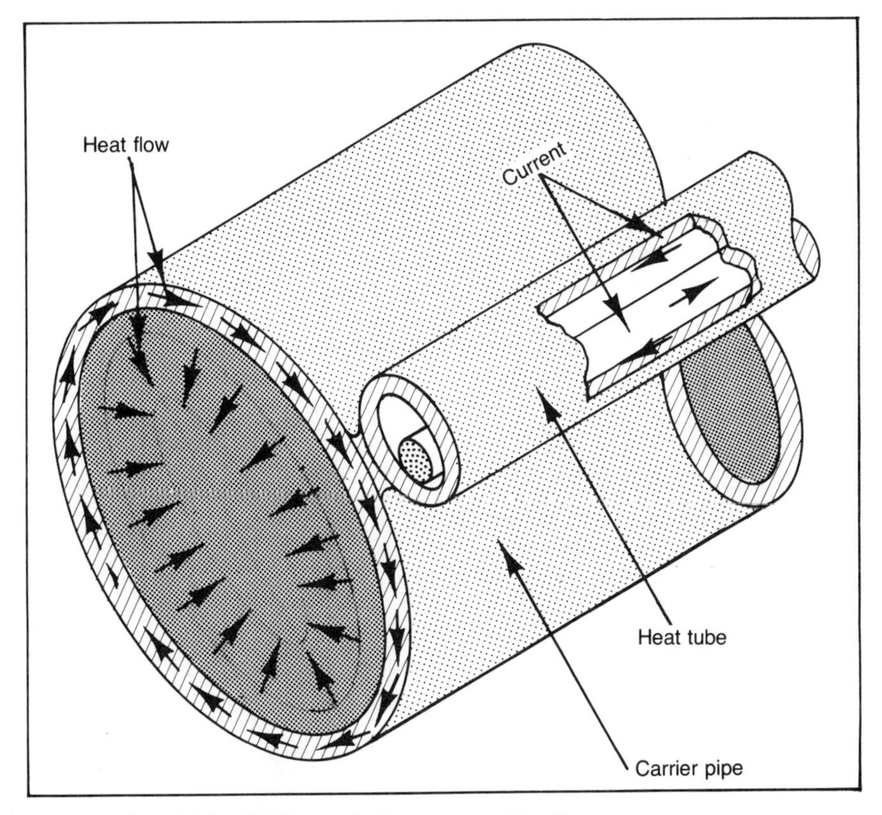


**FIG. 12. Electrically SECT-traced pipe** *(courtesy Ricwil).*

of an insulated copper wire as the return conductor inside the tube. The relatively high resistance of this reduced cross section permits a high heat energy release with practical current values. The absence of appreciable leakage voltage or current on the outer surface allows the heating tube to be welded to the fluid-carrying pipe without affecting the tracing circuit. This welded contact between the heat tube and the carrier pipe provides an excellent heat flow path, ensuring good thermal efficiency and a low operating temperature for the heat tube. This temperature is normally not more than 15 to 20°F higher than that of the fluid (Fig. 12).

For most applications, one heat tube is used for fluid pipes up to 12 in. in diameter, two tubes in the midrange, and three or more tubes on pipes larger than 30 in. in size. Grounding can be applied to the pipeline to whatever extent other factors (such as static electricity and induction from nearby power lines) dictate.

## MISCELLANEOUS HEAT TRANSFER

Tracing or jacketing of piping systems, while initially only used for heating, is now in universal use, it is no longer restricted to that single purpose but often serves for cooling of a process fluid. In many cases, the heating or cooling fluid requires an extensive pumping system operating in parallel with the process system for successful application.

Whenever piping is discussed, the amount of variations as to materials and methods of fabrication is enormous. So it is not unexpected that new and alternate materials or methods of manufacture for the diverse components of both traced and jacketed piping systems become available from time to time to the industry.

# 8

# Fittings and flanges

No piping specification can be considered adequate to cover any project requirement unless it is accompanied by specifications that cover fittings and flanges to be included as part of a given piping system. While flanges are being used only for pipe or equipment connections, fittings can also serve to change the pipe direction, make a branch connection, reduce or increase the continuation in size, or terminate a pipe run.

Fittings and flanges are available to suit the many pipe materials and the various pressure ratings or schedule numbers. Fittings are mostly identified or specified in accordance with the method of connection, such as threading, butt welding, or socket welding since different materials or manufacturing methods are applicable to each. Flanges are generally made from one material, regardless of connecting method, and are initially identified by pressure ratings. The various fittings that are available for all connecting methods are the following:

Elbow (90 or 45 degrees)
Return bend (180 degrees)
Reducing elbow (90 or 45 degrees)
Tee or reducing tee
Lateral or reducing lateral
Concentric or eccentric reducer
Cap
Cross or reducing cross

## THREADED FITTINGS

Threads for fittings and pipe are all machined to meet the NPS (nominal pipe size) standards, and various specifications govern the manufacture of threaded pipe fittings. For threads, ANSI Specification B2.1 applies to most fittings, while different specifications apply to dimensions and materials. For instance, ANSI Specification B16.3 applies to dimensions of malleable iron fittings, while the material has to meet ANSI-ASTM Specification A-197.

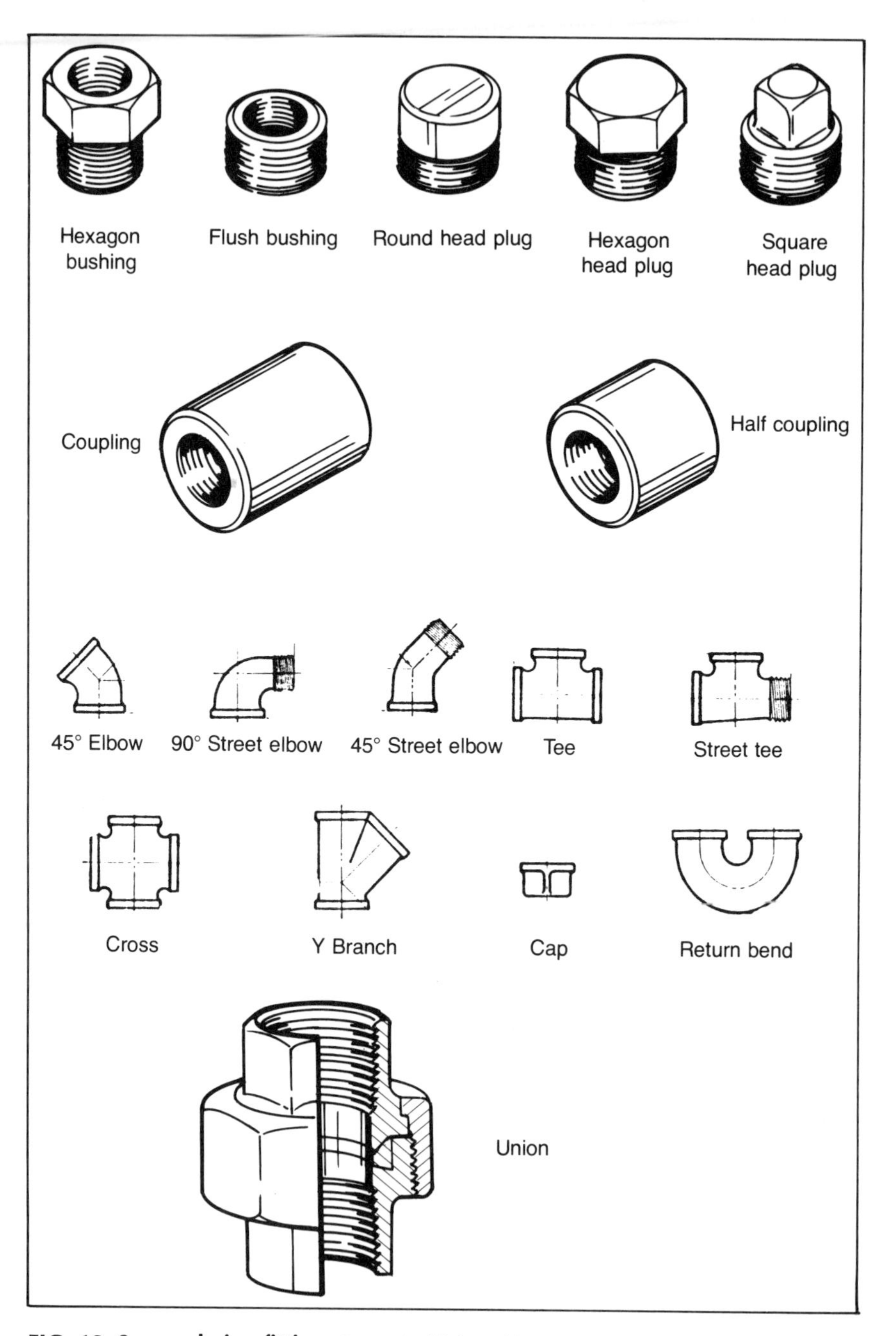


**FIG. 13. Screwed pipe fittings** *(courtesy Walworth).*

While threaded fittings in similar configurations are available for most materials and pressure classes and are listed below, specialty fittings are available in miscellaneous shapes for drainage systems made of threaded pipe (Fig. 13). In addition to the standard fittings listed in the previous paragraph, the items listed below are available as threaded fittings:

Plugs—hexagon, round, square or flush
Bushings—hexagon or flush
Street elbow, tee, or union
Unions
Couplings and half couplings

## WELDING FITTINGS

Welding fittings, as the name implies, are specially manufactured to provide fittings, that are, by means of material composition and end preparation, suitable for welding. The material composition of these fittings is mostly similar to that of the pipe to which they are being connected. Since the manufacturing process for fittings is different from that being used for pipe, different specifications apply.

Metallic welding fittings are normally furnished with 37½ degree bevelled ends so that a V-shaped groove is provided for depositing weld metal wherever a welded connection is being used. When pipe or fittings with a wall thickness of ⅞ in. or more are being used in a piping system, a double-V or U-shaped bevel is normally provided. Wrought steel welding-fitting dimensions and standards can be found in specification ANSI B16.9, and a separate ANSI standard deals exclusively with butt welding ends. Material specifications are contained in ANSI/ASTM A-234.

Besides wrought steel, welding fittings are available for most metallic materials. Some fittings are available only for inception in welded piping systems and they are listed below (Fig. 14).

Long radius or long tangent elbows
Lap joint stub ends
Saddles
Sleeves
Shaped nipples
Welding pads and other branch connections

## SOCKET WELDING FITTINGS

Socket welding for metallic piping is mostly restricted to sizes 4 in. and smaller, and fittings for these sizes are readily available. The fittings for this type of connection are provided with a bell-shaped end that is internally machined so it can encompass the external diameter of the pipe that

**FIG. 14. Welding fittings** *(courtesy Tube Turns, Chemetron Corp.)*

fits into it. Actual I.D. of the fitting matches the internal diameter of the pipe to which it connects.

During construction, care should be taken that the pipe does not butt against the internal shoulder of the fitting but rather leaves a miniscule space for expansion during the welding process.

Various plastic piping manufacturers also furnish their pipe and fittings with bell ends and usually have no restrictions as to pipe size. While connections may be made through heat fusing, they do not really represent socket welding.

Forged-steel socket-welding fitting dimensions are incorporated in ANSI Specification B16.11, while material should conform to ANSI/ASTM A-181. Most socket-welding fittings are similar to threaded fittings, and only the nomenclature "reducing insert" is used in lieu of "bushing" (Fig. 15).

## FLANGES

While fittings are being used to either change the direction of flow, reduce or increase the continuation in size, make a branch connection, or just connect some pipes on a more or less permanent basis, flanged connections permit an easy opening of piping systems for repairs or maintenance, and are most often used as connectors from piping to equipment.

Like piping, flanges are manufactured from a multitude of materials and for different pressure ratings in a variety of types. The basic carbon steel flange types are as follows (Fig. 16):

Slip-on flange
Lap-joint flange (Van stone)
Threaded flange
Welding-neck flange
Long welding-neck flange
Socket-welding flange
Blind flange
Reducing flange (threaded, socket weld, or slip-on)
Orifice flange

The various types of carbon steel flanges in the 150- and 300-lb pressure classes are normally furnished with a $^{1}/_{16}$-in. raised face. For higher pressure ratings, the raised face is mostly $^{1}/_{4}$ in. When requested, a number of different flange facings, such as large or small male and female facing, oval or octagonal ring joint, or large or small tongue and groove are commercially available and have been standardized (Fig. 17).

Lap-joint flanges and other steel flanges used for connection to 125-lb cast-iron flanges are normally flat faced. Gasket surface of all raised-face flanges are normally machined with a spiral serrated finish; however, other

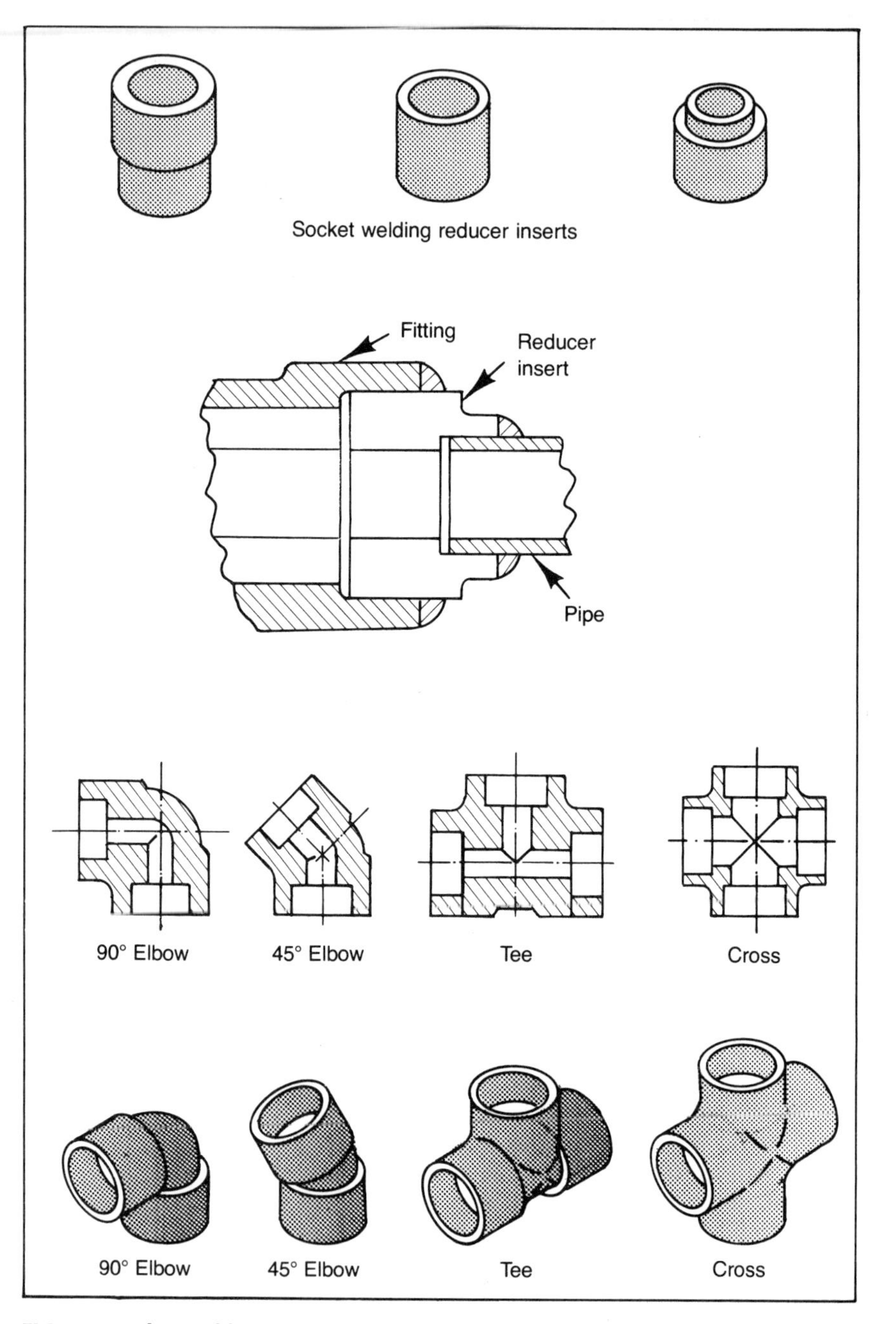


**FIG. 15. Socket-welding fittings** *(courtesy Walworth).*

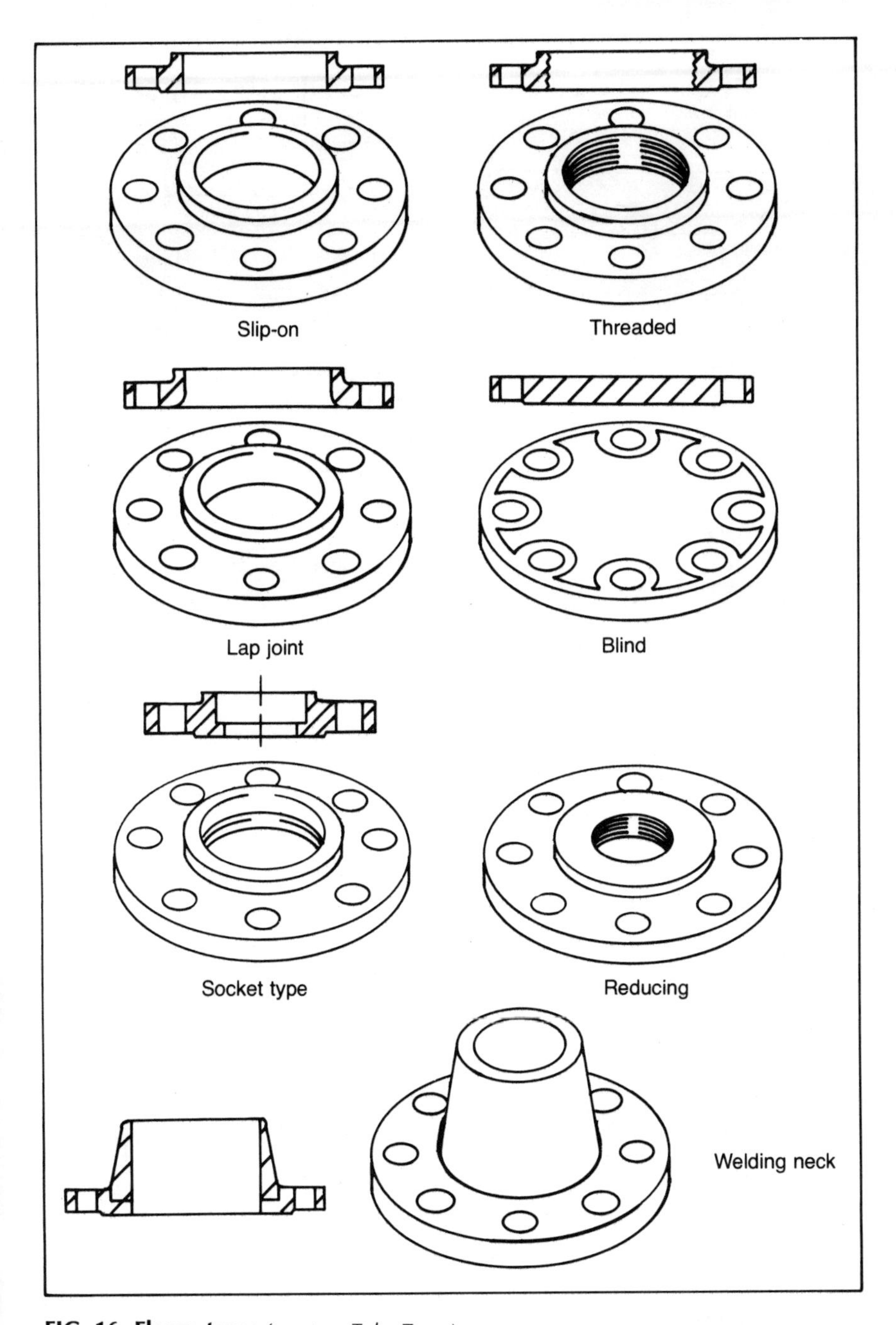


**FIG. 16. Flange types** *(courtesy Tube Turns).*

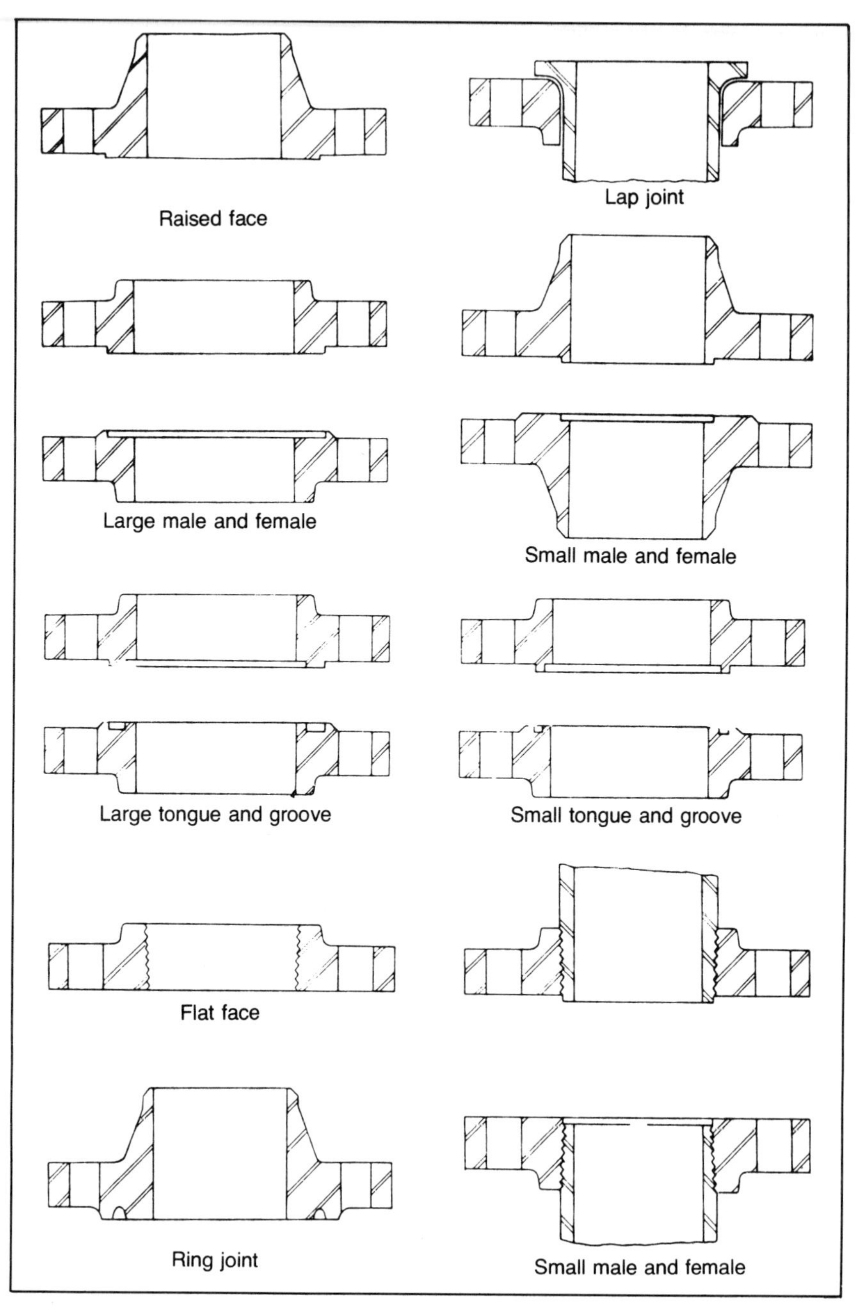


**FIG. 17. Flange facings** *(courtesy Taylor Forge, Gulf and Western Mfg.)*

finishes, such as concentric serrated finish, phonographic finish, smooth or plane finish, and cold-water finish, can be furnished by the various manufacturers.

ANSI has established Specification B16.5 to give dimensional data and operating pressure ratings for seven flange classes 150 through 2500 for various steel and alloy flanges, while other specifications govern material compositions.

It should be noted that ANSI Specification B16.5 covers only sizes up to and including 24 in. Steel flanges for sizes 26 in. and larger are mostly designed in accordance with the dimensions provided in MSS Standard Practices SP-44 or in ANSI Specification B16.1, which covers cast-iron flanges.

API Specification 605 also covers large-diameter flanges (26 in. through 60 in.). But the use of this specification is mostly restricted to the petroleum industry.

Cast or ductile iron flanges are manufactured for threaded connections only. The ANSI Standard B16.1 for cast-iron flanges and flanged fittings lists Classes 25, 125, 250, and 800 as standard classifications. These classifications relate to various working pressures. Thus, the more often used Classes 125 and 250 are good for saturated-steam service at their respective class rating, while Class 25 and Class 800 relate more to lower temperature ratings. Class 25 and 125 flanges are flat faced, while Class 250 flanges have $^1/_{16}$-in. raised face and Class 800 have a $^1/_4$-in. raised face.

## FLANGED FITTINGS

Flanged fittings are being manufactured from metallic and nonmetallic materials, however, the use of cast steel and cast iron flanged fittings in process piping has diminished in direct proportion to the ascendance of modern welding methods. In years gone by, flanged fittings were used largely in conjunction with cast iron pipe with integral or screwed-on flanges or van stone flanges on steel piping.

Standard sizes for cast iron flanged fittings in Classes 25, 125, 250, and 800 are listed in ANSI B16.1, while cast steel fittings run the whole gamut of pressure ratings from 150 lbs to 1500 lbs and comply with ANSI B16.5 (see Appendix 9).

While the use of cast iron flanged fittings is still continuing in association with cast-iron piping in water systems application, new uses have been found also as structural framework in lined piping systems.

Modern plastic piping manufacturers have also found flanged plastic fittings to be preferable in various applications.

## MISCELLANEOUS

Besides the various fittings and flanges that have been mentioned before and for which general standards apply, other standardized items have been designed and are commercially available for special applications. For instance, fittings for drainage systems in various materials and with different connecting methods are available in a number of special shapes for that particular application.

Some manufacturers have developed special piping connecting methods and their trade names have already become industry standards. They also manufacture couplings for plain-end pipe or grooved-end pipe connection. Various plastics manufacturers fabricate not only straight piping connectors but also fittings, which are applicable for their special connecting method. Since these are special end connections, special adapters to fit other pipe connections are also being made available to the various users.

By the same token, manufacturers of piping-system components of what might be called special materials, such as glass or carbon, often use special connecting methods. These connecting methods are then incorporated into the manufacturing process of fittings, flanges, and piping itself.

# 9

# Piping connections

Regardless of what piping materials are involved, there are four basically different methods of pipe connection in use today.

1. Welding, brazing or otherwise, permanently bonding together two pipes
2. Threaded connections
3. Mechanical connectors
4. Bell and spigot-type connections

## BONDED CONNECTIONS

For metallic pipe, the best-known method of bonding piping together is through the welding process. Through welding, two components are being fused together, with or without the supplement of additional filler materials. The heat generated during the welding process is so intense that all affected materials combine into one single material of equal strength.

Another method is brazing or soldering, which is mostly applicable to brass, copper, lead, etc. Brazing or soldering is accomplished through the addition of a filler material different in composition from the piping materials to be bonded. This supplementary material has a lower melting point than the piping materials involved. While the bonding is permanent, the strength of the joint is less than the strength of the piping material.

When bonding plastic piping together, the term *welding* is often used since this bonding process also shows some similarity to a metallic welding process when it involves the application of heat, although the word *fusing* might be more applicable.

## WELDING

One of the best methods to provide a permanent connection between metallic piping materials and prevent leakage is through one of various welding processes. Information on welding and operator specification is contained in ASME Boiler and Pressure Vessel Code Section IX, "Welding and Brazing Qualifications."

Welding provides a permanent bond between the two connected parts through the fusion of the adjoining metals while adding a filler rod of similar material composition. Such a connection cannot be severed without damage to one of the adjoining pipe materials. There is a wide choice of welding methods and equipment available, which is only limited to the extent that some automatic and semiautomatic equipment can only be employed during a fabricating process and do not lend themselves for use at a construction site.

*Oxyacetylene welding* uses a combination of oxygen and acetylene to produce a strong flame, which then melts the adjoining piping metals. At the same time filler rod is being added, thus resulting in a homogeneous fused metal. This welding process has been largely superseded through the use of electric arc-welding methods. However, the oxyacetylene flame is still used extensively for cutting metal.

*Shielded metal-arc welding* is commonly used for welding steel piping and is the method mostly used in construction work. In this process, the arc between a fluxcoated electrode and the material to be welded heats both the electrode and the work while simultaneous filler metal from the electrode is deposited. The filler metal is deposited in overlapping layers after removal of the slag, which is the residue of the burnt flux.

*Gas metal-arc welding* (MIG) is a process used extensively in fabricating facilities because of its easy application for automatic use. This method employs a bare electrode that is automatically fed from a wire-feeder apparatus. The arc is shielded by helium, argon, or a mixture of the two inert gases, and the welding gun is protected from overheating through a continuous flow of cooling water.

*Gas tungsten-arc welding* (TIG) also has features that permit successful automatic application. Welding heat is produced from an arc between a nonconsumable tungsten electrode and the weld area. Like MIG welding, the arc is shielded by an envelope of inert gas. Uncoated filler rod requirements are provided separately in manual or automatic operations.

This process is often used in a critical rootpass because there is less chance of a burnthrough. This welding method is also often employed at construction sites for initial welding passes only where such passes are critical and is then followed by metal-arc welding.

## HEAT TREATMENT

The ANSI Code for Pressure Piping B.31 specifies conditions under which carbon steel and alloy steel pipe have to be heat treated. Such treatment may consist of preheating the area where a weld has to be performed or postweld heat treatment (stress relieving) of the welded area.

Among the various types of heat-treatment equipment, exothermic kits have long been preferred in oil refinery construction. The kit is comprised of combustible aluminum powder material in molded form that accurately fits the welding configuration. The kit is designed to provide sufficient heat for the required temperature and soaking period. However, no accurate temperature measurements can be performed during the treatment cycle.

In the power industry, the requirement for continuous monitoring of material temperatures gave rise to the development of an induction heating process with thermocouples attached to the pipe for transmitting and recording pipe temperatures. Lately, induction heating has been replaced by electrical resistance heating through the use of finger elements that can be provided to fit any given contour. Both induction heating and electric resistance heating can be easily monitored and controlled and provide temperature charts that can be stored and made available at a later date when required for an audit.

## BRAZING

Brazing also represents a type of permanent connection between two metals. After heating the materials to be brazed, a filler is added, whose melting point is slightly below that of the base metal. Close temperature control is necessary since there is little difference between the various melting points. An oxyacetylene torch is mostly used to perform brazing operations, which were at one time largely used in copper and red-brass piping systems.

## SOLDERING

The soldering process uses a filler metal whose melting point is well below that of the material being soldered. A typical soldering application in piping systems is connecting lead piping with a tin-alloy solder.

## BUTT WELDING

This consists of butting together two pipe ends that have been previously prepared. The end preparation of metallic piping consists mostly of a V-bevel shaped at 37½ degrees. When the pipe wall thickness of steel pipe exceeds ⅞ in., the end preparation is changed to a double-V or a U-bevel. For materials with a thin wall thickness, often no other preparation than straight cutting is necessary so that the pipe ends can be butted against each other, with filler material being introduced only atop the adjoining pipes and not in between them. Standards of welding-end bevels are contained in ANSI Specification B16.25 (Fig. 18 and 19).

In order to facilitate welding to have a standardized gap between pipes and to prevent weld spatter from entering the pipe, backing rings are often inserted between two adjoining bevelled pipe ends. These backing rings are manufactured in various shapes and are available as standard commercial items or as specially machined products. Some have protruding nubs

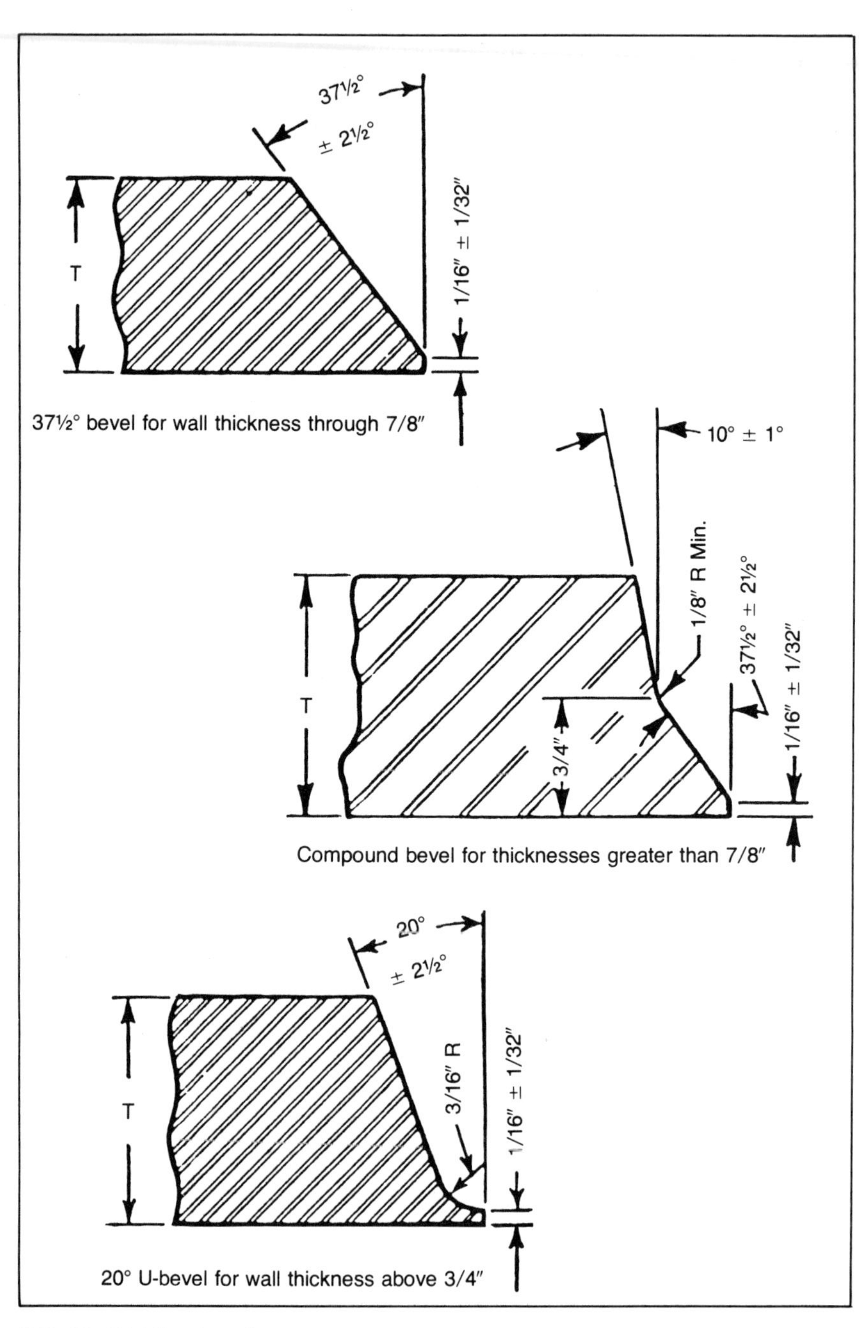


**FIG. 18. Welding bevels** *(courtesy Tube Turns).*

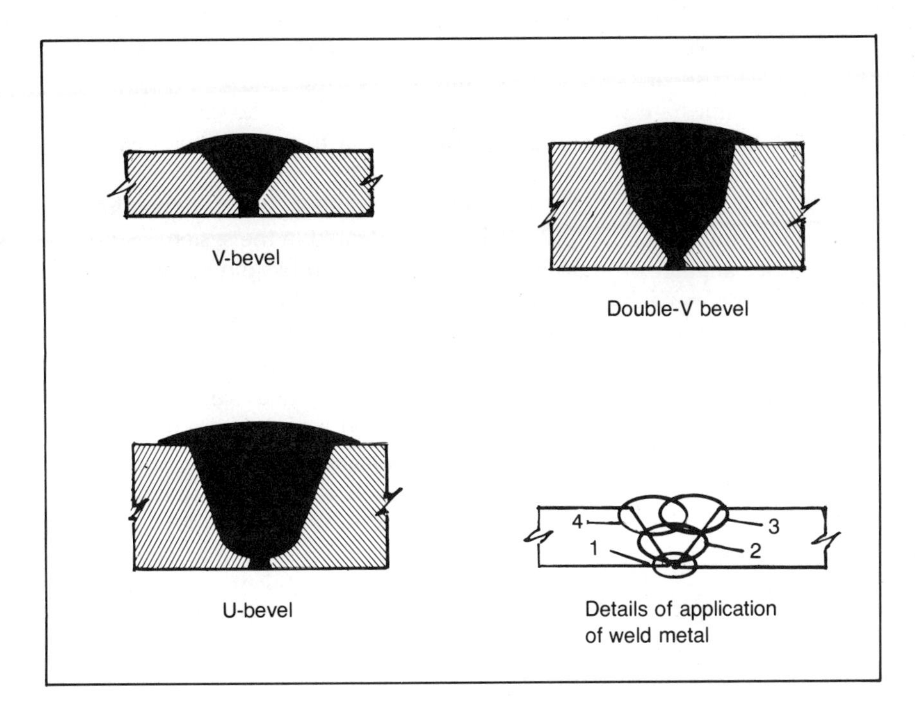


**FIG. 19. Butt welds.**

that can be removed after an initial weld pass has been completed; others have a protruding ridge that serves to keep the pipe ends equidistant. The backing rings mostly remain as a protrusion inside the pipe; however, sometimes the pipe is machined inside to allow for a more or less flush fit of equally machined backing rings.

Another type of backing ring is the consumable backing ring, which is also used in place of filler rod during the welding of the root pass or first pass and normally suffices for this purpose. Various manufacturers provide different consumable backing rings. Some are being furnished in rolls of up to 1,000 ft long, and the mechanic has to cut approximately a long strip and roll it into the required shape on a small rolling machine if the size is less than 12 in. If the pipe I.D. is more than 12 in., the backing ring can normally be shaped by hand. Another type is already manufactured in the round form and can be easily inserted into the piping to be welded.

Great care must be taken not only during the initial but also during the following two to three weld passes in order to guard against burnthrough. The initial weld or rootpass is often made with the TIG method before switching to the more standard shielded-metal arc-welding method. Backing rings are always fabricated from materials equal in composition to piping or compatible weld filler. While piping systems in power plants almost always make use of backing rings during welding operations, gas or oil transmission lines are welded without such assistance. So-called buttwelding of plastic piping is mostly performed by a fusion method, which in some aspects is similar to the welding of metallic piping.

*Branch connecting welds* have different nomenclatures, depending on the type of piping material used to be incorporated in the system. A nozzle weld, either angular or at 90 degrees, is performed by joining pipe to pipe with the contours of the adjoining pipe so shaped that a perfect fit results. The main piping is then cut out to meet the exact dimensions of the pipe nozzle; thus, flow impediments at the point of juncture are held to a minimum. Since this type of connection may result in a weakening of the piping system, a reinforcing pad is often added to regain the full material strength. During the actual welding operation, it may be necessary to restrain the main piping with structural members in order to avoid any distortion (Figs. 20 and 20A).

For instrument or small piping connections, a coupling weld is employed. A full or half coupling, which may be of the threaded or socket welding type, is welded onto the pipe. This is best performed by cutting a hole into the pipe main, which provides a good fit for the coupling, providing a bevel, and filling the resulting angular space with weldmetal (Fig. 21).

One manufacturer is marketing special fittings for small outlet connections that have sufficient body material so no additional reinforcing is required. These Weld-O-Let, Thread-O-Let, or Sock-O-Lets are available for 90- or 45-degree connections and are also contoured to fit onto an elbow.

*Miter welds* are often used in the shop or field fabrication of elbows. Two or more mitered pipe sections are used to fabricate an elbow when such a product is desirable or in an emergency, when a manufactured product is not available. Miter welds are similar to butt welds, except that

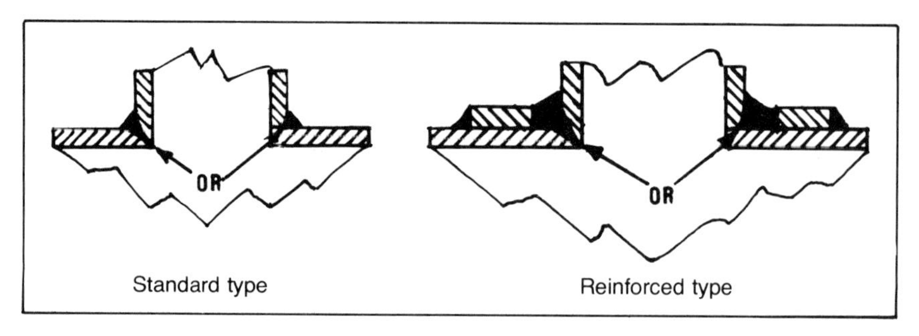


**FIG. 20. 90% nozzle weld.**

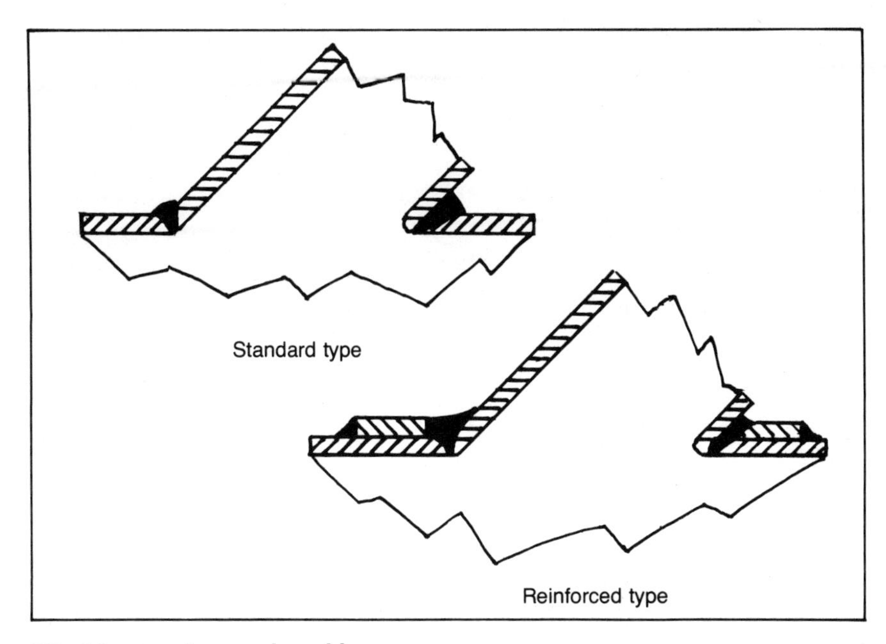


FIG. 20a. Angular nozzle weld.

the bevels employed may not necessarily be the standard 37½ degrees (Fig. 22).

*Socket Welding*—Metallic piping in sizes of 2 in. and smaller is most often connected through socket welding. This means that the fittings or couplings that are used as part of the piping system have their ends internally machined so the pipe can be accommodated. The weld bead will then be placed in the 90° opening provided by the outer rim of the fitting and the exterior wall of the piping (Fig. 23).

A similar type of socket is also used when brazing or soldering is employed. One type of brazing connection named after its various manu-

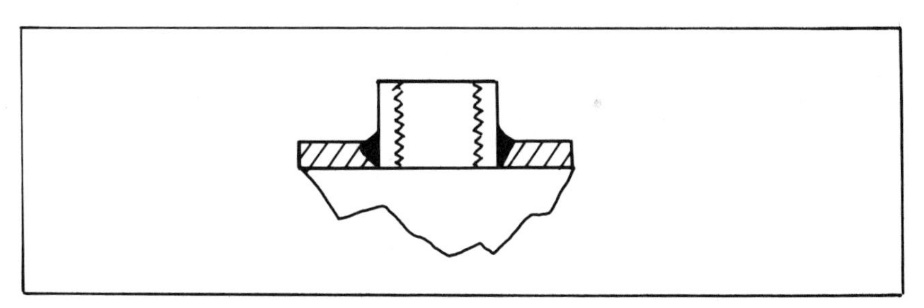

FIG. 21. Welded couling.

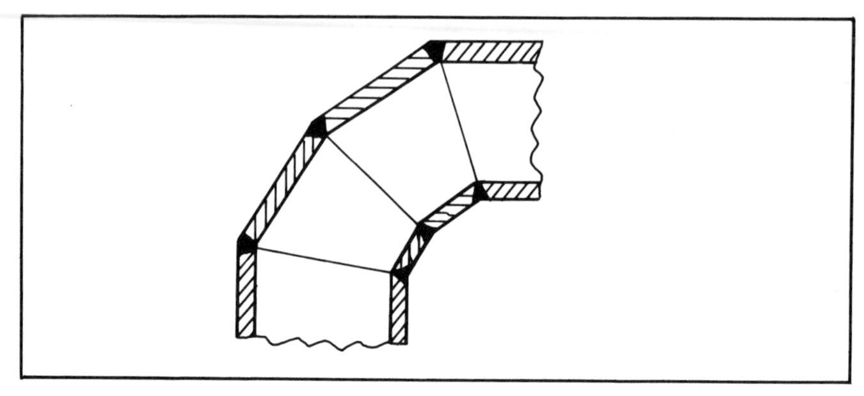

FIG. 22. Miter weld.

facturers has the brazing material located inside a groove in the fittings socket, and it requires only to heat pipe and fitting to the specified temperature in order to achieve a bond.

Plastic piping systems also often use socket-type connections, even for large-size piping, because they are a relative easy way to provide a good bonding between pipe and fittings. Like other connections that are prepared for soldering or brazing, both the piping ends and the internal socket that are going to be joined must be thoroughly cleaned prior to the joining process. However, unlike socket welding of metallic piping, once the exterior of the pipe end has been cleaned, it is then coated with a special resin that is also applied to the interior surface of the socket before joining the two parts.

It is a process very often similar to that of gluing two pieces together. Depending on the type of plastic involved, there might also be some heating involved. On the other hand, this bonding process cannot be per-

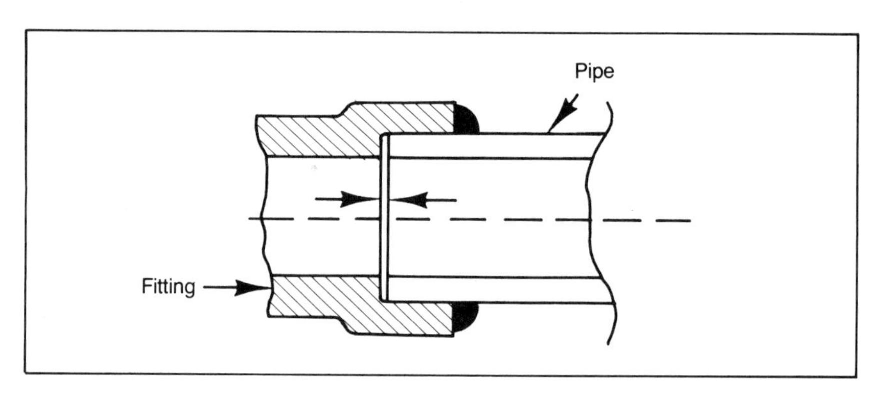


FIG. 23. Socket weld.

formed when the temperature of the pipe and the surrounding environment is below a certain limit.

## THREADED CONNECTIONS

Pipe threads for all types of materials, steel, brass, stainless steel, copper, plastics, and others normally utilize American standard pipe threads. These are the same Briggs threads that date back to more than a century and whose details are listed in ANSI Standard B2.1. Threaded connections are a type of joining that also permits opening up some piping for inspection or maintenance without damage to the system by removing a cap or plug or opening a union. All threads are slightly tapered, which helps in making a leak-proof joint.

Threaded piping lends itself ideally to field fabrication of piping systems since it permits taking the necessary measurements at the place of installation. Pipe threading is performed by hand or machine tools, and motorized pipe threaders are in use at construction sites wherever the size of piping to be threaded favors their use.

Pipe threads for the various pipe sizes should normally be of a prescribed length for each size. Some materials like plastic or copper lend themselves to easier threading, while stainless steel is very hard on pipe dies and requires especially hardened tool steels. Even then, it may require new dies at not too great intervals.

The leak-proof sealing of threaded pipe joints is mostly achieved by covering the threaded pipe part with a sealing compound before starting the connecting operations. In lieu of a sealing compound or in addition to it, various materials, such as plastic tape, hemp, or string are also often used to cover the pipe thread and thus achieve the desired results.

Threaded hose connection standards at fire hydrants differ considerably from location to location. When adding fire hydrants to some existing facility, it behooves the designer to get detailed information regarding existing hose connecting threads.

## MECHANICAL CONNECTORS

Mechanical connectors are flanges or other piping specialties or pipe fittings that permit two pieces of pipe to be joined together with or without special end preparation and that mostly can be performed by using standard tools. Many of these connectors are known by their trade names.

### Flanges

Flanges of the various types are discussed in a separate chapter as well as their diverse pressure ratings, their facings, and their end connections to pipe. At this time, we review the joining of two flanges together and necessary bolt and gasket requirements.

Whenever two flanges are bolted together, a gasket is inserted in between which assists in making a leak-proof connection. When the flanges are flat faced, the gaskets are mostly full faced; that means the I.D. and O.D. of the gasket are equal to the flange dimensions. Gaskets for this purpose are made from a variety of materials (mostly nonmetallic) with the bolt holes punched out. When flanges with raised faces are used, the I.D. of the gasket should be equal to the I.D. of the raised face, while the O.D. of the gasket should be equal to the I.D. of the bolt-hole circle of the flange, thus assuring a good centering of the gasket after bolts have been inserted. Gaskets for raised-face flanges are often of special metallic composition, depending on the fluid being transmitted through the pipe.

Tongue-and-groove flanges also use mostly metallic gaskets since the use of this particular type of flange is generally restricted to high pressure/temperature service.

There are two types of gaskets being used when connecting ring joint flanges: either an oval or an octagonal-shaped ring. This is another flange type reserved for high-pressure, high-temperature duty. Bolting for the various flange types depends on temperature and pressure rating involved. It may vary from run of the mill square-head bolts to high-strength alloy stud bolts or stainless steel bolts for corrosive service.

## Compression-sleeve couplings

These couplings consist of a sleeve, two follower rings or flanges, and two wedge-shaped gaskets and the necessary connecting bolts. Compres-

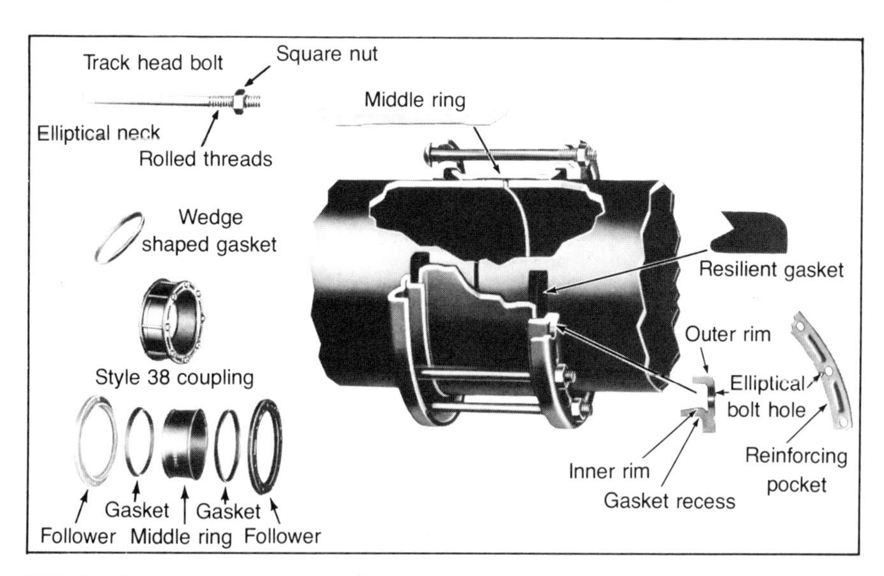


**FIG. 24. Compression-type coupling** *(courtesy Dresser).*

sion-sleeve couplings permit joining of two pipes without end preparation and also permit a certain flexibility at the point of connection (Fig. 24).

## Grooved couplings

Special end preparation of the pipes to be connected is required. Pipe ends must be square cut and a groove has to be prepared. This groove can easily be machined in the field with a tool similar to a pipe threader. The coupling itself consists of two halves or several sections, depending on pipe size and a special U-shaped gasket that spans the two pipe ends and

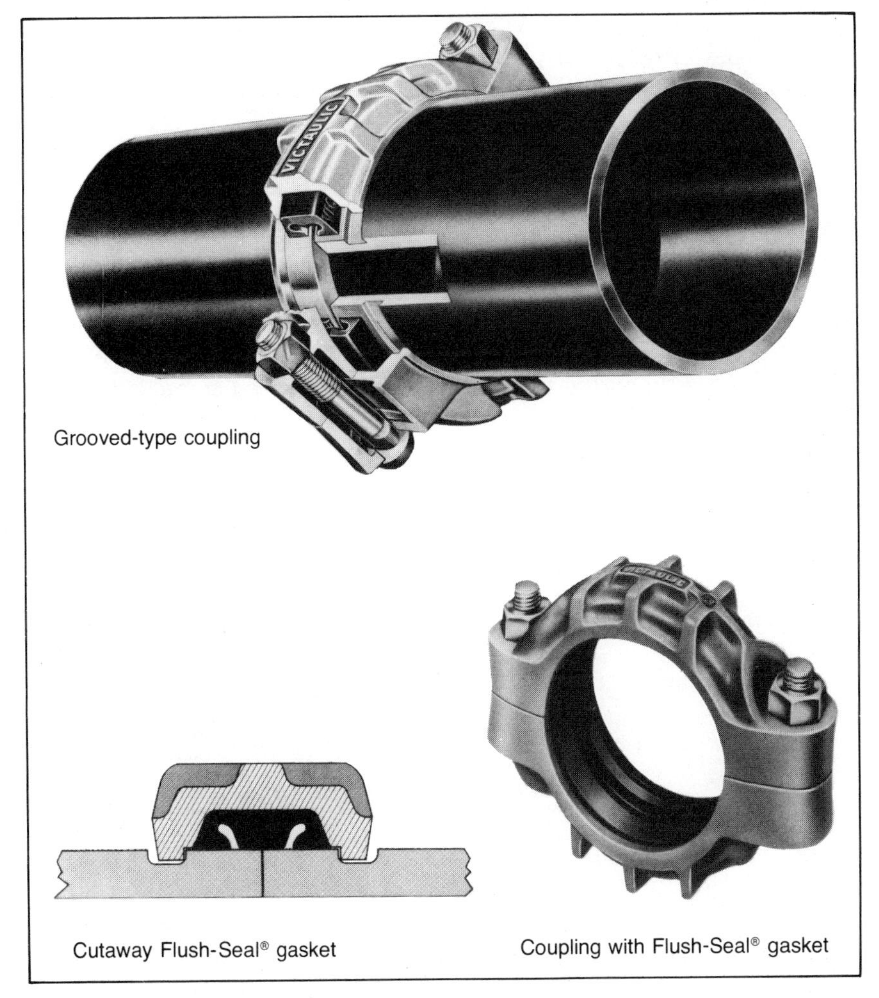


**FIG. 25. Victaulic® coupling** *(courtesy Victaulic Co. of America).*

fits into the two machined grooves. The two coupling halves are then bolted together over the gasket (Fig. 25).

## Plain-end couplings

These couplings are in shape and application similar to grooved couplings; however, they require no special grooves or other end preparation except square-cut pipe ends.

## Pipe clamps

Some cast-iron soil pipe, vitreous clay, and other pipe for nonpressure application is manufactured without any hub, just as plain length of pipe. Various types of rubber or plastic sleeves are used to join the two pipes, and this assembly is then held together by a pipe clamp.

## Ring couplings

For asbestos-cement pressure-pipe connections, the manufacturers have produced couplings with grooves inside that contain rubber or plastic rings. The ends of each piece of pipe are machined flat, as compared with the normal rough texture of asbestos-cement. The coupling is factory applied or initially forced over one end of the pipe, and then the adjoining pipe is pulled into the open end of the coupling. Pipe ends have to be marked in order to ensure that the coupling is centered correctly between two pipes (Fig. 26).

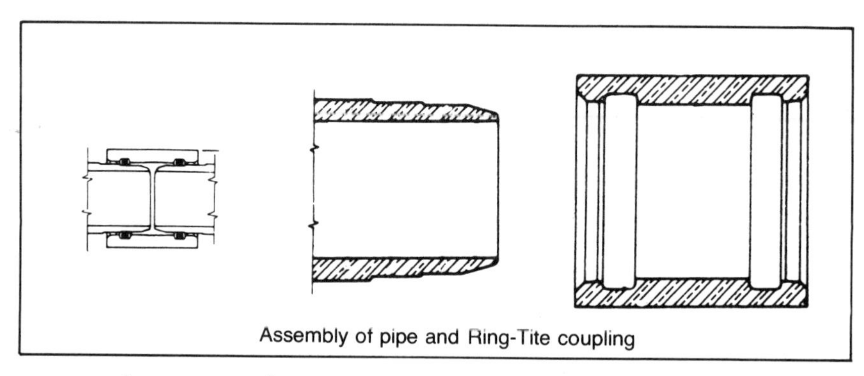


**FIG. 26. Ring-type coupling** *(courtesy (Johns-Manville).*

## Miscellaneous

There are many other types of mechanical connectors, some being manufactured only for special applications such as couplings for high pressure service or pipe sleeves for repair work, while others are being manufactured for use only with special materials such as split flanges for duriron piping. It is impossible here to mention each and every one of the various special applications.

## BELL AND SPIGOT-TYPE CONNECTIONS

This type of pipe connection was initially used for cast iron, concrete, and vitreous clay piping but is now also in use with a variety of other materials. Many variations of this originally very simple connection have been created.

### Basic bell & spigot

Under the original concept, the cast-iron or clay pipe was manufactured in varying length with a bell or hub at one end and a small protrusion or lip at the other end. The terminology *spigot* was probably initiated because one pipe end fits into the bell of the other pipe like a spigot or peg.

This type of piping is mostly installed by using a single starting point and progressing from there piece by piece. One pipe end is being fitted into the bell of the next pipe. The cavity between inner bell and outer pipe is then filled partially with jute or some like material. This is then followed with lead that is then compacted and thus ensures a stable and leakproof connection in cast iron piping.

Clay and concrete piping bells are often completed by filling them with a soft cement mix. Steel piping is also sometimes manufactured with bell ends that fit tightly around the adjoining pipe and the pipe connection is then sealed by welding the outer rim of the bell to the adjoining pipe (Fig. 27).

Among the large variety of plastic piping available today, various different types of piping with bell ends are also encountered.

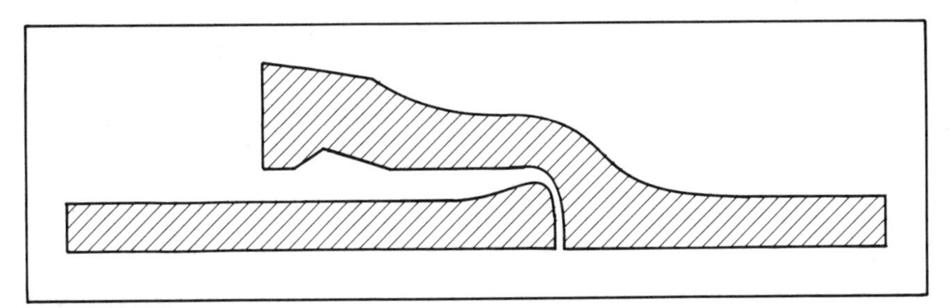

**FIG. 27. Bell and spigot joint** *(courtesy CI Pipe Research).*

## Push-on type

This is a more refined type of connection in use mainly in cast iron and vitreous clay piping systems. A seal ring of rubber or plastic material is factory installed in the bell and the connecting pipe is pushed into the bell up to an indicated length and thus completes the connection without any further adjustments (Fig. 28).

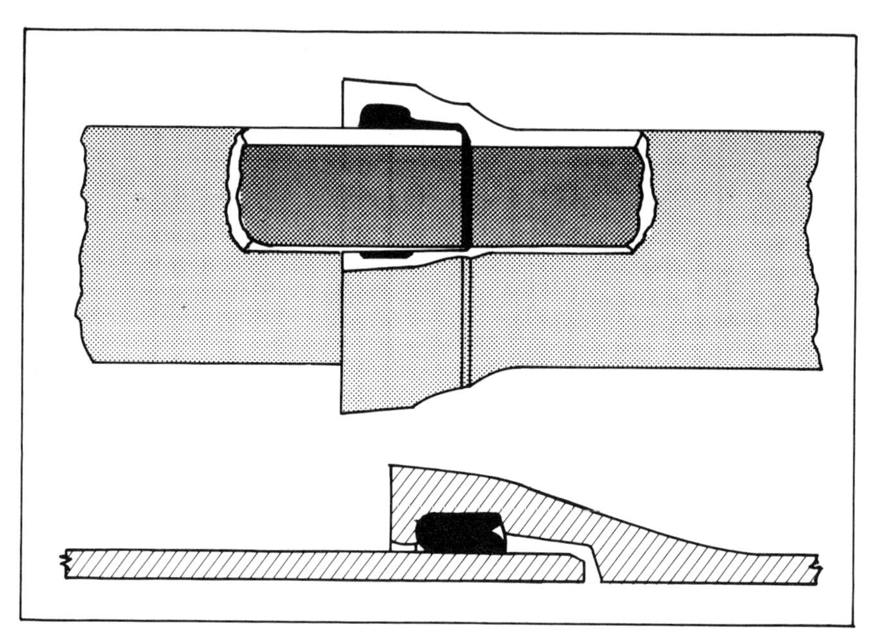

**FIG. 28. Push-on joint** *(courtesy Clow).*

## Mechanical joint

The mechanical joint is a bolted joint of the stuffing box type and is manufactured only in cast-iron piping. The joint consists of an integral bell with flange and is made up with a rubber or plastic ring gasket, follower gland, nuts, and bolts. The assembled joint provides normal expansion or contraction and also permits some flexibility (Fig. 29).

## Ball or river crossing joint

This is a ball and socket-type joint that permits a great deal of flexibility. The joint has no bolts, and restraint is provided by a bayonet-type locking of the retainer over the lugs on the bell. To prevent rotation of the retainer after assembly, a retainer lock is inserted between the lugs and

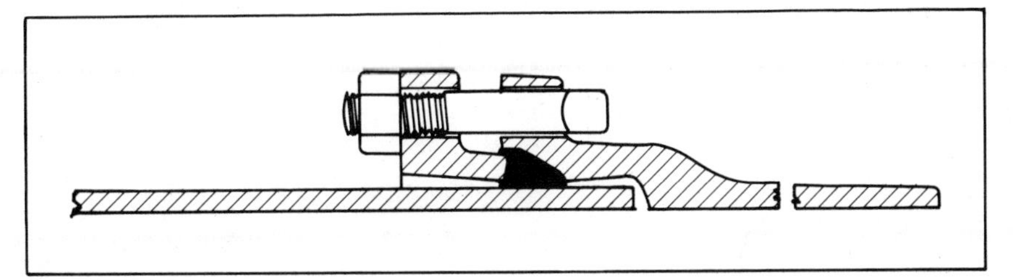

**FIG. 29. Mechanical joint** *(courtesy Clow).*

held in place by a roll pin. Piping cannot be field cut and must be provided in exact length.

## BOLTS AND GASKETS

Almost all mechanical pipe connections require the insertion of a gasket to act as a retaining seal between the rigid connecting surfaces. In many instances, bolts are required to produce sufficient pressure and provide a leak-proof seal. The ANSI Power Piping Code B31.1 spells out specific requirements for bolts and gaskets if the installed piping is to comply with that code (Table 12).

### Gaskets

The gasket is mostly a compressible material that will permit the leak-proof coupling of flanges or other surfaces, even if they contain irregularities. However for high-pressure applications, gaskets are machined from steel in such a design that they fit into prescribed sealing cavities. Through the application of pressure, either by the bolting together of flanges or the stabbing of pipe into a bell-shaped opening, or through similar means, the friction relative to the sealing surface becomes so great that no leakage will occur.

Flat gaskets for insertion between flanges are either full face or ring type (Fig. 30). The full-face gasket, which is used mostly in conjunction

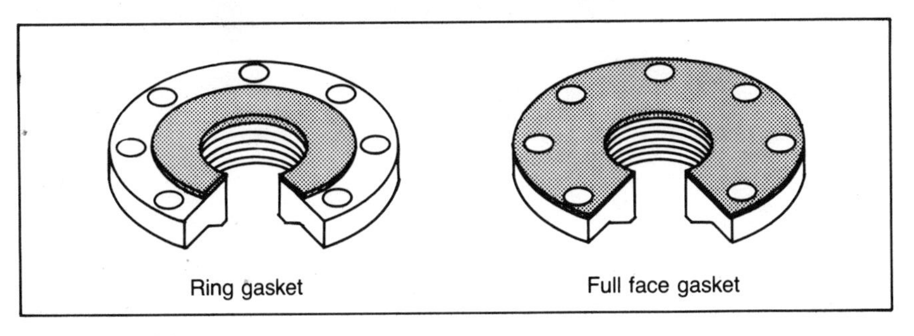


**FIG. 30. Gaskets** *(courtesy Clow).*

**TABLE 12** Piping flange bolting, facing and gasket requirements *(Reprinted from Power Piping Code ASME/ANSI B31.1-1980 with permission of American Society of Mechanical Engineers)*

| Item | Mating With | | Bolting | Flange Facings | Gasket |
|---|---|---|---|---|---|
| | Flange A | Flange B | | | |
| a | Class 150 and Class 300 steel and stainless steel (except MSS SP-42 and SP-51), or Class 150 and Class 300 ductile iron | Class 150 and Class 300 steel and stainless steel (except MSS SP-42 and SP-51), or Class 150 and Class 300 ductile iron | Alloy steel; with carbon steel to ASTM A307 optional to 500°F (260°C) max. | Raised or flat | Full face or ring type |
| b | Class 125 cast iron, Class 150 bronze, MSS SP-42 and SP-51 stainless steel, "Large diameter" steel, or nonmetallic | Class 125 cast iron, Class 150 bronze, Class 150 steel and stainless steel including MSS SP-42 and SP-51, Class 150 ductile iron, "Large diameter" steel, or nonmetallic | Carbon steel to ASTM A307 or alloy steel | Flat | Full face nonmetallic |
| c | Class 300 bronze | Class 300 bronze, Class 300 steel and stainless steel, Class 300 ductile iron, or Class 250 cast iron | Carbon steel to ASTM A307 or alloy steel | Flat | Full face nonmetallic |
| d | Cast 250 cast iron | Class 250 cast iron, Class 300 steel and stainless steel, or Class 300 ductile iron | Carbon steel to ASTM A307 Grade B | Raised | Ring type; nonmetallic or spiral wound (metal winding with nonmetallic filler) |
| e | Class 400 and higher steel and stainless steel, or Class 400 and Class 600 ductile iron | Class 400 and higher steel and stainless steel, or Class 400 and Class 600 ductile iron | Alloy steel | Raised or flat | Full face or ring type |

with flat faced flanges, covers the entire flange face and O.D. and I.D. of flange and gasket are the same. The ring-type gasket may differ in size depending on the type of flange; however, in most instances the I.D. is equal to the I.D. of the raised-face flange facing, while the O.D. may equal the inner bolt circle and thus facilitate installation. Numerous gasket materials are available, and most of the materials for flat-type gaskets, used for insertion between flanges, are listed in the ANSI standard for steel pipe flanges, flanged valves, and fittings B.16.6 (Table 13).

Other types of gaskets include O-ring gaskets that are used in couplings for asbestos cement pipe or as an insert where pipes are connected through bell type ends. Gasket shapes of *V* or *U* form are often used in conjunction with mechanical connectors.

Flanged joints for high pressure/high temperature are often sealed with metallic gaskets whose shape conforms to the particular sealing indentations, such as ring-joint gaskets or lens gaskets that have been especially developed for high-pressure service in the chemical industry. ANSI Specification B16.20 covers ring-joint gaskets, while ANSI B16.21 covers non-metallic gaskets for flanges.

## Bolts

Bolts can be generally classified as machine bolts, bolt studs, or stud bolts (Fig. 31), and very often there is confusion as to what is a bolt stud as opposed to a stud bolt. The bold stud is fully threaded and has a nut affixed at each end, while the thread of the stud bolt is not continuous, thereby permitting the end with the short thread to be affixed permanently into any machined surface, which might be used as an alternate to a flange facing. Material standards for bolting material have been published by

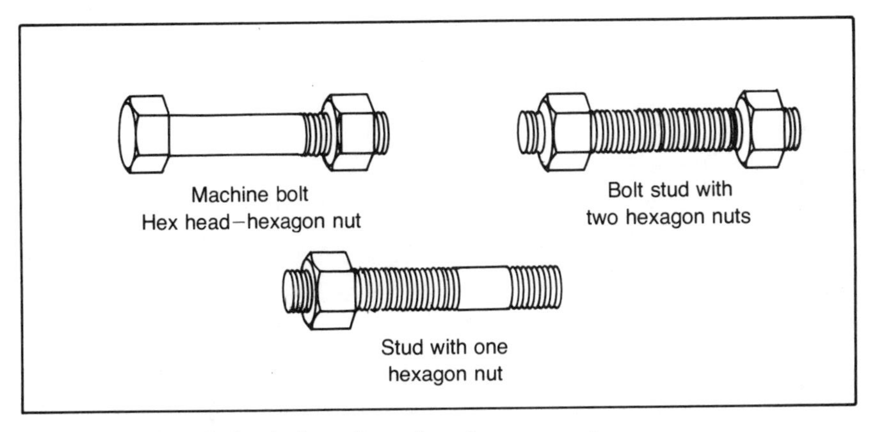


**FIG. 31. Machine bolts, bolt studs, and studs** *(courtesy Clow).*

**TABLE 13** Gasket materials for steel flanges *(Based upon the ASME Unfired Pressure Vessels Code, Section VIII, Division 1. Reprinted from ANSI B16.5–1973 with permission of American Society of Mechanical Engineers)*

| Gasket Group Number | Gasket Material | | Sketches |
|---|---|---|---|
| | Elastomer without Fabric or a High Percentage of Asbestos Fiber:<br>Below 75 Shore Durometer<br>75 or Higher Shore Durometer | | |
| | Asbestos with a Suitable Binder for the Operating Conditions | 0.12″ Thick<br>0.06″ Thick | |
| | Elastomer with Cotton Fabric Insertion | | |
| | Elastomer with Asbestos Fabric Insertion, with or without Wire Reinforcement | 3-Ply<br>2-Ply<br>1-Ply | |
| | Vegetable Fiber | | |
| | Spiral-Wound Metal, with Asbestos or other Nonmetallic Filler | Carbon Steel<br>Stainless Steel or Monel | |
| | Corrugated Metal, Asbestos Inserted<br>—or<br>Corrugated Metal Double Jacketed Asbestos Filled | Soft Aluminum<br>Soft Copper or Brass<br>Iron or Soft Steel | |

| | | | |
|---|---|---|---|
| | Corrugated Metal | Soft Aluminum<br>Soft Copper or Brass |  |
| IIa & IIb | Asbestos with a Suitable Binder for the operating conditions | 0.03" Thick | 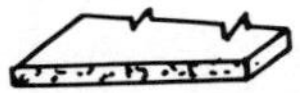 |
| | Corrugated Metal, Asbestos Inserted<br>—or<br>Corrugated Metal Double Jacketed<br>Asbestos Filled | Monel or 4–6% Chrome<br>Stainless Steels | 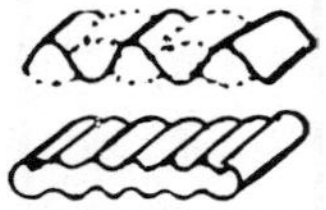 |
| | Corrugated Metal | Iron or Soft Steel<br>Monel or 4–6% Chrome<br>Stainless Steels |  |
| | Flat Metal Jacketed<br>Asbestos Filled | Soft Aluminum<br>Soft Copper or Brass<br>Iron or Soft Steel<br>Monel<br>4–6% Chrome<br>Stainless Steels | 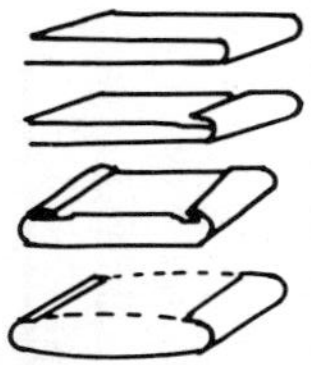 |
| | Grooved Metal | Soft Aluminum<br>Soft Copper or Brass<br>Iron or Soft Steel<br>Monel or 4–6% Chrome<br>Stainless Steels |  |
| | Solid Flat Metal | Soft Aluminum | |
| IIIa & IIIb | Solid Flat Metal | Soft Copper or Brass<br>Iron or Soft Steel<br>Monel or 4–6% Chrome<br>Stainless Steels |  |

ASTM, and the Code for Pressure Piping B31.1 refers to the following specifications:

| | |
|---|---|
| A-193 | Alloy and S.S. bolts for H.T. service |
| A-194 | Carbon and alloy nuts for H.T. service |
| A-307 | Low carbon steel threaded fasteners |
| A-320 | Alloy steel bolting material for L.T. service |
| A-354 | Quenched and tempered alloy steel bolts and studs |
| A-437 | Alloy steel turbine-type material for H.T. service |

While installing bolts, it is important that bolt tightening follows the right procedure and that correct torque be applied. Bolts should first be installed hand tight, then tightened in sequential order, and finally tightened in rotational order until all bolts have reached final torque level (Table 14).

**TABLE 14** Bold torque procedures *(Reprinted with permission of Tube Turns Division, Chemetron Corp.)*

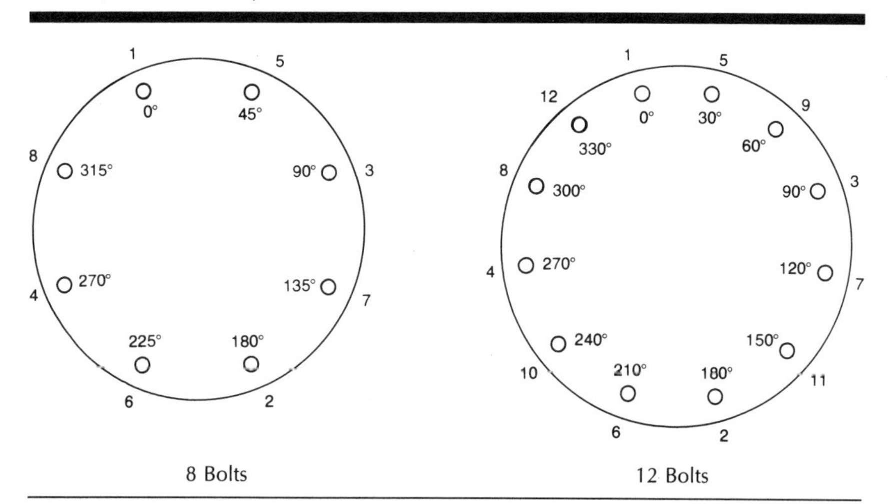


| 8 Bolts: Sequencial Order | 8 Bolts: Rotational Order | 12 Bolts: Sequencial Order | 12 Bolts: Rotational Order |
|---|---|---|---|
| 1–2 | 1 | 1–2 | 1 |
| 3–4 | 5 | 3–4 | 5 |
| 5–6 | 3 | 5–6 | 9 |
| 7–8 | 7 | 7–8 | 3 |
| | 2 | 9–10 | 7 |
| | 6 | 11–12 | 11 |
| | 4 | | 2 |
| | 8 | | 6 |
| | | | 10 |
| | | | 4 |
| | | | 8 |
| | | | 12 |

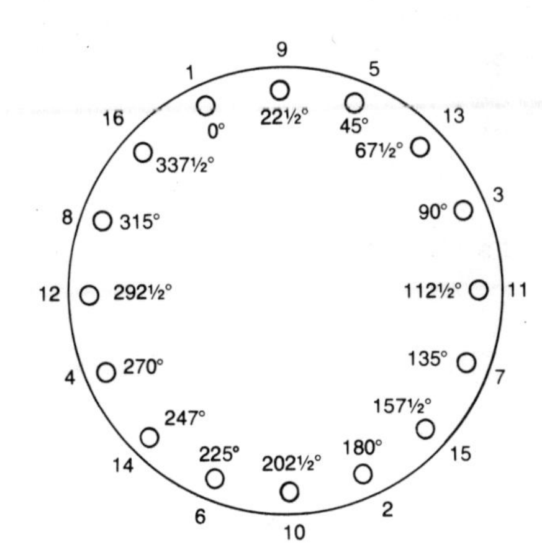


16 Bolts

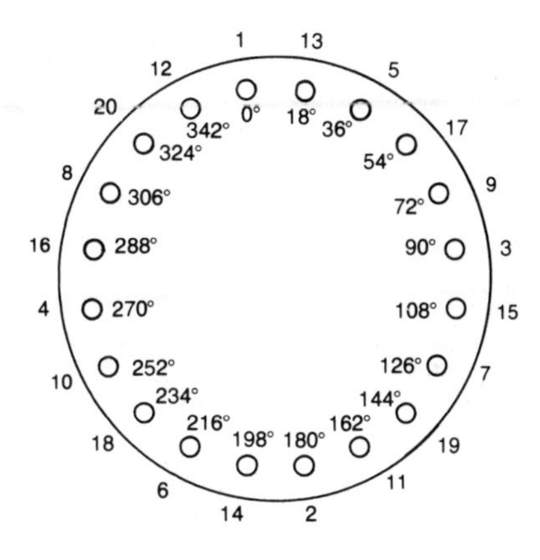


20 Bolts

| Sequential Order (16 Bolts) | Rotational Order | | Sequential Order (20 Bolts) | Rotational Order | |
|---|---|---|---|---|---|
| 1–2 | 1 | 2 | 1–2 | 1 | 2 |
| 3–4 | 9 | 10 | 3–4 | 13 | 14 |
| 5–6 | 5 | 6 | 5–6 | 5 | 6 |
| 7–8 | 13 | 14 | 7–8 | 17 | 18 |
| 9–10 | 3 | 4 | 9–10 | 9 | 10 |
| 11–12 | 11 | 12 | 11–12 | 3 | 4 |
| 13–14 | 7 | 8 | 13–14 | 15 | 16 |
| 15–16 | 15 | 16 | 15–16 | 7 | 8 |
| | | | 17–18 | 19 | 20 |
| | | | 19–20 | 11 | 12 |

# 10

## Valves

Valves in any piping system serve three elementary functions:

1. Shut off or open a system to fluid flow
2. Regulate or throttle any fluid flow
3. Prevent backflow

Valve installation may be in the normally open position or in the normally closed position, depending on the particular requirements. An unloading valve at a storage tank is normally closed and is only opened when the tank's contents are required, while a pump discharge valve is normally open. For functions that require only a fully open or closed position during operation, the gate valve is well suited. When the flow has to be regulated or throttled through only partial opening of the valve, the globe or angle valve is much preferred. Check valves have only one function to perform, and that is the prevention of backflow in any system. A particular application for check valves is on the discharge side of a pump.

Besides these three categories, there are many other valve types available, and each type may have its own particular application. Often, two or more valve styles can fulfill the same service requirements. For instance, ball valves and plug valves have different body styles but are suitable for similar use.

Not only do we look at this assortment of the various valve types, but there is also the choice of material of construction, various pressure ratings, and finally the type of connection in a given piping system. There is also a choice of the various types of manual, mechanical, or electrical operators. The number of possible selections seems endless. Let us first introduce the main valve categories, based on the different body styles:

Gate valves
Globe valves & angle valves
Check valves
Plug valves
Ball valves

Diaphragm valves
Butterfly valves
Pinch valves
Blow-off valves

Besides these nine main categories, various other types of valves are available for more-or-less special purposes. Control valves, which may use basic design features of any of these main valve categories but which serve specific operating functions, are described separately.

## STANDARDS

Specifications for metallic valves and their components have been promulgated by ASTM and are mostly incorporated in ANSI, while the American Petroleum Institute (API) has developed various valve specifications for its industry. In addition, other standards of beneficial interest to the valve user are available, particularly those published by the Manufacturers' Standardization Society of the Valve and Fittings Industry (MSS).

Some of the most applicable standards follow:

| | |
|---|---|
| ANSI B2.0 | Pipe threads |
| ANSI B16.1 | Cast iron pipe flanges and flanged fittings |
| ANSI B16.5 | Steel pipe flanges, flanged valves and fittings |
| ANSI B16.10 | Face-to-face and end-to-end dimensions of ferrous valves |
| ANSI B16.20 | Ring-joint gaskets and grooves for steel pipe flanges |
| ANSI B16.21 | Nonmetallic gaskets for pipe flanges |
| ANSI B16.34 | Steel valves, flanged, and butt welding ends |
| ANSI/ASTM A181 | Forged or rolled-steel pipe flanges, forged fittings, and valves and parts for general service |
| ANSI/ASTM A182 | Forged or rolled alloy-steel pipe flanges, forged fittings, and valves and parts for high-temperature service |
| MSS DS-13 | Corrosion-resistant cast flanged valves |
| MSS SP-25 | Standard marking system for valves, fittings, and flanges |
| MSS SP-45 | Bypass and drain-connection standard |
| API 593 | Ductile iron plug valves |
| API 594 | Wafer-type check valves |
| API 595 | Cast-iron gate valves |
| API 597 | Steel venturi gate valves |
| API 599 | Steel plug valves |
| API 600 | Steel gate valves |
| API 602 | Compact cast-steel gate valves |

| | |
|---|---|
| API 603 | Class 150 corrosion-resistant gate valves |
| API 604 | Ductile iron gate valves |
| API 606 | Compact carbon-steel gate valves (extended bodies) |
| API 609 | Butterfly valves to 150 psig and 150 F. |
| API 6D | Pipeline valves |

## IDENTIFICATION

Valves are fully described in manufacturers' catalogs and are identified through a numbering system that, by implication and sometimes through the use of prefixes or suffixes, contains a full description of the valve in terms of type, pressure rating, connecting end preparation, and the details of the various trim materials.

Such catalog items are therefore easy to identify or specify without going into the details of materials and manufacture, which otherwise are a prerequisite for a purchase requisition. Since there are a number of manufacturers that produce similar or equal valves in any given category, most of them have prepared a cross index that lists comparative valve numbers assigned by other manufacturers. These are equal to the valves produced by the originator of the index.

Most valves are identified by the manufacturer's name or symbol embossed or otherwise presented on the valve body. The applicable valve number is also shown on a disc mounted on the handwheel or a small plate fastened to the valve body.

Many valves and particularly those for high-pressure applications normally have plates attached that give the valve number and show size, pressure, temperature rating, and material composition of basic parts.

## GATE VALVES

Gate valves are mostly multiturn valves that, in their basic construction, consist of valve body, seat and disc, spindle or stem, gland, and rotating wheel, The seat is located at the bottom of the valve and may be fixed or removed together with the disc to provide the actual valve components which regulate the flow of any fluid through the valve body (Fig. 32).

In order to actuate a gate valve, the disc is either lowered or lifted by means of a stem that projects outside of the valve body and is crowned by a handwheel, which activates the stem. The protrusion of the stem to the outside atmosphere requires some method of retaining the fluid in the pipeline, which is accomplished by installing a gland packed with a fluid-resisting barrier to prevent leakage.

Gate valves are mostly used as stop valves, that is to fully shut off or provide full flow. They are ideally suited for wide-open service, such as at an outlet of a storage tank for liquids in oil and gas pipelines and firelines.

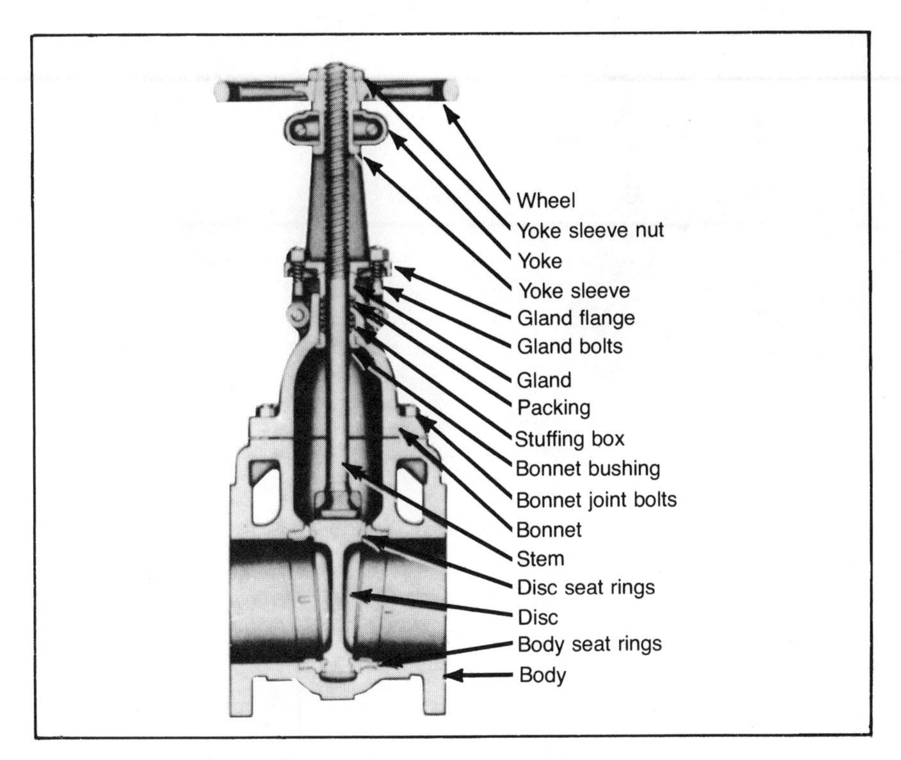


**FIG. 32. Gate valve, wedge type** *(courtesy Crane).*

The flow can move in a straight line—practically without resistance when the disc is fully raised. Seating in a gate valve is at a right angle to the line of flow, which makes the valve impractical for throttling operations and makes close regulation a near impossibility.

While many of the valve accessories have been improved throughout the years, and a great many variations of gate-valve design are commercially available today, the basic components have not changed. The valve body is still surmounted by a bonnet. In small valves, it is common practice to use a U-bolt, a screwed joint or union-type joint to connect the bonnet to the valve body (Fig. 33) that suffices for low pressure applications. For larger-sized valves and for more demanding applications with higher pressure and temperature ratings, other means must be utilized to retain the fluid. The obvious solution is a flanged and gasketed joint.

Large, flanged, bonnet-type valves with many variations of gasket types and materials proved satisfactory for a long period of time. These gasketed joints were always improved upon through the development of better bolting and gasketing. With the introduction of satisfactory welding tech-

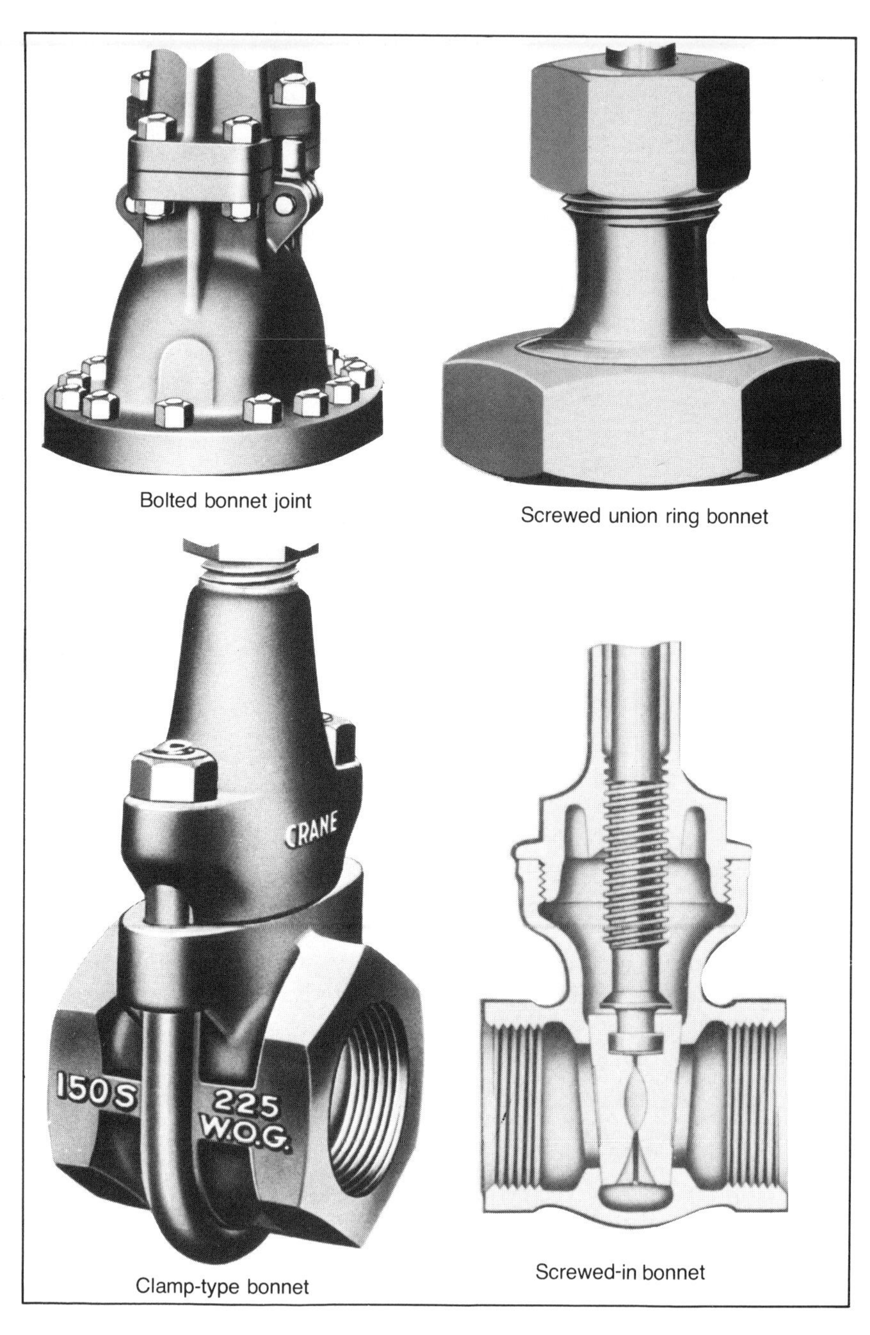


**FIG. 33. Valve bonnet types** *(courtesy Crane).*

niques, a welded bonnet joint made its appearance. Still later, the pressure-seal joint has been applied to gate-valve bonnets. High-pressure valves today are mostly of the pressure-seal bonnet construction (Fig. 34). For nuclear applications, where any leakage would be disastrous, a hermetically sealed valve has made its debut.

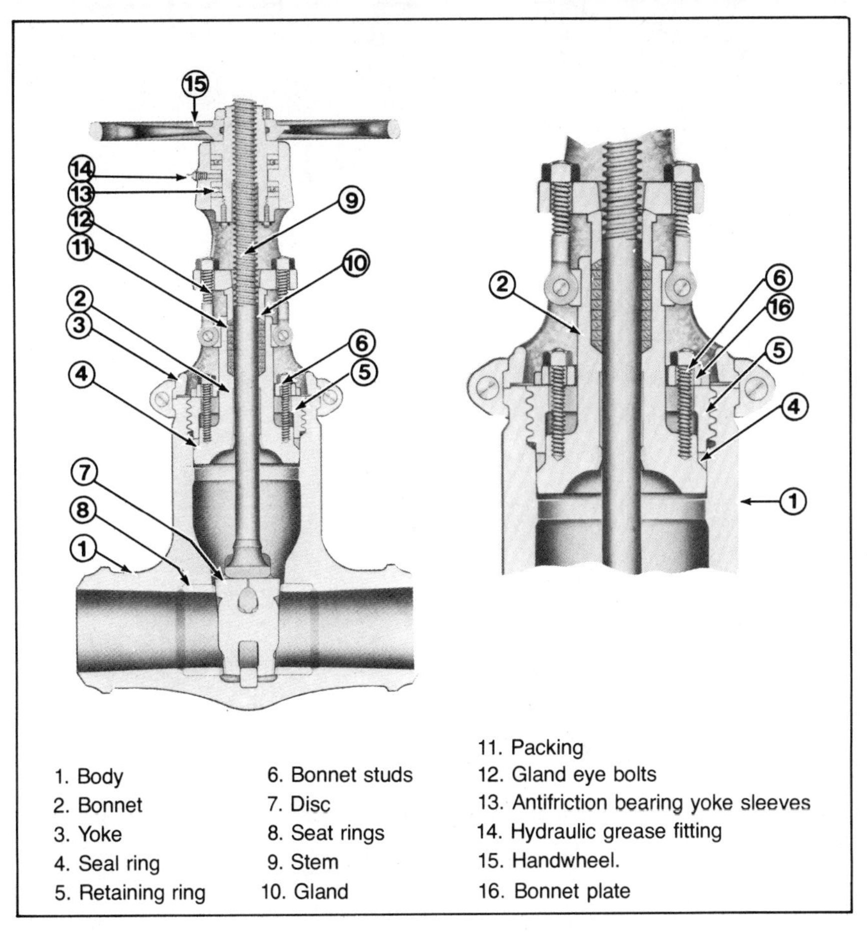


**FIG. 34. Pressure seal valve** *(courtesy Crane).*

Packing glands, which prevent leakage around the valve stem, are often of simple construction with a threaded gland follower and a graphited packing material, in particular for smaller valves. However, more sophisticated designs make use of the lantern-type or bellow seals for packings,

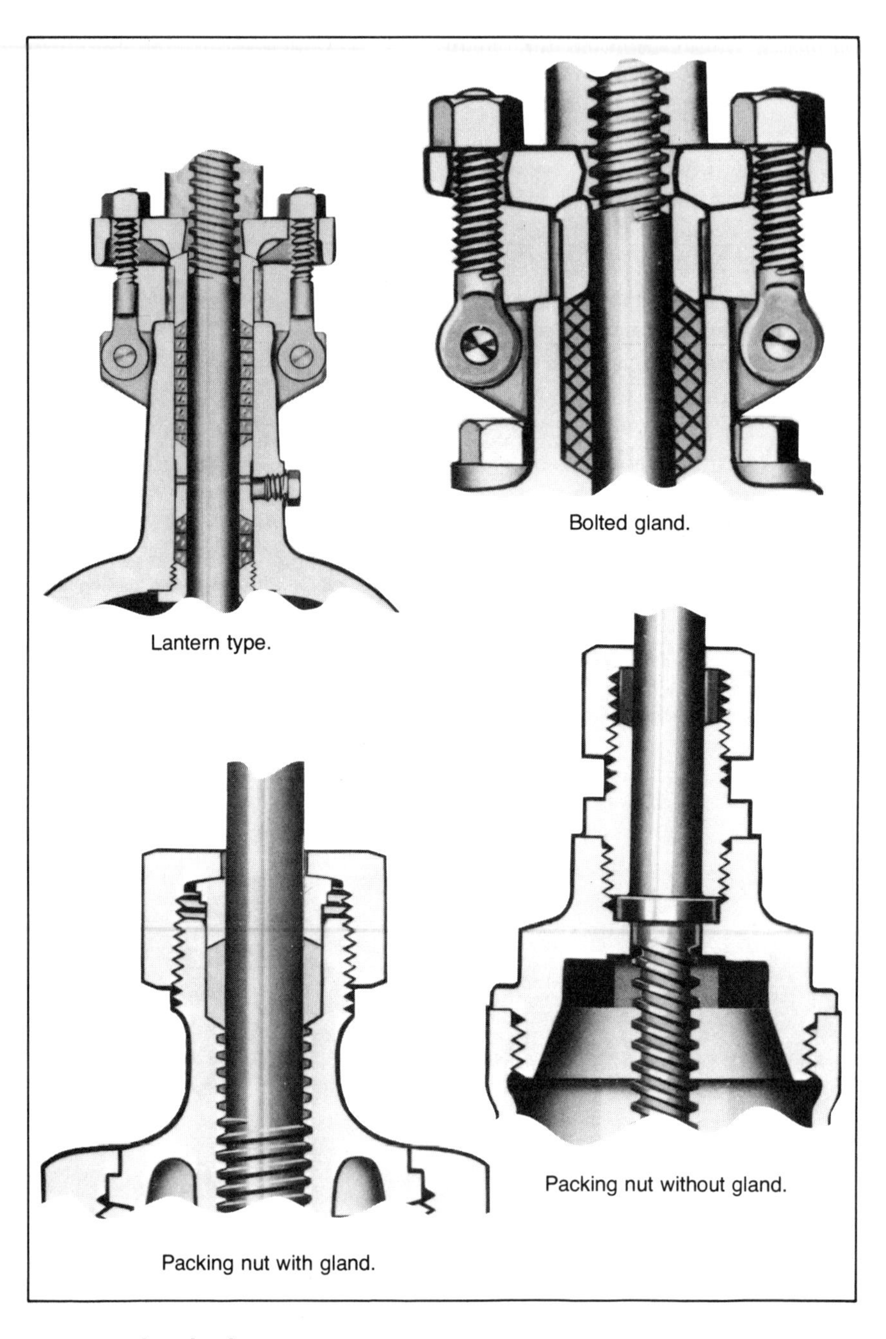


**FIG. 35. Valve gland types** *(courtesy Crane).*

and an outside screw and yoke (OS&Y) to hold the gland follower have gained wide acceptance (Fig. 35).

Besides the differences in wedge, bonnet, and gland design, there are more differences. Some valves are designed with a rising stem, where an indicator riding on the valve spindle can show if and to what degree the valve is open. As opposed to the rising stem, there is, of course, the valve with the nonrising stem (Fig. 36).

Depending on what type of gate valve or for that matter any valve referred to, there always crops up the question of repairs and replacement parts. Repacking gland boxes can be accomplished on many valves, even when under pressure and in the open position, because of a feature called *back-pressure seating.* This effectively seals the packing retaining chamber from the pressure of the fluid flow.

One major difference in gate valve design is the various different types of wedges available.

### Single-wedge disc gate valve

The single-wedge disc is usually solid and fits into tapered valve seats, which may be replaceable, or into an internal part of the valve body. This single-wedge design is particularly well suited to overcome misalignment and dimensional changes within the valve body due to temperature variations. The tapered construction of the disc and seats with the resulting large seating area normally result in a wide and true contact between the disc and the corresponding faces of the valve seat (See Fig. 32).

A variation of the solid wedge is the so-called flexible disc, which is solid only throught the center so that some movement of the faces relative to each other is possible. This flexibility, which is attained without adding additional parts, can assist greatly in ease of operation and guaranteeing valve tightness, not only on the inlet seat but also on the outlet seat. Another design that is used in conjunction with tapered valve seats is the split wedge (Fig. 37).

Large-sized valves are often provided with a bypass around the valve seat, which may assist in pressure equalization or warm-up of a steam-carrying pipeline.

### Double disc gate valve

In double-disc gate valves, the two seating surfaces can move relative to each other. This gives good shut-off even with misaligned seats or different seat angles. In one design, the two discs are held together by a ball and socket that permit movement of the discs when they mate against tapered seats. Parallel seating double-disc designs use spreaders or wedges

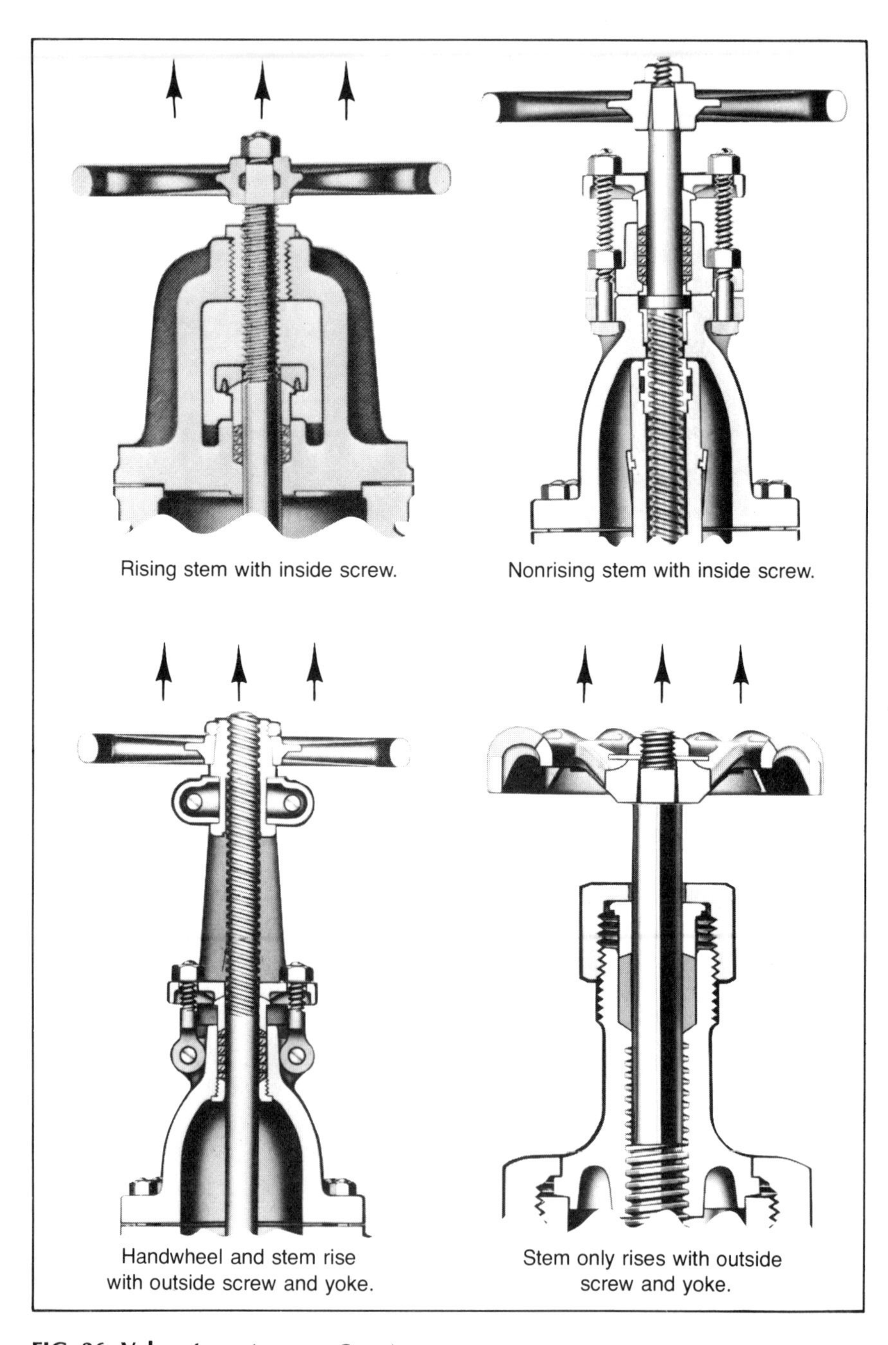


**FIG. 36. Valve stems** *(courtesy Crane).*

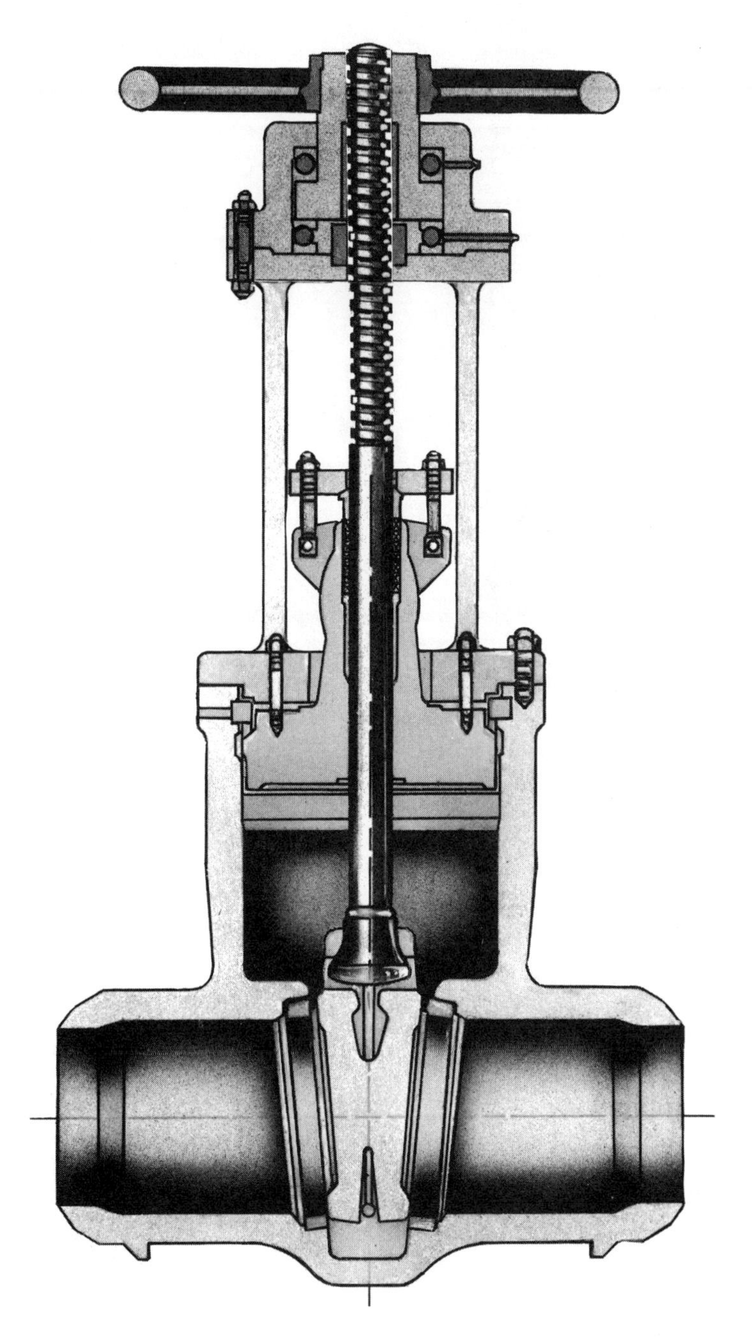

**FIG. 37. Typical gate valve, flexible wedge** *(courtesy Crane).*

to force the discs against the seating surface. Seat wear is minimized because the discs contact the seat without sliding motion (Fig. 38).

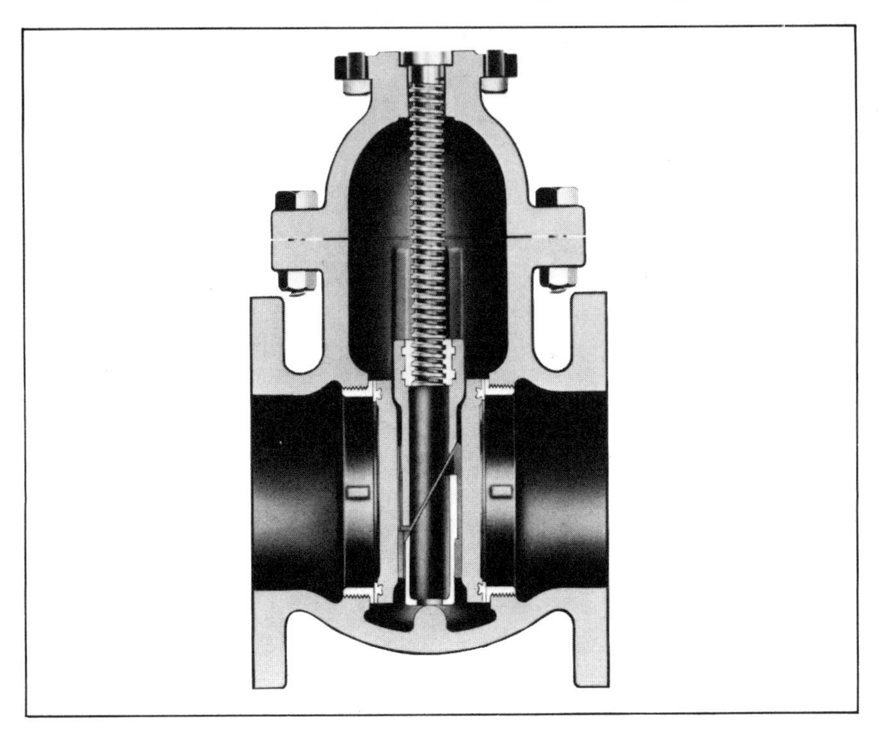

**FIG. 38. Double-disc gate valve** *(courtesy Crane).*

Another design with soft inserts in the discs and vent or drain connections in the valve body provides tight shutoff and permits the valve to be used for double block and bleed applications.

Large-sized valves when installed in a horizontal position often are required to have an internal track and rollers that assist in opening and closing operations.

Flanged cleanouts can also be provided to assist in the cleaning of valve seats without dismantling the entire valve during maintenance operations.

## Knife gate valve

This variation of a single-wedge gate valve has a very thin, knife-like wedge and is mostly used in specialty applications, like in the paper industry or for slurry services. Because of special design features, the valve is clog-proof, and materials that otherwise might cause an obstruction in the valve port are sheared off.

This valve is also furnished as a quick-action valve, where the knife-wedge is operated through a lever arm that eases operation with slurry-like substances.

### Conduit gate valves

This nomenclature has been adapted for large gate valves. The main feature of these valves, which are mostly used for oil pipeline service, is a smooth, round bore that permits passage of pigs or scrapers. The gate itself may be a single gate or a parallel double-gate assembly. As an additional feature, most conduit valves have a base that permits easy valve support from any supporting structure.

## Globe valves and angle valves

Both globe and angle valves are covered under one heading because the seat and disc arrangements in both types are similar. The only difference is that the globe valve provides for continuous straight flow while the angle valve provides for a 90-degree change of flow direction through the particular design of its valve body.

There is also a Y-shaped globe valve in wide use today, particularly in high-pressure systems. Because of its particular body formation, it has less flow resistance than the regular T-shaped globe valve. There is no adaption of this valve body to an angle valve since the 90-degree flow change cannot be lessened (See Fig. 39).

Bonnet, gland, and stem design of globe or angle valves is in many respects similar to gate valves, but the valve internals are markedly different. While gate valves are primarily designed to completely shut off fluid flow or provide an unimpeded port opening when fully open, the globe or angle valve is better suited to perform a throttling operation or to permit flow through only a partially open port opening.

Globe-valve seating is parallel to the line of flow, and there is no contact between seat and disc as soon as the valve is opened. Because of its obvious advantages of service in the partially opened or throttled condition, globe valves are often used in conjunction with home appliances. In industrial applications such as the feeding of liquid or steam to equipment with differing flow requirements at various times, i.e., water supply to a boiler or steam supply to a heater, the globe valve is also preferred.

As with gate valves, there are also various globe or angle valve designs, and there is practically no limit to the materials of construction. Bronze, iron, and steel are preferred valve body materials, but valves in stainless steel, aluminum, PVC, and other materials are also abundantly available. The various valve designs are generally identified through the variations in internal or, more specifically, their disc design.

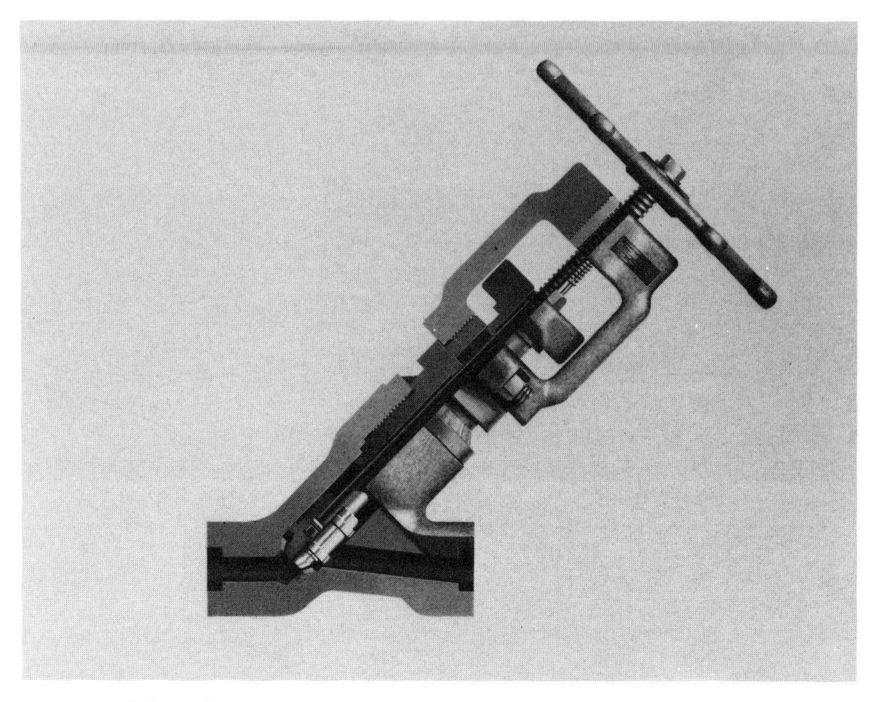

**FIG. 39. Globe valve, Y type** *(courtesy Rockwell International).*

## Conventional or ball-type disc

The original ball-type disc is the oldest kind of globe valve. As of late, it is also being replaced by the conventional type, which has retained most features of the ball type with the exception of the convex shape of the disc (Fig. 40). The basic design feature is a flat-surfaced though internally slightly tapered valve seat that is fitted with a disc of convex configuration that uses the taper in the seat for closing. This type of seating has only a narrow line of contact that normally assists an easy pressure-tight closure; however, deposit of foreign particles on the narrow seat ledge, which is a not uncommon condition, often prevents such tight closure. The valve should therefore be used more in undemanding applications and low-pressure service. Seat and disc are mostly metallic, and the disc is normally replaceable. The seat can be reground without removing the valve from service.

## Composition or renewable disc

The renewable or composition disc got its name from the material rather than from the configuration of the disc (Fig. 41). The disc is normally a circular shaped, approximately $^3/_{16}$-in. thick flat piece of material that

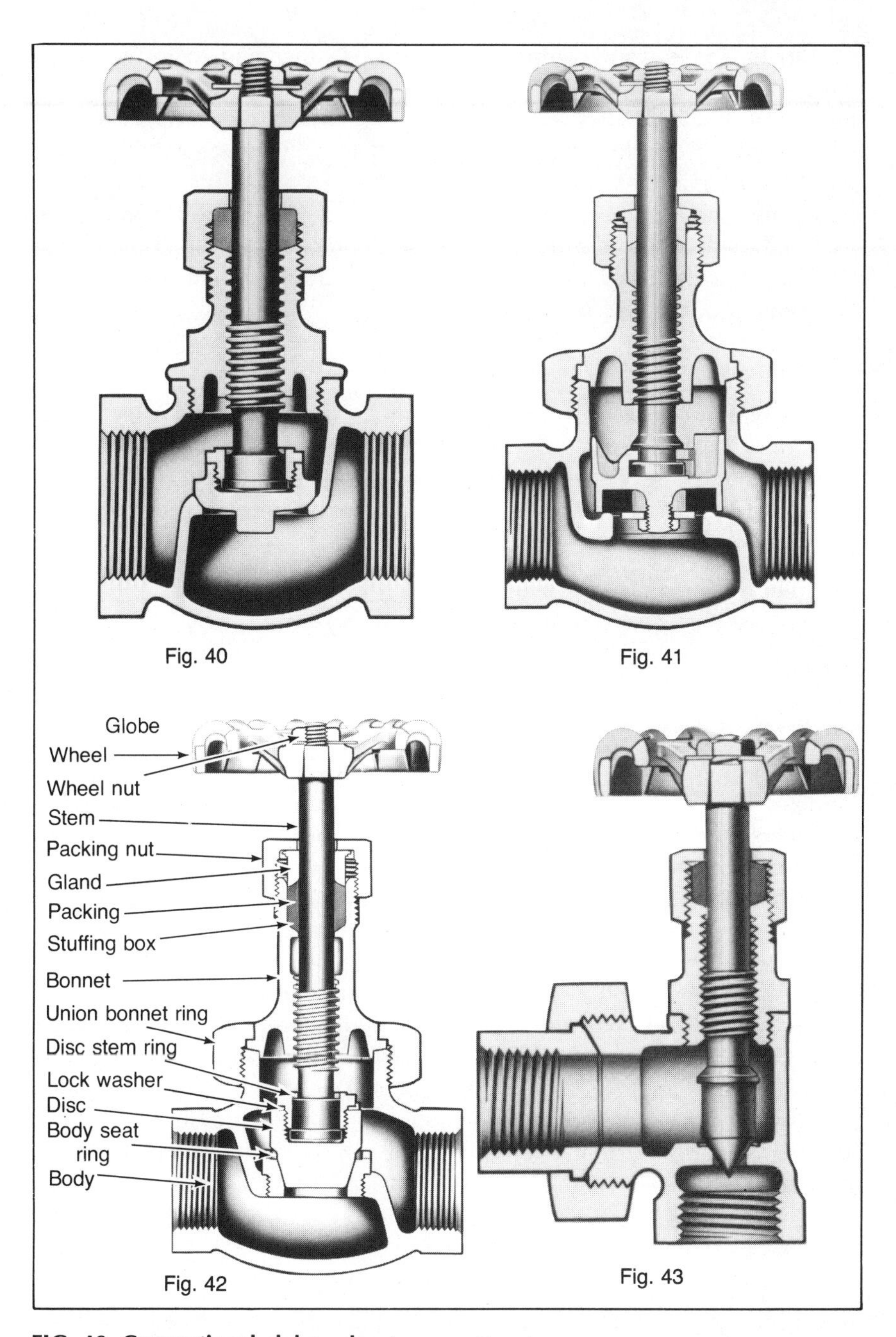


**FIG. 40. Conventional globe valve** *(courtesy Crane).*
**FIG. 41. Globe valve, composition type** *(courtesy Crane).*
**FIG. 42. Globe valve, plug type** *(Courtesy Crane).*
**FIG. 43. Needle valve** *(courtesy Crane).*

used to be made from compressed fiber or leather and today is mostly plastic but varies, depending on application. This renewable disc is fitted into a disc holder and is retained there by a small screw. Closure is effected against a thin lip protruding from and actually constituting the valve seat. Foreign particles, which so easily can prevent closure of a balltype glove valve, do not affect this valve since they can easily be absorbed into or imbed themselves into the relatively soft composition disc. This type of valve is most often found in homes. For the do-it-yourself homeowner, replacement of the disc is a familiar chore.

### Plug-type disc

This type of globe valve is best suited for its avowed throttling application and is also best able to withstand the rigors of high pressure and high-temperature service (Fig. 42). A long, tapered metallic plug that is fitted into a corresponding seat provides a wide area of seating contact combined with a proper selection of metals. This is most effective in resisting erosive effects of close throttling. Because of the wide seat-bearing area, foreign matter in flow can seldom damage a seat or plug area large enough to cause leakage. Both seat and disc can be replaced in most plug-type valves.

### Needle-point disc

Needle-point valves are designed to give fine control of flow in small-diameter piping (Fig. 43). The name is derived from the sharp-pointed elongated plug that is provided in lieu of a disc and which fits into a matching orifice-like seat area. Even when fully open, the needle-point valve does not permit a full flow, since the open seat area is only a fraction of the piping flow area. The stem threads are finer than in any other comparable valve, so that considerably more rotations of the stem are required in order to permit full flow through or full closing of the valve.

This specialized construction permits application of the valve in many flow-control situations, which require a close regulation as is often necessary when calibrating instruments. For such a purpose, needle-point valves are also available outfitted with an indicator, which lists the number of stem turns made, so that fine adjustments can be achieved and can be repeated after valve opening or closing for whatever reason.

## GENERAL

While globe valves normally impede the flow of fluids because of their varied seating construction, angle valves reduce the number of pipe fittings required (Fig. 44). By their very construction, they perform the functions of

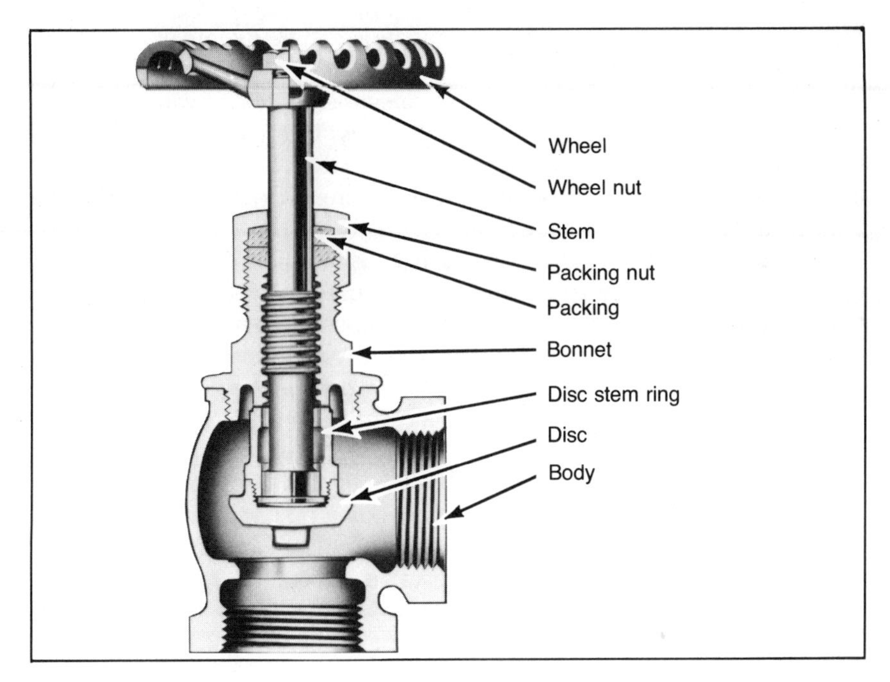


**FIG. 44. Angle valve** *(courtesy Crane).*

both a valve and an elbow. Small globe valves have a major advantage over gate valves in that it is very easy to replace a disc or seat or even regrind an integral seat without removing the valve from the system. This makes the globe valve ideally suited in applications where ease of repair and maintenance are prime considerations and minor flow restrictions are of no importance.

Installing globe valves should always be such that line of flow exerts pressure from below the disc and assists in lifting the disc when opening a valve. Such a direction of flow is mostly indicated through an arrow embossed on the valve body.

## CHECK VALVES

Check valves are normally closed and open automatically when pressure is applied to the valve by the flow of a fluid. The valve also closes automatically if a pressure builds up against the direction of flow, thus preventing any backflow or pressure buildup beyond the design requirement of a given component of a piping system. Flow of fluid is permitted in only one direction and prime importance is directed toward the correct valve installation. An arrow on the outside of the valve body indicates direction of flow, and counterflow installation can easily impair or damage delicate parts of a piping system.

Although there are only two basic categories of check valves, namely swing and lift checks, each has many variants. Like with all other valves, body materials and end preparation of check valves can be made available to suit any given piping system, and there are two basic types of valve bonnet: a flanged or a threaded bonnet.

*Swing check valves* are so named because of their particular mode of operations. They are by far the most widely used in general industry since they offer little flow resistance and are virtually foolproof in operation. Like gate and globe valves, check valves have a valve seat and a disc or an aberration thereof that, in this case, is the only moving part. The disc, which is hinged at the top, seats against a machined seat in the tilted bridge wall opening (Fig. 45). The disc swings freely in an arc from fully closed position to one providing unobstructed flow. Discs can be furnished with metallic or nonmetallic facings, depending on operational or maintenance requirements. To increase sensitivity to flow, an outside lever and weight can be attached to assist in the valve's operation.

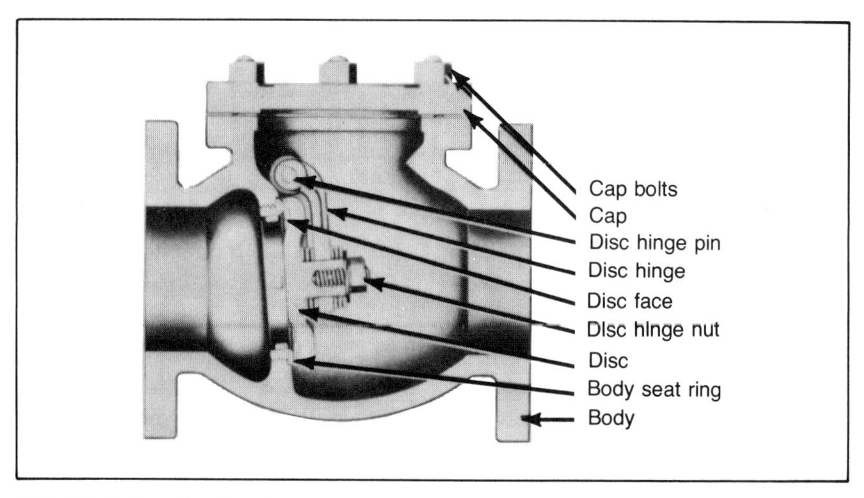


**FIG. 45. Swing check valve** *(courtesy Crane).*

A variation of the regular swing check valve is the tilting-disc check valve. The hinges that support the disc are located just above the center of the disc, and this different pivot location is instrumental in minimizing slamming. This type of valve is normally not furnished in sizes smaller than 2 in. (Fig. 46).

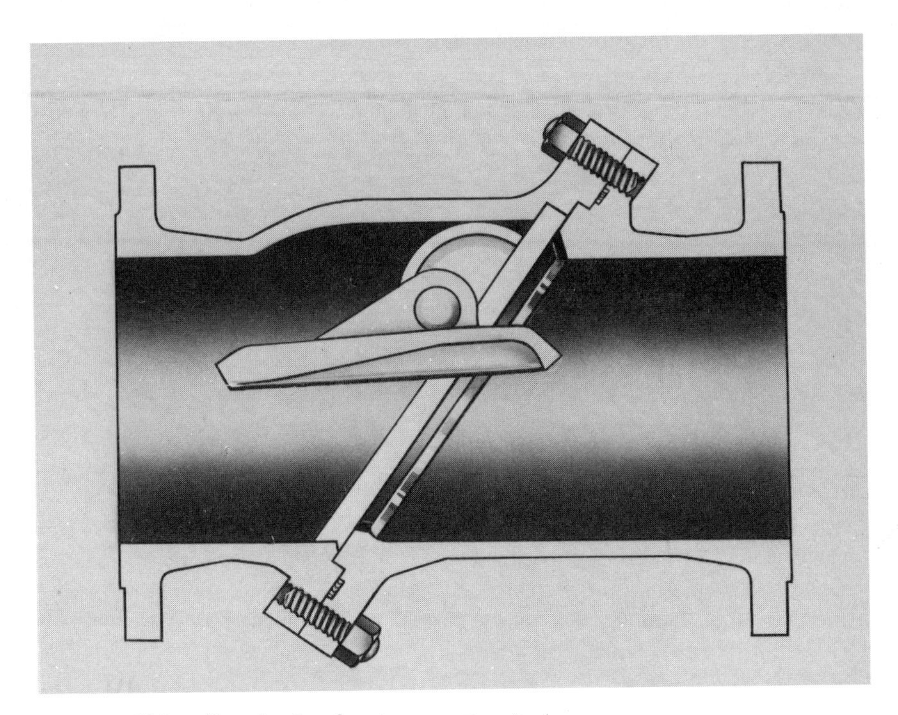

**FIG. 46. Tilting disc check valve** *(courtesy Crane).*

*Lift check valves* can be divided into three different types that differ in construction and application.

1. *Horizontal-lift check valves* have an internal construction similar to glove valves, and the same body casting is often used in both applications. The disc, which is seated on a horizontal seat, is equipped with guides above and/or below the seat and is guided in its vertical movement by integral guides in the seat bridge or the valve bonnet. Both seat and disc are metallic and can be easily replaced after removal of the valve bonnet. As the name implies, the valve is specially designed for installation in horizontal position and should not be used otherwise (Fig. 47).
2. *Vertical-lift check valves* have the same guiding principle as horizontal lift check valves, namely a free-floating guided disc that rests when inoperative on the seat. Because of the very nature of their operative system, vertical-lift check valves are of practical use only when installed in a vertical piping system with an upward-directed flow.

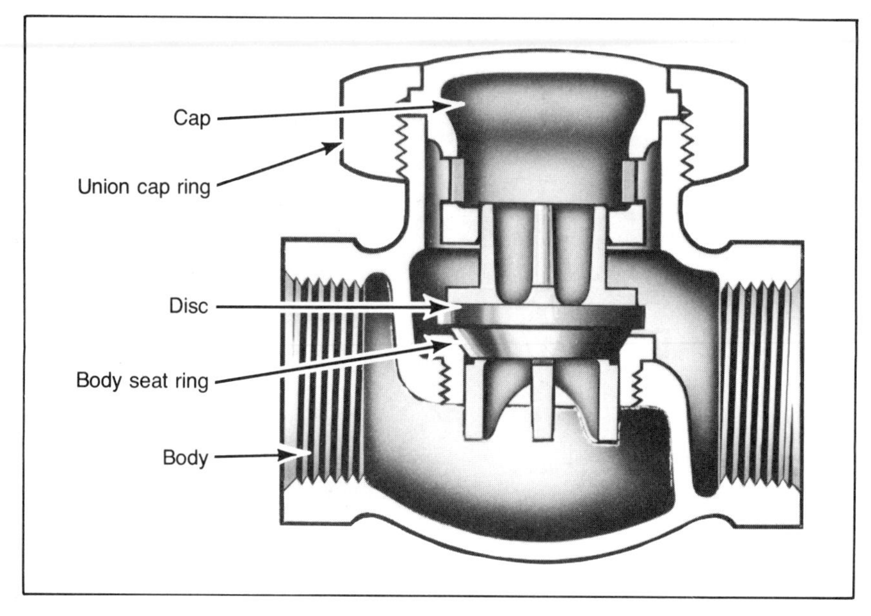


**FIG. 47. Lift check valve** *(courtesy Crane).*

3. *Ball check valves* are of similar construction as horizontal or vertical lift check valves. Instead of a guided disc, a ball serves as the flow-control medium. When operating, the ball is constantly in motion, reducing the effects of wear on any particular area of its sphere. While ball valves are manufactured from the various metallic materials, this type of check valve design has been found well suited for manufacture and operation in plastics materials.
4. *Miscellaneous check valves.* Besides the variations of lift check valves that are indicated above, a variety of check valves are readily available that represent a single manufacturer's effort to introduce a speciality item.
   a. A wafer-type check valve, for installation between two existing pipe flanges, has a helical spring loaded disc and is of very compact design. Because of special design considerations, it is considered to be of a nonslam nature and lends itself to horizontal or vertical installation (Fig. 48).
   b. Another wafer-type check valve for installation between existing pipe flanges is of different design. The disc is composed of two separate half discs that are mounted through hinges on one pin. For flowthrough, the two disc halves fold back and are side by side in the center of the pipe. The closing is performed with

**FIG. 48. Wafer-type check valve** *(courtesy Daniel).*

the assistance of separate spring arrangements for each half disc.

This valve type also lends itself to horizontal or vertical installation. A prerequisite during horizontal installation is that the holding pin must be in a vertical position.

c. A flanged check valve that is composed of two halves flanged together with a disc and seat assembly inserted between the two flanged halves is suitable only for air or gas service. A slotted arrangement permits flowthrough. Even in the fully open position, flow is severely restricted.
d. There are also available other check valves that are either piston or spring operated, as well as other variations of the three basic designs.

### General

Check valves, like gate and globe valves, rely mostly on iron, steel, and bronze as materials of body construction, while seats and discs are furnished in a variety of metallic alloys and nonmetallic materials. However, the availability of check valves in other materials for incorporation in piping systems that are based on other than steel piping is practically unlimited. Most check valves have removable seat rings, and discs are also easy to replace after opening of valve.

Valves in horizontal service with the exception of the tilting disc check valve have either a screwed cap or bonnet or a flanged top that can be opened without removing the valve from the line. The resultant opening is big enough to lift out the valve internals. Valve seat removal mostly requires special tools and is not a job for the unitiated, while the disc can be changed quite easily. The disc on swing check valves is supported by a hinge, which in most valves can be removed by loosening the screws on both sides of the valve body. Check valves should be inspected or tested periodically, since an inoperative or leaky check valve may otherwise not be detected until a possibly serious occurrence, when malfunction might have serious consequences.

## PLUG VALVES

The plug valve or cock may very well be the oldest member of the valve family since it can trace its origin to the simple peg. Like all other valves, plug valves come in a variety of types and materials. The two main groupings are lubricated and nonlubricated valves. There are also differences in the shape of the plug and its opening for fluid flow, as well as the general valve housing.

Plug valves are at first identified with positive closure since there is no possibility of seat leakage.

Another outstanding feature of the plug valve is its quick-opening and closing operation. A quarter turn fully opens or closes the valve. With the help of a stop collar on the valve, the operator does not have to rely on the feel of a wheel's resistance to tell if the valve is closed. Valves may be wrench, wheel, or gear operated.

Almost all wrench-operated plug valves have a flat or a square wrench head on the plug stem and a stop-and-indicating collar that also serves as a weather shield for the stem clearance space.

The plug valve in its basic form is of very simple design and consists of only three parts: body, plug, and cover. Of these three parts, only the plug is nonstationary and is used during valve operations to open or close the flow of fluid through a 90-degree rotation. This movement either aligns the porthole that is part of the plug with the direction of flow or shuts off any flow. The porthole or flow opening in the plug may be round, oblong, or diamond shaped and may permit either full or restricted flow through the valve.

### Nonlubricated plug valves or cocks

These range in size from the very small to the very large, and applications may vary from the small drip cock at a gauge glass or the shut-off cock at a gas range through various industrial applications to the large so-called cone valves that are used in water-supply installations (Fig. 49).

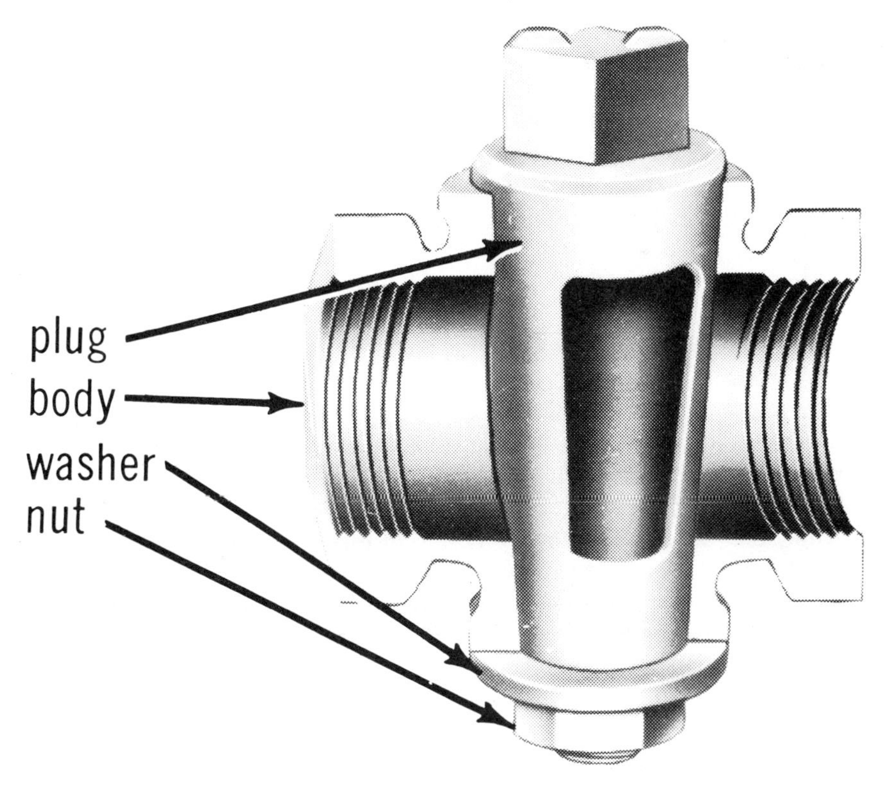


**FIG. 49. Basic plug valve or cock** *(courtesy Crane Co.)*

Depending on the manufacturer, the plug may be inserted from the top or the bottom into the valve body. The use of cylindrical plugs is often preferred in nonlubricated plug valves since they are less likely to experience galling or freezing than conical plugs. In the various designs, plastic seals are often molded into grooves of the plug to provide better seals, and bottom springs assist in the operation.

An aberration of the standard plug is the eccentric plug, which is only about one-third of a full plug in area. It permits a full flowthrough in the open position and closes with less contact of the seat on body walls. Another manufacturer encapsulates all parts of the valve that come in contact with the fluid in plastic materials so that a corrosion-resistant valve is created. Most manufacturers use their own longitudinal size so that interchangeability of these valves is practically nonexistent. Similarly, design details for most valve parts vary so that design and purchase requisitions have to be dovetailed toward one particular valve model.

### Lubricated plug valves

These come in even more variations than the nonlubricated plug valves, but the main difference is, of course, already indicated in the nomenclature. A lubricant is forced into various grooves in the plug body to minimize friction and thereby prevent sticking and also assists in sealing surfaces and valve stem. The lubricant pressure is also used to unseat the plug from its position and through this action nullifies any adhesion which may have taken place, and thus adds another positive factor in favor of lubrication.

The plug itself may be cylindrical or conical in shape with manufacturers of both types extolling the virtues of their valves. Proponents of the tapered form make reference to quicker and simpler valve operation because of the tapered rotary principle (Fig. 49). Another advantage of the tapered form is the combination of the tight-fit, wedge-like seating together with the sliding movement of plug rotation. Proponents of the cylindrical shape point to special seating construction, such as spring-loading or teflon bearings, that also seems to guarantee an ease of operation (Fig. 50). Lubricants can be forced into its various distribution channels either by a special lubricant gun that fits a buttonhead fitting on top of the plug or they may be inserted as stick lubricants into the lubrication opening of the plug shank and then have pressure applied through the turning down of an elongated lubricant screw.

A lubricant check valve normally prevents any lubricant from leaking to the outside. Most other accessories vary, depending on manufacturers design. No standardization of face-to-face dimensions of flanged valves prevails, although some valves are manufactured in accordance with the

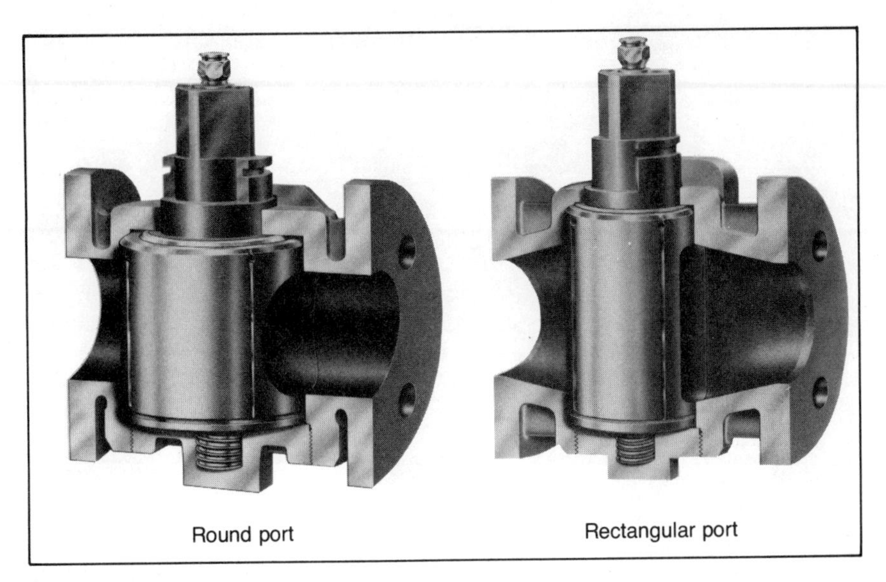


**FIG. 50. Cylindrical plug valves** *(courtesy WKM, ACF Industries).*

standards of ANSI B16.10 (face-to-face dimensions of ferrous flanged end valves, Class 125).

Because of the lubrication required, a valve maintenance program must be maintained that provides for the adequacy of the lubricant level in each valve at all times when the valve is in operation.

### Multiport valves

These are yet another variation of both the lubricated or nonlubricated plug valve. The basic structure of the plug valve permits a very simple adaption of multiport construction. In the three-way form, a single L-shaped passage connects any of two ports; with a T-shaped opening, any of three ports can be simultaneously connected. The four-way valve type may have two cutouts on opposite sides that connect any two adjacent ports of four ports or may otherwise connect two out of four available piping systems.

Multiport valves provide for a wide selection of interconnection of multiple piping systems. In lieu of having the plug available with many port openings, an eccentric-type plug can also be provided, which either seals off only one of the various piping systems or leaves all valve ports open. Multiport valves are adaptable to numerous valve port possibilities, but purchase requisition of the required valve has to give exact details of open and close port requirements (Fig. 51).

**FIG. 51. Multiport plug valve** *(courtesy Rockwell International).*

## General

As previously stated, most plug valves are lever operated since only a 90-degree turn of a lever is sufficient to open or close a valve. However, in larger-sized valves even that little movement is often difficult to achieve, and valves of six-in. and larger size can mostly be furnished with a gear-type operator that facilitates operation.

Plug valves are normally furnished with flanged or threaded ends: however, some valves are also furnished with weld ends. Welding connections to such valves must be made under controlled conditions. If too much heat is dissipated from the welding area, the body or plug of the valve might become affected and the valve seating could become impaired.

If a plug valve is disassembled for repairs or maintenance, this is usually easy to achieve since in most valves the plug can be removed from the valve body without difficulty. However, cleanliness before reassembly is absolutely necessary, and a cleaning solvent is the best answer.

## BALL VALVES

When unique features of plug valves are mentioned, we should not forget that ball valves have similar features, such as quarter-turn valve operation and complete flow shut-off. As a matter of fact, it might be said that the ball valve has evolved from the plug valve. The ball-shaped internals provide good seating, especially when used in conjunction with plastic surfaces, and thereby reduce the possibility of leakage through the valve when in the closed position.

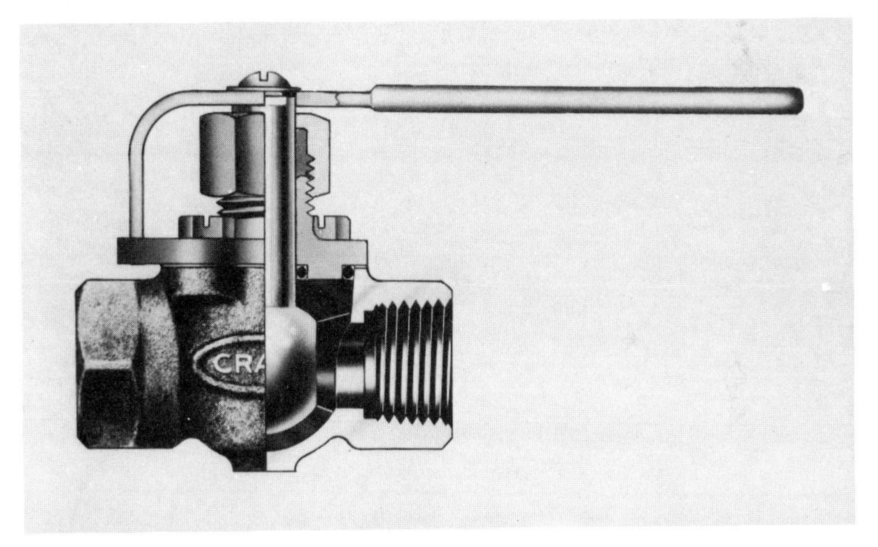


**FIG. 52. Ball valve, top entry** *(courtesy Crane).*

Ball valves are also manufactured to a large extent with easy disconnect features. That means that the actual valve body can be removed from the piping system by just opening a few bolts. Such valves sometimes have end fittings that remain part of the piping system, while the valve body is either flanged or clamped in between these end fittings and thus is easily removed for repair or replacement. Some of the other classic designs permit removal of the ball-type disc through the top and are top entry valves, as opposed to the end or side entry valve that must be disconnected to remove the ball disc (Fig. 52).

While ball valves are manufactured with metallic valve bodies of various composition, this particular type of valve has been found very successful in a plastic valve body adaption in small sizes for low-pressure applications. At the other end of the spectrum, ball valves are being manufactured in sizes of 36 in. and larger for special applications, in particular for high-pressure service in the oil and gas industry.

Some of the advantages of ball valves, such as unimpeded flow when fully open and fast 90-degree closure, can be of prime importance in specific installations where closure might be affected and speeded up through remote control in case of emergency.

## DIAPHRAGM VALVES

The diaphragm valve takes its name from the importance that the diaphragm plays in the valve design. Like the plug valve, the valve consists of only three basic elements, namely valve body, valve diaphragm, and the operating mechanism, which might be referred to as valve bonnet (Fig. 53).

**FIG. 53. Diaphragm valve** *(courtesy Hills-McCanna).*

The valve body is so designed that flow of fluid is directed over an indented weir, which closes the flow when the diaphragm is brought down and compressed against it by the operating mechanism. The operating mechanism is a convex compressor disc that is raised or lowered by a handwheel-operated stem or spindle. In lieu of this handwheel-operated stem, an air actuator can also be used to open or close valves by applying compressed air with or without the assistance of helical springs (Fig. 54).

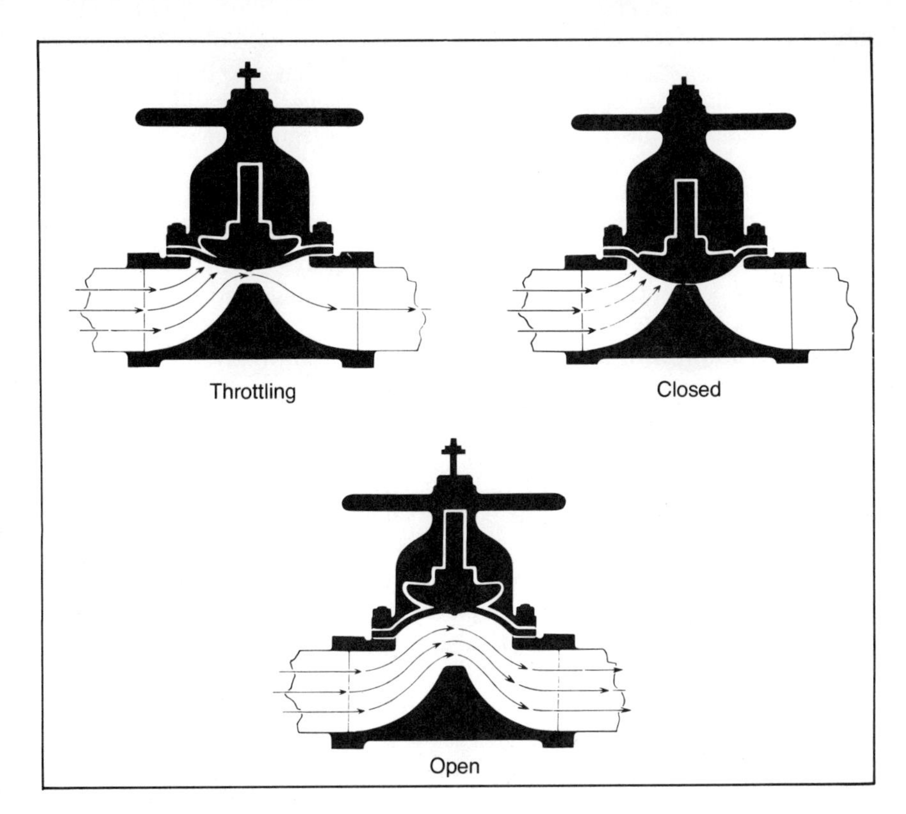


**FIG. 54. Diaphragm valve flow** *(courtesy Hills-McCanna).*

The main valve feature, the diaphragm itself, can be furnished in a variety of elastomeric materials or rubber, which should be chosen depending on the valve service requirements. The resilient diaphragm provides a cushioned, leak-tight closure and is designed so the fluid cannot penetrate it. It thereby isolates bonnet and operating mechanism from the fluid being handled by the valve. This simplistic construction eliminates the need for glands or valve-stem packing. Thus the flanged connection between bonnet and valve body and, of course, the actual valve connections into the piping system are the only weak spots where a leak might occur. Because of its elastomeric composition, the diaphragm and the valve itself are limited as to pressure and temperature ratings of fluids being handled.

The diaphragm valve can be manufactured in the various metallic materials and is yet another valve that is much sought after when manufactured in one of several plastic materials. Through its ease of assembly from the three basic components, the diaphragm valve has also been developed as a basic valve that can be lined with nearly every pipe-lining material available today. It has been established as an integral part of many lined piping systems where valving is required.

Valve endings are normally either threaded or flanged, but other specialized endings such as socket-bonding end for PVC and butt weld for stainless steel are available. Welding a diaphragm valve must be approached in a cautious manner and with the bonnet and diaphragm removed so as not to warp the valve body.

In order to prevent diaphragm rupture due to an inadvertently applied overbearing force during closing, a special indicator can be furnished to avoid such an occurrence.

The diaphragm valve has found a multitude of applications, and the various lining materials and their corrosion or abrasion resistance have greatly attributed to its widespread acceptance in the process industries. It is equally useful as a shut-off valve or can serve as a flow control or throttling valve. Sizes of diaphragm valves normally range from 1/2 in. to 16 in. Maintenance work or diaphragm replacement is easily accomplished without removing the valve from its location in a piping system by simply removing the valve bonnet.

Diaphragm valves are being manufactured in an angle-type valve configuration and also as straight flow through valves without a weir. The valve without weir has a diaphragm that is formed in a shape similar to the compressor. When fully opened, the valve has no flow restrictions.

## BUTTERFLY VALVE

The origin of the butterfly valve goes back to the shutter-like damper that was initially in use in applications where no tight shutoff was required but rather served more as flow restriction. Today's valves, which are mostly outfitted with rubber or elastomeric seats, provide a shut-off service like any other valve. The valve design is particularly suitable for installations where space considerations are important and makes this type of valve a favorite also for very large piping systems since there is practically no size limitation.

The valve consists basically of the valve body, shaft and butterfly disc, sealing gland, and valve actuator. The simple valve design has been diversified by introducing three different valve bodies without variations in the interaction between seat and disc.

*The flanged butterfly valve* has a short valve body and is flanged at both ends. If necessary, welding ends, in lieu of flanges, can be provided. But welding butterfly valves is not a standard connecting method and is not desirable because of possible damage to the seating surfaces. AWWA Specification 504, "Rubber-Seated Butterfly Valve," lays down requirements for such valves in water service (Fig. 55).

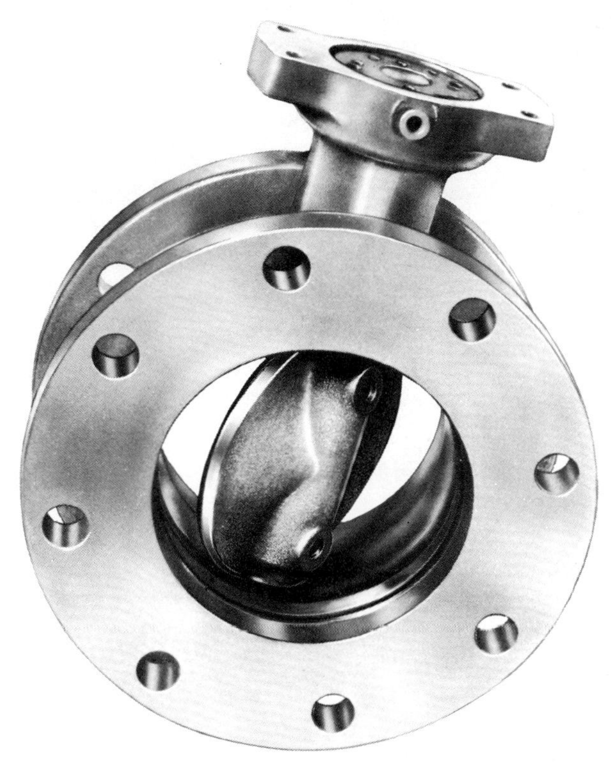

**FIG. 55. Flanged butterfly valve** *(courtesy Crane).*

*The lug-wafer butterfly valve* has a shortened valve body with protruding lugs whose bolt circle matches adjoining flanges. Lugs can be connected to the mating flanges either through full-length bolts that squeeze the wafer between two pipe flanges. Tapped holes can be provided and cap screws can be used to fasten the lugs individually to each flange, thus permitting the valve to be used as a dead-end valve also (Fig. 56).

**FIG. 56. Butterfly valve, lug wafer type** *(courtesy Crane).*

*The wafer butterfly valve* consists of a short body like the lug wafer but without the lugs. This valve can be inserted between two adjoining flanges but must be centered exactly since there is no guidance provided for its location. Gaskets may be molded onto the body or may have to be inserted for satisfactory flanged joint (Fig. 57).

**FIG. 57. Butterfly valve, wafer type** *(courtesy Crane).*

## General

Most butterfly-valve component parts are of metallic materials with stem and disc often furnished in a higher-grade material than that of the body because of service requirements. The valve body, which is also the valve seat when the butterfly disc reaches a perpendicular position, is often lined with rubber or plastic materials to provide a pressure-tight shutoff. Where the stem protrudes the valve body, a gland sealing is provided to eliminate fluid loss at this point. The valve actuator for the simple quarter-turn operation may be manual like a simple lever or gear operated for a large valve or may consist of a motor or cylinder operator.

Torque characteristics play an important part in the operation of butterfly valves, and the various friction effects, as applicable to bearing, seal, and seating, should be determined when selecting a butterfly valve.

## PINCH VALVES

Pinch valves consist of only two working parts, a rubber or plastic tube, which is the fluid-conveying part of the valve, and an outer casing or pinching mechanism, which is the operating part of the valve. The tube may be fully enclosed in a metal body or can be just encased in a clamp-like device that provides the pressure to interrupt the fluid flow. Automatic operation can also be provided by using an air-operated actuator to provide the necessary power to operate the pinching mechanism.

Since there are no internal operating parts, this valve represents a unique nonclogging variation that is also resistant to a variety of abrasives, depending on the inner tubing material. End connections are mostly of the flanged type when the fluid-conveying tube is fully contained, or the tube may be directly fastened to the adjacent piping. Because of the simplistic valve construction, no maintenance of fluid-bearing components is required: however, the inner tube may call for replacement periodically, depending on the fluid being conveyed (Fig. 58).

Pinch valves are favored for flow control of slurries and other abrasive liquids or semiliquids.

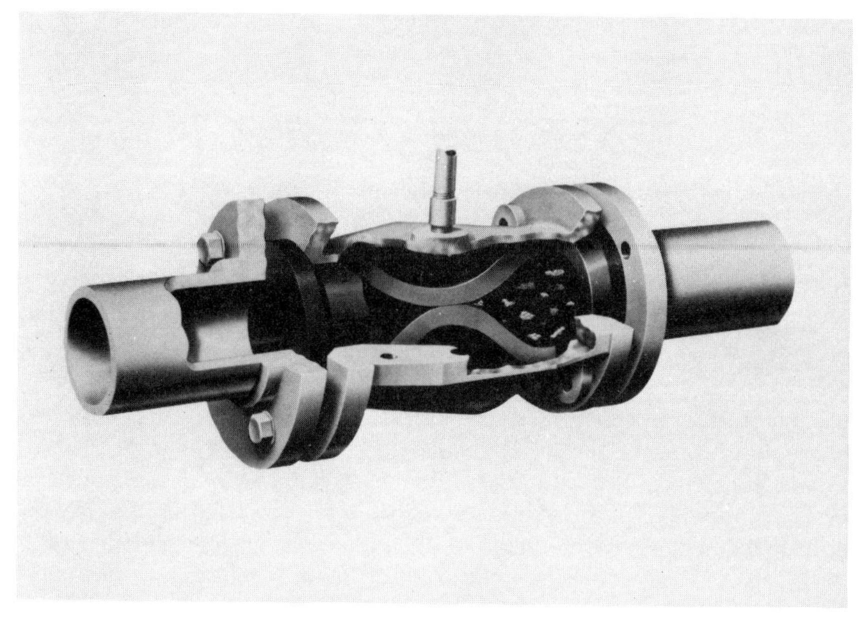

**FIG. 58. Pinch valve, air operated** *(courtesy Flexible).*

## BLOW-OFF VALVES

Blow-off valves are a necessary adjunct of steam generators or boilers, and their function as well as general requirements are spelled out in the ASME Boiler and Pressure Vessel Code, Section I, for power boilers. These specifications permit the installation of straightway-Y-type globe valves or angle valves for this special function under certain conditions. However, the valve industry has developed a valve especially designed to be applicable to this need and to meet the code requirements.

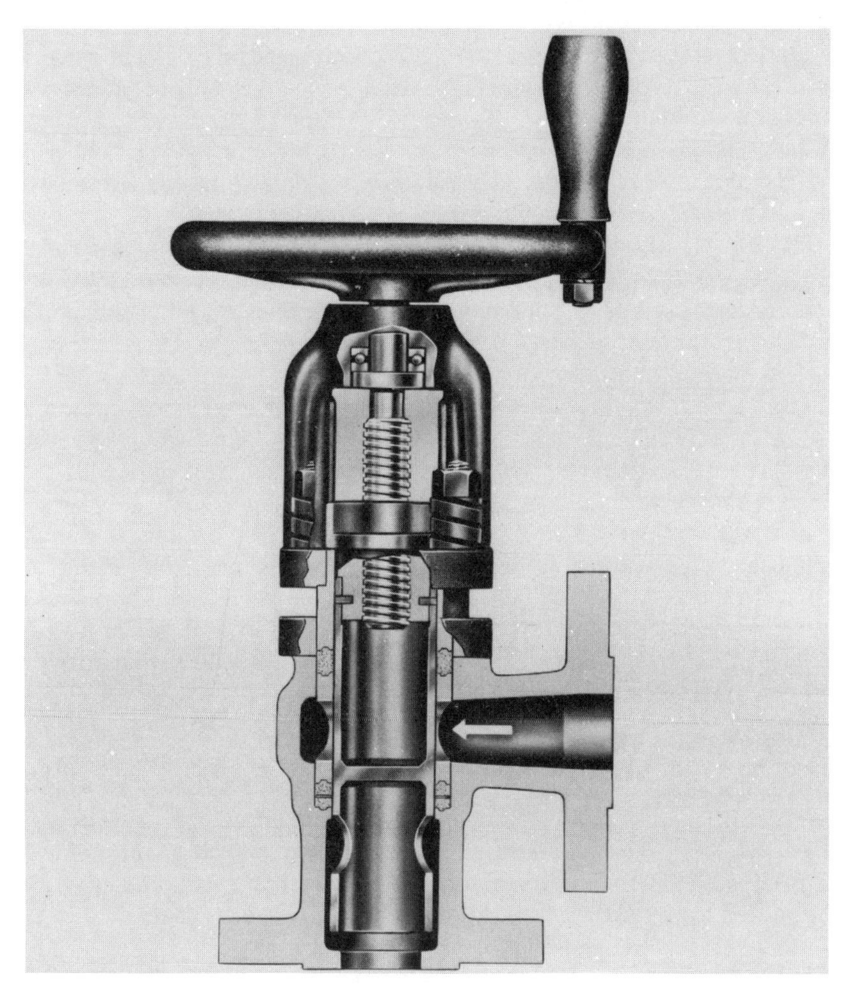

**FIG. 59. Blow-off valve** *(courtesy Yarway).*

The function of the blow-off valve is to assist in blowing down or reducing the water level in a boiler (Fig. 59). Its prescribed size ranges from $^3/_4$ in. to $2^1/_2$ in., and it has to be installed so that it is below the lowest waterspace in a boiler. Since the Boiler Code Section I also calls for double valving of all boiler drain piping above minimum pressure conditions, it is clear that blow-off valves have to be paired or that a tandem valve that incorporates two valves in one valve body has to be utilized.

Valves used in blow-off service normally must withstand heavy service requirement, and the special blow-off valves are intended to fill all these special needs. There are three basic types of blow-off valves manufactured, and they are classified by their internal construction.

The valve that differs most strikingly from other gate or globe valves is the sliding-plunger-operated seatless type. The plunger or piston is incorporated in a straight flow or angle-type valve body and is raised or lowered through a handwheel-operated nonrising stem. The plunger has port openings that correspond to the valve inlet when fully open and discharge through the plunger outlet (Fig. 59).

Another type of blow-off valve is a variation of the globe valve, which is classified as hard seat or seat and disc type. Yet another blow-off valve is based on the gate-valve principle and is better known as the sliding-disc type. It incorporates a lever-operated spectacle disc that opens or closes the flow of fluid in a fast action.

The tandem valve, which incorporates two valves in one body and thereby provides for reduced field installation through elimination of a flanged or welded connection, normally incorporates two different types of valves in one body.

Most blow-off valves are manually operated, but nonmanual operators, such as electrical, hydraulic, or air-activated attachments, are available for local or remote-control operation. Since blow-off valves may be inactive for a prolonged period of time but when called upon must perform flawlessly, preventative maintenance to keep the valve in good condition is a necessity and cannot be neglected.

## CONTROL VALVES

By implication of its very name, the control valve performs an important function in any piping system where nonmanual control is a desired system function. Plant modernization tends to require more and more reliance on automatic control from a centralized control room, and the same can be said for pipeline and tank-farm operations (which also rely heavily on remote-controlled valves).

While any manually operated valve can control fluid flow in any piping system, most control valves have been specifically designed to provide a variable resistance in a piping system, an operation that is better known as throttling. However, the indicating characteristic of a control valve is that its operating function is being guided by the condition of the fluid that is being handled.

Control valves mostly operate with a control loop that consists of two other important elements besides the control valve. A sensing element or transmitter measures temperature, pressure, or level of the fluid that is to be controlled. Finally, there is the controller itself, which remits the necessary signal either electronically or pneumatically that will actuate the operating mechanism of the control valve. The actual opening or closing of the control valve is then being performed by an actuator, which is part of the control valve. Since the control valve positively reacts to the command by the controller, it lends itself also to an application for fail-safe action and will open or close if the power supply to the actuator should fail.

Control valves are being manufactured in the same wide range of body styles, materials of construction, pressure ratings, and end connections that are available for hand valves. There are, however, differences, and certain types of valve bodies are more preferred than others. The most often used valve body is a globe-type unbalanced contoured plug valve, which lends itself best for throttling actions. Variations of the plug type globe valve are the balanced tight shut-off plug (Fig. 60), which is also available with a balanced trim. A double-seated valve normally has top and bottom guidance and thus lends itself easily to be either opened or closed by air pressure through reversal of the valve body.

Diaphragm valves are often used in piping systems that have rubber, plastic, glass, or other liners because the bodies of these valves can easily accommodate most commercial lining materials. A modification of the gate-type valve body is the sliding-gate valve, which mostly has a stationary gate with slots in it, serving as the valve seat, while a sliding seal covers or uncovers these slots depending on fluid pressure.

Rotary valves, which can be opened or closed through a quarter turn of the valve (ball valves, plug valves, or butterfly valves) are also often used as control valves, particularly in situations that require quick-opening or closing response. A variation of the plug valve called a cone valve often serves in large water-supply systems where elapsed closing time is a critical requirement. Similar requirements are served in petroleum-product tank farms through remote-controlled valves that may have to be closed because of events not involving the fluid that is being transported, such as fire or leakage at a nearby location.

While it was previously stated that actuators are part of the control valve, such actuators differ markedly from those that are normally in use

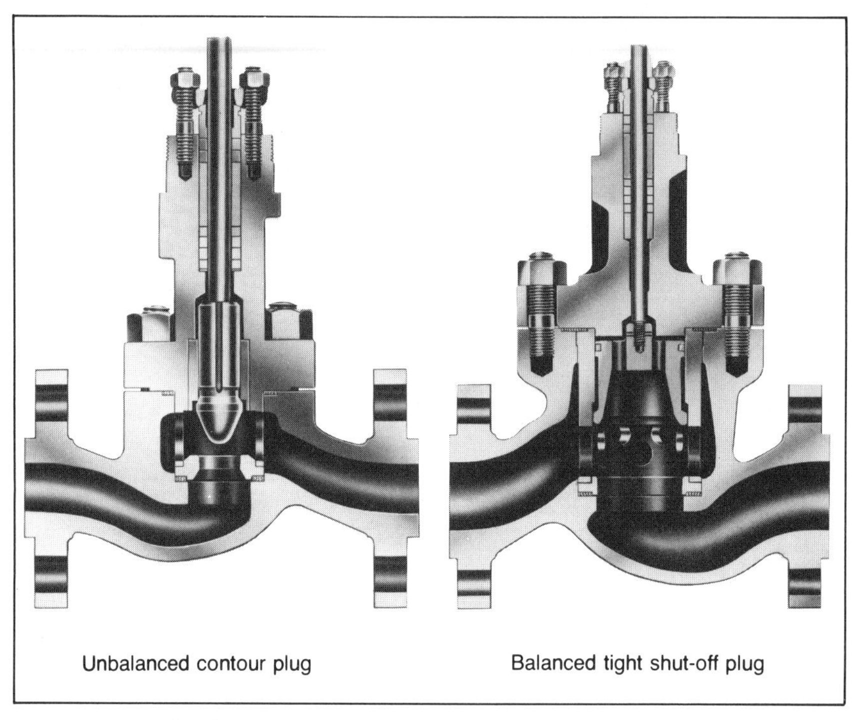


**FIG. 60. Control valves,** *(courtesy Masoneilan).*

with a manually controlled valve but are like those used for a remote-control hand valve. Power supply of the actuator may come from three different sources: compressed air, hydraulic fluid, or electric current. Depending on the power source, various actuators have been developed, sometimes combining electric power with another power source.

a. Pneumatically operated actuators often operate through a diaphragm, which is connected to the valve stem of the control valve.
b. Hydraulic pressure is mostly transmitted through a piston-type actuator. However, the piston may also be actuated by pneumatic pressure; this type of actuator is often preferred for rotary valves.
c. Electromechanical actuators use a motorized gear assembly that drives the handwheel of the control valve.
d. Electrohydraulic actuators combining the electrical control signal with the actual operation of a hydraulic piston.

In addition to these most widely used remote-control actuators, there are numerous variations available that are deemed better suited by the manufacturer for a particular application. Manually operated control valves are also in use and serve where fluid flow through a valve must be

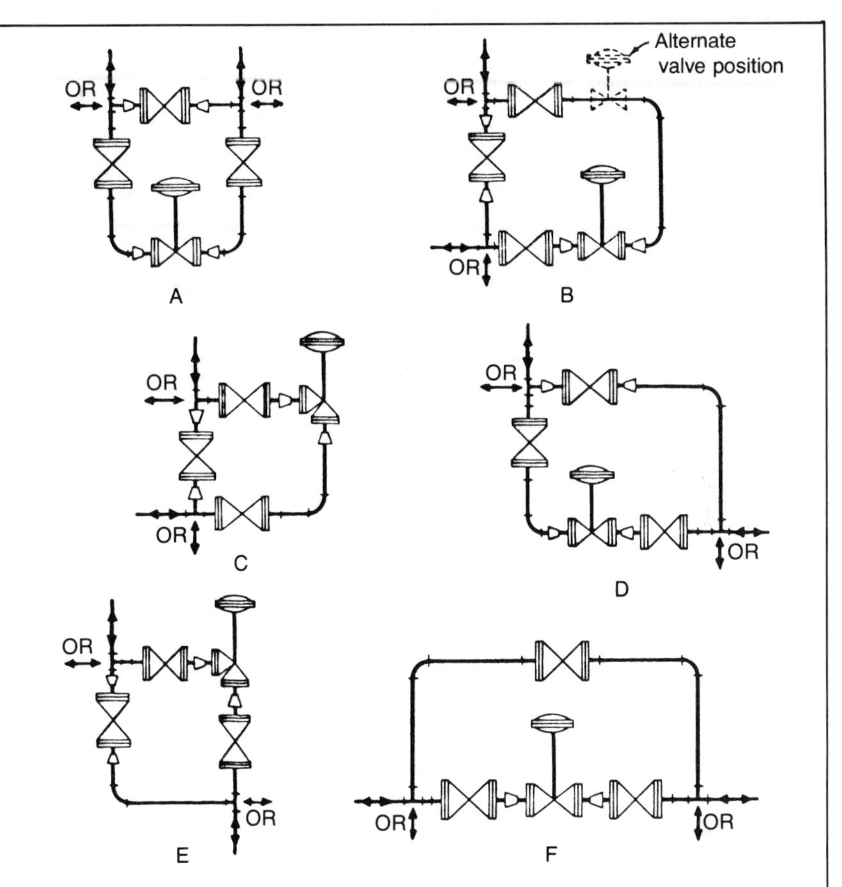


Arrangement A (ISA RP 4.2, Type 1) is preferred because manifold is compact, control valve is readily accessible for maintenance, and the assembly is easily drained.

Arrangement B is preferred because control valve is more readily accessible.

Arrangement C is often used with angle valves. Control valve is self-draining and valve is inaccessible.

Arrangement D (ISA RP 4.2, Type 3) is preferred because the control valve is readily accessible. Bypass is self-draining.

Arrangement E results in a compact manifold, but control valve may not be too accessible.

Arrangement F is preferred because bypass is self-draining; however, requires greater space.

NOTE: Block and bypass valves should be installed close to tees, as shown, to minimize pockets. Drain and vents on either side of the control valve are not shown, nor are the supports. Many manifolds can be rotated into any plane, keeping control valve vertical.

**FIG. 61. Control valve manifold arrangements** *(from Manual on Installation of Refinery Instruments and Control Systems, courtesy API).*

adjusted manually to comply with exacting flow requirements. Manual actuators are also often available in addition to the remote control and thus permit operating personnel to override the control system, or operate the control valve in case of power or signal failure.

With the advent of the Occupational Safety and Health Act (OSHA), noise problems have become a subject of intense concern in any operating facility. Control valves are often a cause of concern in this respect because of their contribution to noise pollution and the difficulty in containing it. Since OSHA requirements are very specific and operating-valve noise levels are not always predictable, even those operators who try to comply with the law may find that they have a problem on their hands and may have to replace an installation at great cost.

Control valves are mostly installed in conjunction with a manifold system that permits the piping system to operate with or without the control valve in service (Fig. 61). The control valve is located between two manually operated valves that when closed, isolate the control valve and thus permit maintenance while the system continues to operate after a third manually operated valve (which is normally closed) has been opened to permit the bypass of the control valve.

Pressure regulators (which might be classified as control valves, since their function is control of the pressure aspects of fluid flow) deserve specific mention. Here is an aberration of a control valve that performs a specific function but sometimes operates under different conditions than previously outlined. Pressure regulators are installed in piping systems and reduce any existing line pressure to a desired level. This pressure reduction can be achieved through manual operation, as is being done with oxygen or other gases that are furnished in bottled form. However, the regulating function can also be performed through remote operation. This is one type of control valve that is not installed with a bypass system.

## MISCELLANEOUS VALVES

There are numerous other types of valves manufactured for special applications and just a few are mentioned here:

*Cone valves* are large-sized plug-type valves that are mostly used in waterworks or water-treatment plants.

*Flush valves* have a seat and disc design similar to angle valves and are mostly installed at the lowest point of a tank to permit easy and fast draining.

*Quick-opening valves* are either gate or globe valves that are operated by a quick-acting lever instead of having a multiturn stem. There are no size restrictions and applications vary, but a good example for its use is the valving in an emergency shower (Fig. 62).

**FIG. 62. Quick-opening valve, knife gate type** *(courtesy Crane).*

*Jacketed valves.* Similar to jacketed piping and fittings, gate, globe, diaphragm, and plug valves are being manufactured for use in jacketed piping systems. The actual valve is encased in a second hull with entry and exit ports for the heating fluid at various locations.

*Stop-check valves.* A prevailing adjunct in steam-generating plants with more than one operating boiler, these valves have a cylindrical-shaped disc that is the only moving part. Large valve port areas normally produce only a minimum of pressure drop, which is of utmost importance in their intended service. These valves can perform several essential functions in the steam piping system of any boiler plant by:

1. Acting as automatic nonreturn valve and thus preventing a backflow of steam flow from the main steam header to the boiler.
2. Assisting in cutting out a boiler when firing ceases or assisting in bringing a boiler into service after a shutdown.

Due to their particular design, correct installation of stop check valves is a prerequisite for proper performance (Fig. 63).

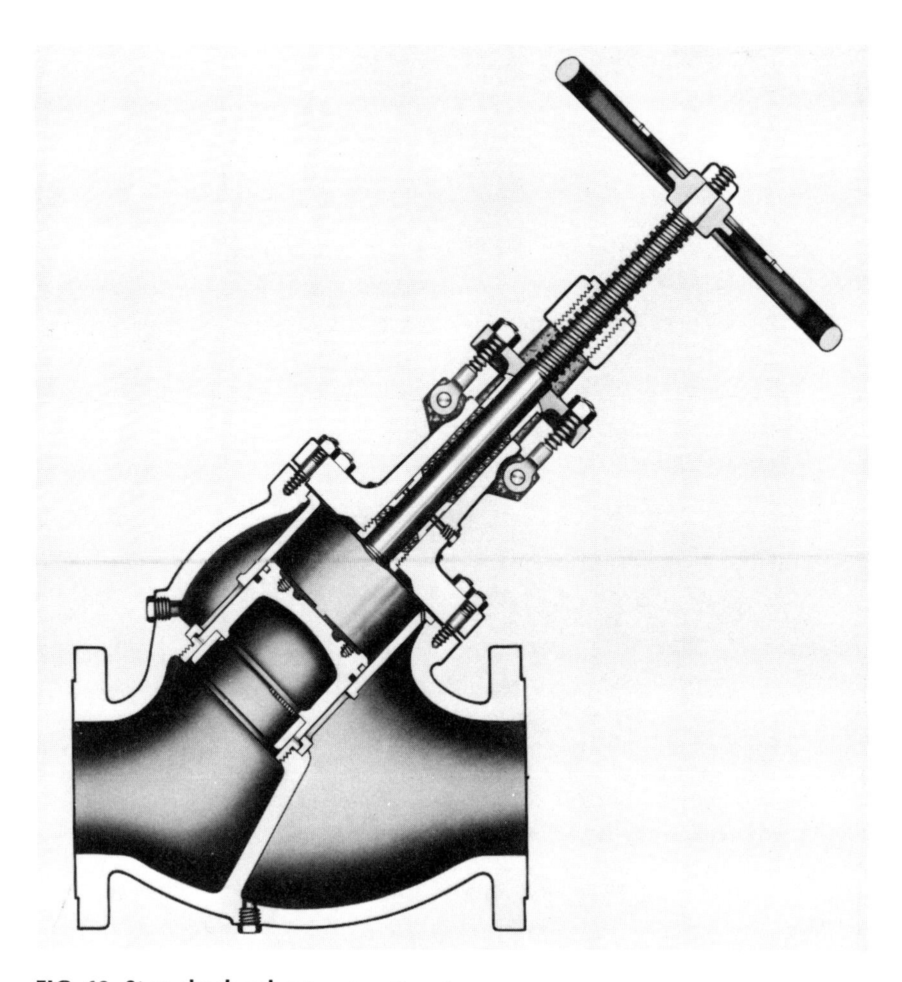

**FIG. 63. Stop check valve** *(courtesy Crane).*

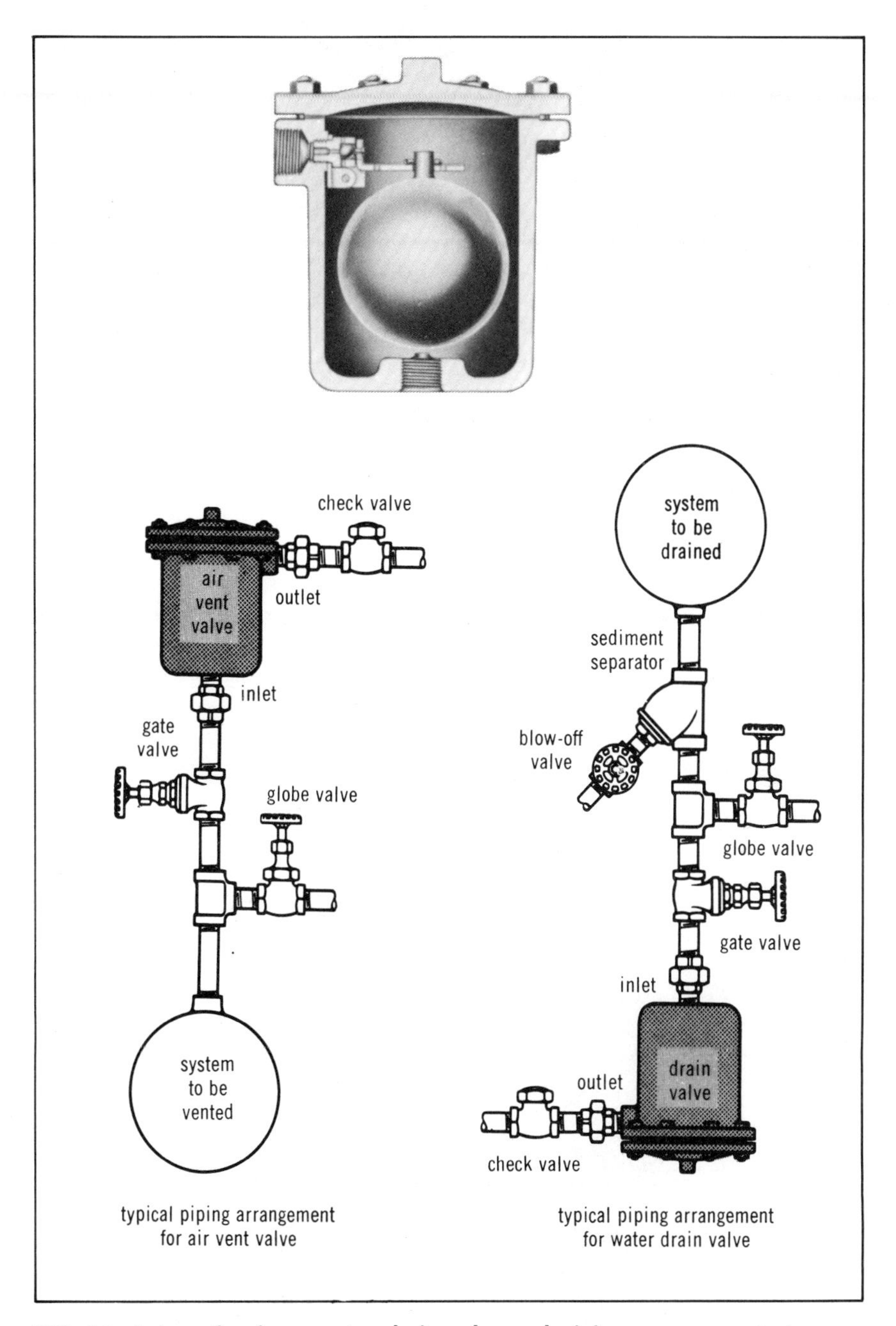


**FIG. 64. Automatic air or water drain valve and piping arrangements** *(courtesy Crane).*

*Sluice gates* are mostly rectangular and have been designed to handle large volumes of water while filling or emptying water reservoirs or for similar purposes. They are not normally found in any piping system but are often used in conjunction with concrete structures where part of the valve is embedded in concrete.

*Flap valves.* This simplest of all valves actually works like a nonreturn valve on lines or as an overflow valve from water reservoirs. Whenever water reaches the flap valve, it will permit the discharge of water but will not permit any reverse flow.

*Mud or plug-drain valves* are for similar service as the flush valve, only the valve construction is based on a plug-valve design with top entry. They are used mostly in concrete water reservoirs and water-treatment plants.

*Air-vent valves* operate automatically. Their purpose is to eliminate air in any piping system that is going to be filled with liquid or with steam. They can differ in size and construction, depending on the amount of air that has to be displaced. They can be found in steam heating systems but are also used in large overland pipelines.

Some air-vent valves are so constructed that they can serve as water drain valves when installed upside down. In either case, a float-operated discharge valve permits the automatic functioning of this type of valve (Fig. 64).

The design of air vent valves is often based on similar engineering prin-

**FIG. 65. Foot valve** *(courtesy Crane).*

ciples that are used for steam traps, and other air vents are reviewed as an adjunct to the discussion of steam traps.

*Foot valves.* As the alternate to the air valve, which ensures elimination of air in pipelines, the foot valve ensures that a pump suction line remains filled with liquid, especially for pump priming. A foot valve operates automatically and has only one moving part, a disc that is guided to its seat when in operation. Most foot valves are manufactured from iron or bronze and have an integral strainer to prevent solids from entering the valve body or the pump (Fig. 65).

*Special systems.* There have been special valves developed for use in particular piping systems, such as heating, ventilation, air conditioning, plumbing, drainage, etc. Their application is then mostly restricted to such a particular system.

## VALVE TRIM

This is the normal designation given to the various working parts of a valve, such as stem, wedge, disc, plug, seat, etc., which are all additives to the basic valve body. While most piping systems are designed around a single material of construction that is used for piping as well as for fittings, valve-trim materials may be composed of half a dozen different materials. In many cases, it is of the utmost importance that the correct material be specified for a given service. All valve fabricators normally designate their product through a numbering system, but this does not always suffice to identify some of the trim materials. It is therefore incumbent upon the purchaser to specify his requirements in detail.

A typical specification for a gate valve might possibly be given as follows:

> 6-in., 600-lb. gate valve, cast steel, flanged ends, rising stem 13% CR, single wedge C.S. stellite faced, seat rings S.S.TP304, XYZ manufacturer #24.

Even such a detailed description assumes that certain trim materials are the manufacturer's standard items, which are shown in a catalog as standard for the indicated valve.

## VALVE ACCESSORIES

These are certain accessories that are applicable in one form or another to both gate and globe valves. One such accessory is a bypass arrangement for larger valves which can assist in equalizing pressure around control valves or warm up valves used in steam-line service. In order to prevent freezing of residual fluids in closed piping systems, many valves can be furnished with a drain valve installed at a low point in the valve. Extended stuffing boxes are a particular requirement of valves in cryogenic service. The long stuffing box provides sufficient insulating area, with the valve operator extending beyond the insulation. There are some special accessories for large valves, such as clean-out windows or internal runners to assist in opening or closing a gate valve.

## VALVE OPERATORS

The simplest valve operator is the handwheel or lever, which is directly attached to an operating stem. Larger valves that require more force for opening or closing can be furnished with a hammer-blow attachment or a gear system. Gears are mostly furnished as either bevel or spur gears. Another alternative for gearing is to provide motor or cylinder operators. The purpose of such operators can also be remote control of valves. Cylinder-operated valves can use water, air, oil, or any fluid as an operating substance, depending on the service requirements and availability of the operating fluid. Like everything else connected with valves, there are also numerous variations of motor operators available.

When hand-operated valves are installed in locations that are not easily accessible, they can be equipped with chain wheels and chains, an extension stem, or even a flexible extension shaft. For easier operation, an extension stem can be supplemented with a floor stand that may also have a position indicator that will indicate to what degree a valve is open or closed. Buried valves may have a simple valve box flush with the ground that permits valve operation or an indicator post may be used.

# 11

# Pipe expansion devices

Temperature differentials in piping systems, either created by the fluid being transmitted or the result of the outside surroundings, must be taken into account during the design phase of any system. Some methods of connection, like bell-and-spigot joints or mechanical couplings, permit a minimum of movement. However, in most piping systems where temperature differentials can occur, some other means have to be found to control the movement of pipe, which is the result of such temperature differentials.

In the Arabian desert where outside temperatures vary widely between day and night, the solution to this problem is very simple. Surface pipelines are not being laid in a straight line but are being installed in a continuous zigzag. Any expansion during the day or contraction during the night is compensated for by increasing or decreasing the angle at these bends. A more sophisticated approach has been standard practice in steam piping systems for a long time. Expansion loops, whose expansion ratios have been the subject of extensive calculations, are installed in such lines.

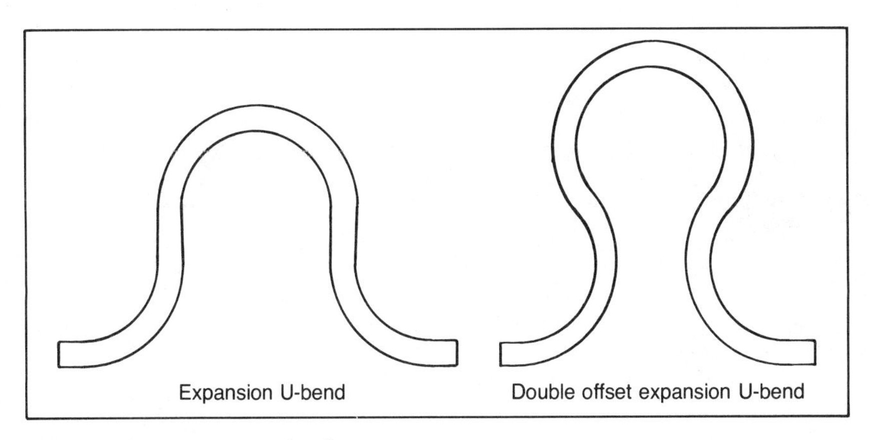


**FIG. 66. Pipe-type expansion loops.**

These loops may be welded using 90° elbows at the four corners or may consist of bended pipe.

Pipe bends are usually fabricated to the expansion U-bend pattern or as a double-offset U-bend. Since these expansion loops require a considerable area for installation, manufactured expansion joints for use in piping systems with space restrictions became a necessity (Fig. 66).

There are two basically different types of expansion joints available. One is the sleeve or slip type and the other the bellows type. It is important that piping systems, which incorporate these manufactured expansion joints, are also restricted in their movements so that any expansion or contraction is not deflected but is restricted to the expansion joint. The best way to achieve these results is through the use of properly located pipe anchors and alignment guides.

## Sleeve or slip-type expansion joints

This type consists of three major parts: an external sleeve that is connected to the piping on one side, an internal slip that is connected to the piping on the other side, and a stuffing box or packing-gland arrangement to hold the pressure. This type of expansion joint is manufactured to allow for expansion or contraction from an anchor point in either one or two directions along its axis. Some of the variations of this device are:

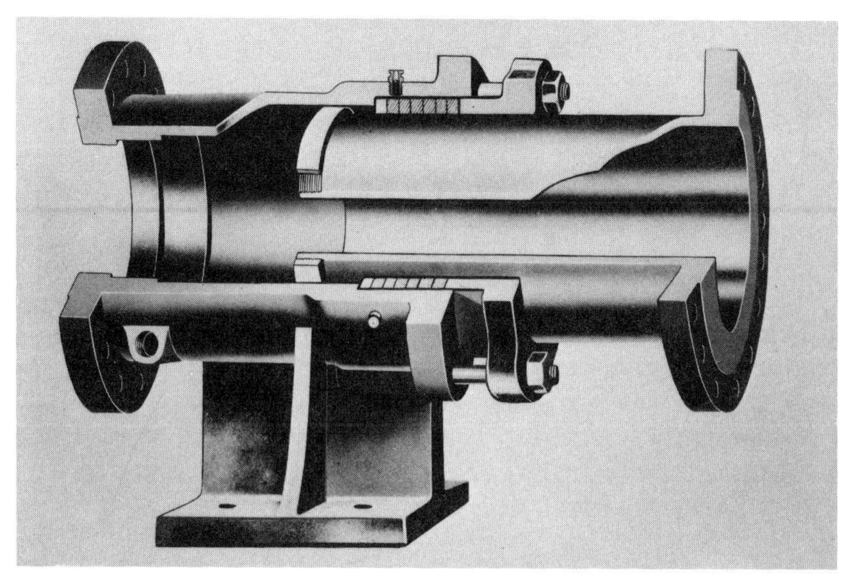

**FIG. 67. Slip-type internally guided expansion joint** *(courtesy Adsco).*

1. An internally guided type where the slip is guided internally and prevented from being pulled out of the sleeve (Fig. 67).
2. An internally externally guided type that permits the slip to travel the entire length of the sleeve since a guide is provided at both ends.
3. A piston-ring-type expansion joint with either an internal or external guide that can be packed under pressure. Piston rings hold the line pressure during packing operation.

## Bellows-type expansion joints

Bellows-type or corrugated expansion joints are manufactured as either equalizing or nonequalizing. The number of bellows incorporated in any expansion joint can vary in accordance with the requirements of any particular system and may range from a single bellows to more than 20.

Most metallic bellows are being fabricated from different materials than the piping system into which they are incorporated. Favorite materials for construction of bellows are copper, rubber, teflon, monel, and stainless steel: however, there is no limitation on material or size.

1. *Nonequalizing type (Fig. 68).* This consists of bellows only, with suitable ends for installation into a piping system. After expansion or contraction, there is no control available to return the bellows to their original configuration.

**FIG. 68. Bellows-type nonequalizing expansion joint** *(courtesy General Rubber Co.).*

2. *Self-equalizing type (Fig. 69).* This design uses control rings to distribute the compression equally among all bellows. It limits the amount of compression to which any one bellows is subjected and thus ensures equalization after expansion is being relaxed.

**FIG. 69. Equalizing-type expansion joint** *(courtesy Zallea Bros.).*

3. *Internal guides and liners.* An internal guide, consisting of an internal sleeve that ensures that only an axial movement will occur, can be provided in both the equalizing and nonequalizing type of expansion joint.

   An internal liner, serving to protect the bellows from early failure due to erosion, corrosion, and/or the exposure to high fluid velocities, may thus sometimes serve as both guide and protection. Liners also protect the bellows from high temperatures of the fluid flow.
4. *External guidance.* External guidance for bellows-type expansion joints can take various forms. Tie rods may be provided that keep the expansion joint in a compressed form for easier installation and allow for more expansion after installation. Limit rods, on the other hand, provide limited expansion during system operation. Hinges are often furnished when a lateral expansion is involved.

5. *Torsional expansion joints*. A variation of the U-shaped bellows has been introduced in the form of helical-constructed bellows that are especially suitable for torsional applications.
6. *General*. The nonequalizing bellows-type expansion joint is preferable in various situations where no axial expansion is contended with, but a lateral offset or angular displacement of the piping may occur. There is, of course, a limit to this type of application for expansion joints. Where more severe displacement is expected, another form of pipe equalization must be utilized—for instance, the use of flexible piping.

### Alignment guides

Keeping correct pipe alignment is of utmost importance for piping systems that make provision for axial expansion by installing expansion joints. Various alignment guides that permit free axial movement of the pipe while restricting angular or lateral displacement are in use today. They may range from a restrictive box fabricated from structural steel members at the point of installation to standardized manufactured pipe guides.

### Installation

The following installation instructions are an excerpt from the Standards of the Expansion Joint Manufacturers' Association and refer to metal bellows expansion joints.

## DO'S*

- Inspect for damage during shipment, i.e., dents, broken hardware, water marks on carton, etc.
- Store in clean dry area where it will not be exposed to heavy traffic or damaging environment.
- Use only designated liftings lugs.
- Make the piping systems fit the expansion joint. By stretching, compressing, or offsetting the joint to fit the piping, it may be overstressed when the system is in service.
- It is good practice to leave one flange loose until the expansion joint has been fitted into position. Make necessary adjustments of loose flanges before welding.
- Install the joint with the arrow pointing in the direction of flow.
- Install single Van Stone liners pointing in the direction of flow. Be

*Reprinted courtesy of Expansion Joint Manufacturers' Association.

sure to install a gasket between the liner and Van Stone flange as well as between the mating flange and liner.

- With telescoping Van Stone liners, install the smallest I.D. liner pointing in the direction of flow.
- Remove all shipping devices after the installation is complete and before any pressure test of the fully installed system.
- Remove any foreign material that may have become lodged between the convolutions.
- Refer to EJMA Standards for proper guide spacing and anchor recommendations.

## DON'TS

- Do not drop or strike carton.
- Do not remove shipping bars until installation is complete.
- Do not remove any moisture-absorbing dessicant bags or protective coatings until ready for installation.
- Do not use hanger lugs as lifting lugs without approval of manufacturer.
- Do not use chains or any lifting device directly on the bellows or bellows cover.
- Do not allow weld splatter to hit unprotected bellows. Protect with wet chloride-free asbestos.
- Do not use cleaning agents that contain chlorides.
- Do not use steel wool or wire brushes on bellows.
- Do not force-rotate one end of an expansion joint for alignment of bolt holes. Ordinary bellows are not capable of absorbing torque.
- Do not hydrostatic pressure test or evacuate the system before proper installation of all guides and anchors.
- Pipe hangers are not adequate guides.
- Do not exceed a pressure test of 1 1/2 times the rated working pressure of the expansion joint.
- Do not use shipping bars to retain the pressure thrust if tested prior to installation.

## FLEXIBLE CONNECTORS

Another way to provide for pipe expansion is to incorporate flexible connectors, better known as swivel joints, into a piping system. These swivel joints also provide flexibility to otherwise rigid piping systems. For instance, an oil-truck loading station that has steel pipe for loading purposes may include several swivel joints to supply the required versatility of the discharge pipe. Swivel joints are mostly manufactured from various metals such as steel, stainless steel, bronze, or similar material and can be connected into any piping system through the use of flanges, or through threaded or welded connections.

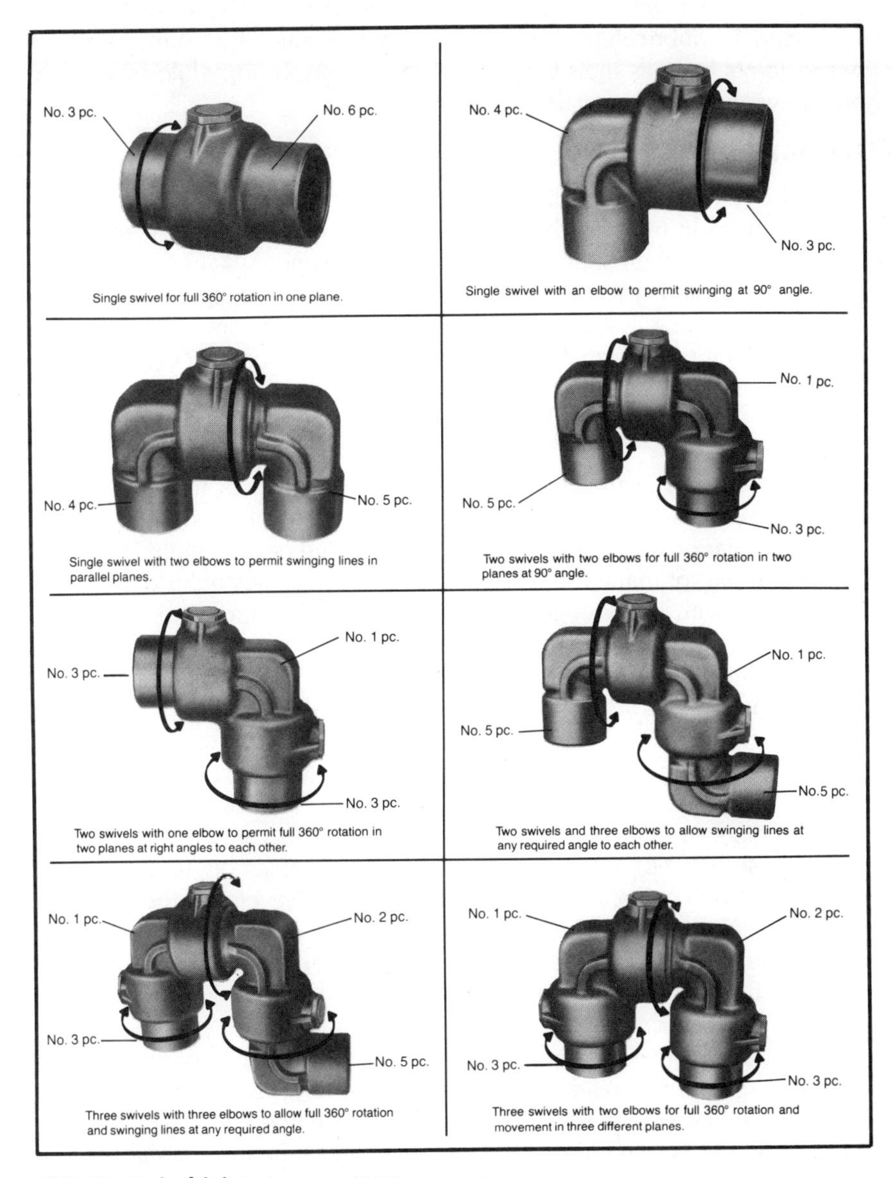


**FIG. 70. Swivel joints** *(courtesy FMC).*

Swivel joints are manufactured from two or more components, depending on the degree of flexibility required, and permit in a simple configuration a rotation of 360 degrees. By adding additional components to the swivel joint, one can increase flexibility from the single plane to additional configurations (Fig. 70).

Various manufacturers have originated different models of swivel joints; some can be lubricated or greased, others operate dry. Some types of swivel joints have as their basis a ball-joint connection, while others rely on a bayonet type connection.

## FLEXIBLE PIPING

While some flexible metallic piping such as copper tubing or lead piping have been in use already for a long time, they can be considered flexible only during their initial installation and are not suitable for applications that demand the ability to withstand frequent flexing or movement. Problems in piping systems or equipment connections are always created through vibration, thermal expansion and contraction, shock, or swing connections. These can best be solved by using flexible piping that is especially designed to withstand the rigors of continuous or frequent movement.

Rubber hose is available in many variations and varies from plain hose to multilayered heavily reinforced (with fabric or steel) hose. Inner liners made from various plastics, depending on the fluid being handled, are also an integral part of many rubber hoses. The various types of hose now suitable for use with increasingly higher pressure and temperature conditions also require the use of suitable hose connectors.

It is relatively easy to find the limits of pressure ratings for various types of hose and connectors through thorough material testing. But the point of connection defies such testing standardization. Therefore, each hose should be individually tested before being put into service, where rupture might have grave consequences. Connectors for rubber hose are mostly metallic pipe, which is grooved at the one end that is inserted into the rubber hose. Clamps are then tightened over this assembly to prevent the slippage of the connector out of the hose. The other end of the pipe will then be available for a standard connection to a piping system or equipment through the use of threaded, welding, or flanged ends.

Hose connectors are often designed as special mechanical couplings for quick connection or release of the flexible hose. Such specific adaptors may be a bayonet-type coupling as used for the unloading of oil trucks or may be of the large-thread type as used on fire hoses and hydrants. While almost everybody imagines small-diameter-type garden hose when rubber hose is referred to, medium-sized (8-in. to 12-in.) hose is a standard equipment item used on loading arms for the loading or unloading of oil tankers.

Rubber or elastomeric hoses are still limited in many applications, and the development of metallic flexible hose is designed to fill the need that exists in modern industry for suitable materials. Two basic types of metallic hose are being manufactured, corrugated and interlocked, that are so designated because of their different construction (Fig. 71). While there are

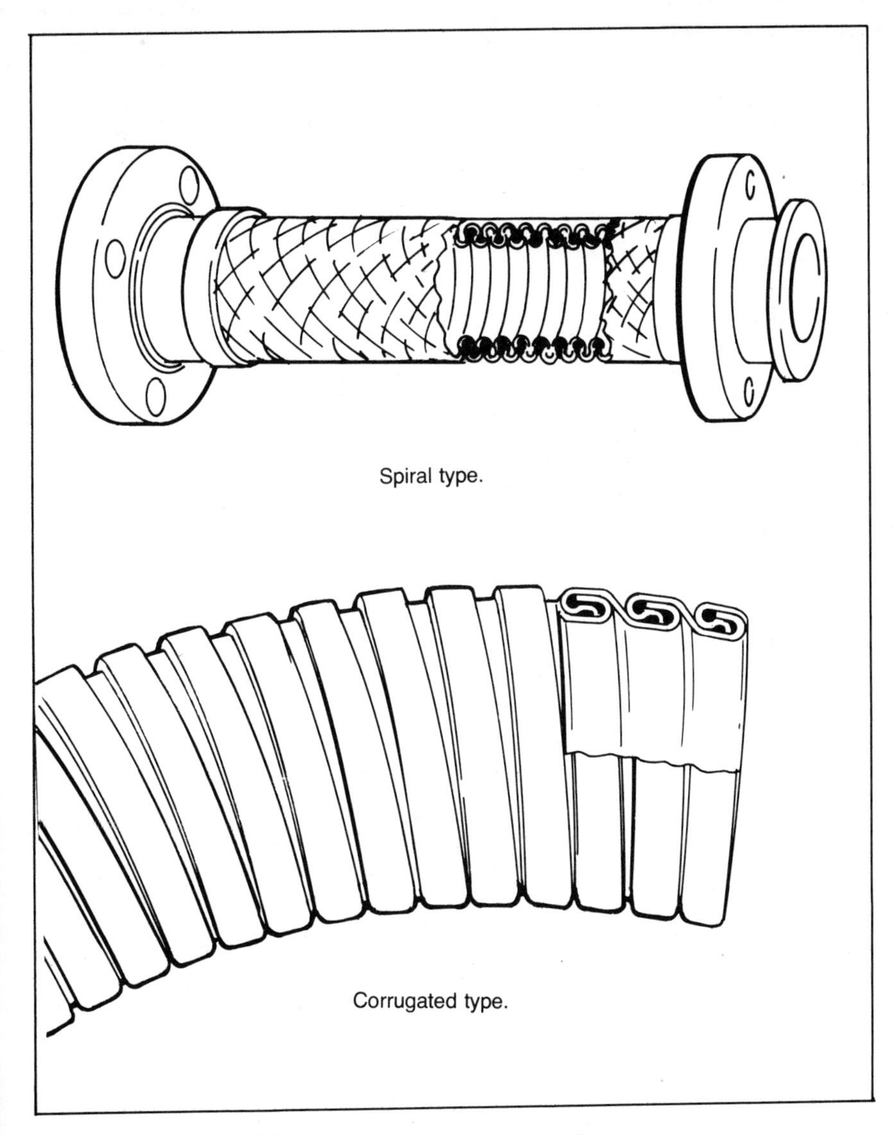


**FIG. 71. Metallic flexible hose** *(courtesy Universal Metal Hose).*

essentially no restrictions on metals being used, stainless steel and bronze seem to be favored by both users and manufacturers.

Metallic flexible piping is often furnished with a protective covering of braided material, which helps to preserve the natural contours of the original corrugations or interlock. Connectors for metallic hose are usually welded or brazed; thus, the point of connection can withstand similar pressures as the pipe itself. Depending on what metallic hose material is being used, applications may vary from the low temperatures of liquid nitrogen to 1,500°F. Pressure ratings in excess of 3,000 psig can also be accommodated in certain flexible hoses. Braided metallic hose is used, for example, to provide isolation from vibration, to allow for movement on burners of steam generators, or to permit a position change of relief lines from safety valves in steam lines.

# 12

# Piping specialties

There are, of course, numerous specialties manufactured that pertain to a special service, such as heating, air conditioning, plumbing, drainage, lubrication, and fire lines. This chapter does not intend to go into details of specialized applications but deals with some of the specialties that are generally applicable in piping systems and also refers to some of the specialized items which are being found in process plant service.

## SPECTACLE BLINDS

Spectacle blinds are being used to ensure 100-percent cutoff to flow in any piping system. During maintenance work on certain equipment, the assurance of total flow cutoff is mandatory and the use of this or a similar type of blind flange is the answer. Spectacle blinds are so-called because their appearance resembles a giant pair of spectacles. One side is the solid blind flange; the other side is cut so full seating of the gasket surface is accomplished while providing full flow through a cut-out inner circle. A

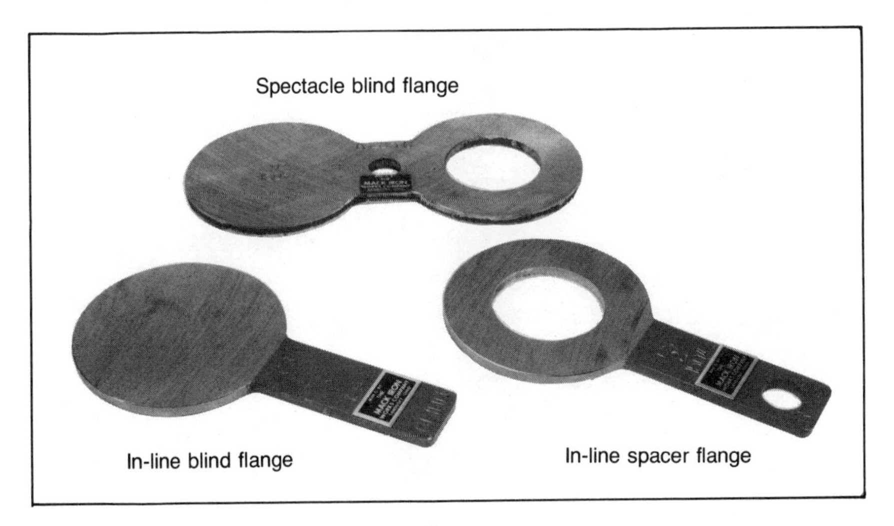


**FIG. 72. Flanges** *(courtesy Mack Iron Works).*

small bridge is between the two sides with a single or double bolt hole. This permits the flange to rotate from the open to the closed position or vice versa without disturbing alignment of the piping system (Fig. 72).

A simpler version is an in-line blind flange with a handle. When being inserted between two flanges, the flanges must be pried apart sufficiently to permit insertion of the blind and associated gasket, unless space has been provided by having a spacer already in place.

## ORIFICE FLANGES

This special type of flange is used in conjunction with the placement of an orifice disc for flow measurement. Orifice flanges are always installed in pairs with the flow-measuring orifice inserted in between. The flange body is increased in thickness compared to regular flanges, and a hole is drilled sideways from the outside to the inside of the flange to permit withdrawal of fluid before and after passing through the orifice. This hole is tapped from the outside to permit threaded connection of a sampling pipe.

End connections for orifice flanges are mostly of the weld-end type; however, socket welding and threading are also being used. In that case, care must be taken to ensure that the side hole is extended through the pipe wall to permit fluid removal (Fig. 73).

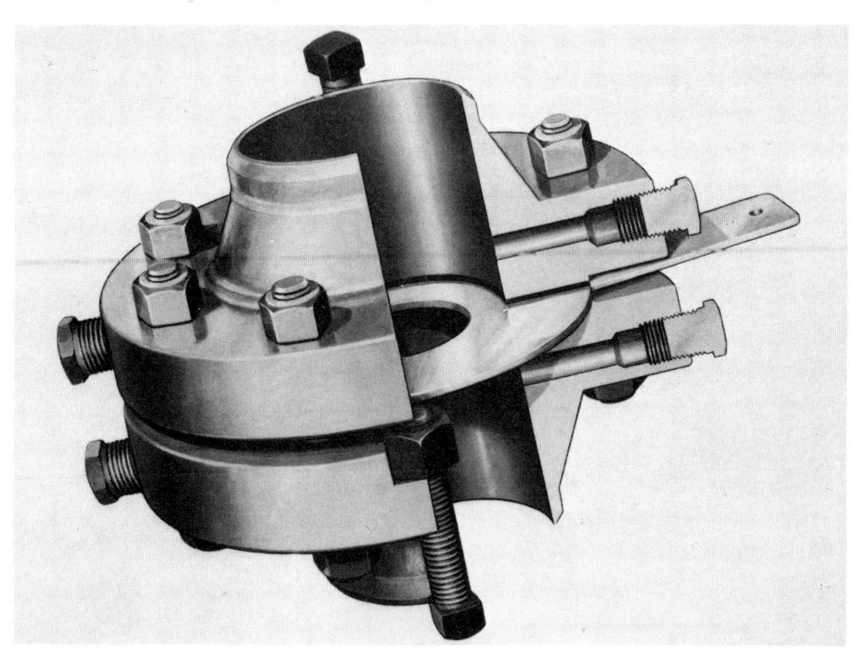

**FIG. 73. Orifice flange** *(courtesy Daniel Industries).*

## ORIFICE FITTINGS

Orifice fittings serve the same purpose as orifice flange assemblies; however, a single fitting is replacing the special flanges. These fittings are commonly used in the oil and gas industry and permit an easier installation than orifice flanges. Another feature of some of these fittings is that they permit removal of the orifice plate even when the piping system is pressurized.

Orifice fittings are commercially available in sizes up to 48 in. I.D.

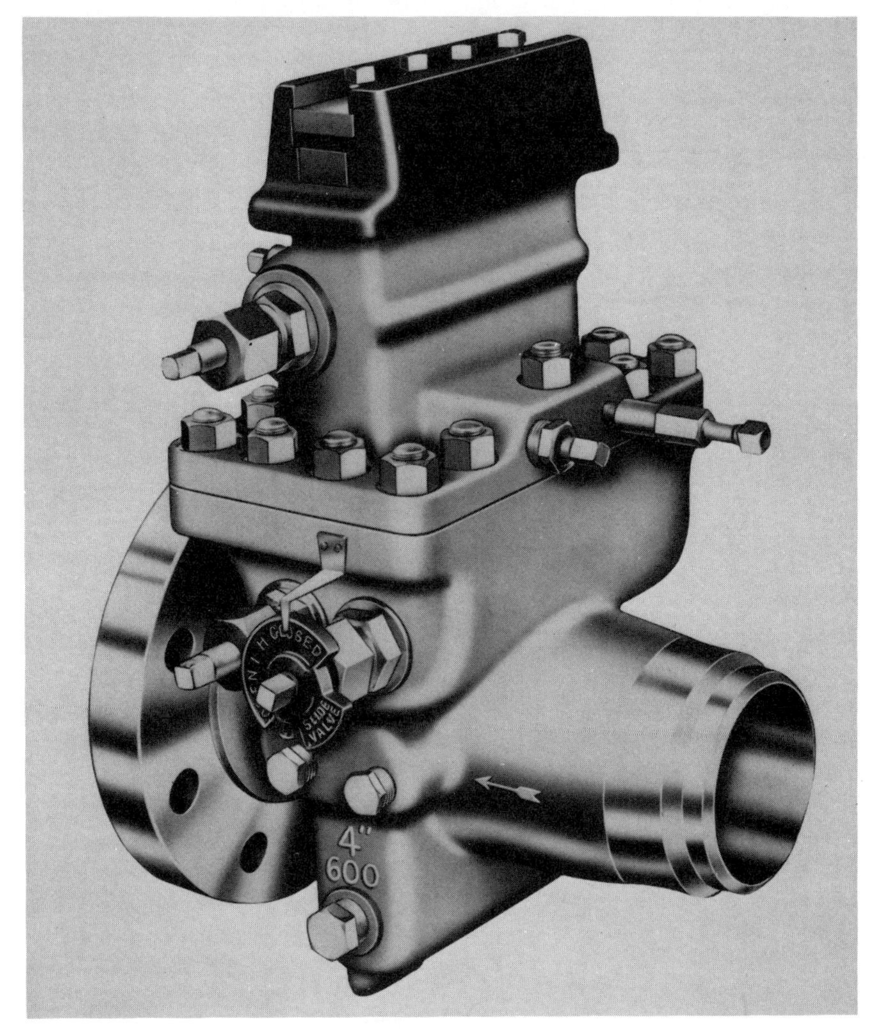


**FIG. 74. Orifice fitting** *(courtesy Daniel Industries).*

and can be furnished either flanged or with a flange at one end and prepared for a welding connection (butt weld or socket weld at the other end (Fig. 74).

## SIGHT GLASSES

During many operations, it is necessary to observe if the flow of fluid in a piping system is unobstructed even while the system operates under pressure. The obvious solution is to provide a window in the piping through which observation can be made. However, even after having come up with a special glass window in a pipe, it is not always possible to observe any liquid flow, yet some kind of an indicator is required to show if there is liquid movement in the pipe. Sight glasses are therefore provided with flappers or rotating wheels that indicate movement.

A different type of sight glass is provided for an automatic lubricating system where there is no pressure in the sight-glass part. A drip glass permits viewing of the lubricating liquid after it has been dispensed from a central reservoir and is entering a point of lubrication.

## SAFETY OR RELIEF VALVE

A safety or relief valve is a major protective device that is designed to avoid accidents through relieving pressure when a malfunction occurs in the system or vessel that is protected. It is one of the few pieces of equipment that has a standby function most of the time, but whose operation requires split-second timing in case of need. From the production point of view, the valve should normally not disrupt any ongoing operation. But from the safety point of view, it is essential that in an emergency the valve must open and discharge at a previously determined flow rate.

The ASME Boiler and Pressure Vessel Codes specify the use of pressure-relief devices in Section VIII for pressure vessels, as well as in the divergent boiler sections. These codes provide detailed requirements as to manufacturing and installation methods of pressure relief valves, which guarantee a safety outlet for fluids, in case higher than working pressure in a fired or unfired pressure vessel or piping system occurs.

In the petroleum industry, API provides several specifications that deal specifically with pressure-relieving systems for refinery service. The most important ones are:

RP 520 Recommended practice for the design and installation of pressure-relieving systems in refineries.

Part I Design

Part II Installation

RP 521 Guide for pressure relief and depressuring systems.

RP 526 Flanged steel safety relief valves.

In addition, the ANSI Code for Power Piping B31.1 specified design parameters for discharge lines from safety devices.

To make a distinction between the various nomenclatures used for similar devices that are pressure-relief valves, the term *safety valve* is used mostly for steam and air. The terms safety-relief or liquid-relief valves are primarily used for petroleum or chemical processing and storage in either vapor or liquid form.

The primary function of any pressure-relief valve is adequately described as the prevention of excessive pressure buildup in any fluid container or system. However, the secondary function is an important one as well and requires reclosing the device immediately after the emergency condition has been alleviated. Such reclosing should occur automatically at a designated pressure.

A pressure relief valve is a fully automatic type of valve that, because of its universal function, deserves to be listed under specialties rather than in a category with standard valves. It is an essential element of construction that presssure relief valves should open and close quickly, and for steam and air service that they can be tested under operating conditions.

The basic elements of a pressure relief valve are a seat and disc as in any standard valve. However, an internal spring holds the disc against the seat. The disc can be lifted for testing through a connection to an external lever. A huddling chamber initiates the automatic operation of the valve by permitting a static pressure to be built up that forces the valve to open wide, after operating pressure has exceeded the permissible limit and has opened (or cracked) the valve slightly (Fig. 75).

An example for installation of a safety valve is on a boiler drum, where it will open when overpressure occurs and will remain open until the pressure inside the drum drops to the reseating pressure. These valves mostly vent to atmosphere, but provisions should exist to drain off any condensate.

As an alternative to the spring-loaded type of direct-acting safety valve, a pilot-operated relief valve that permits remote control and which is piston operated in lieu of the spring operated model is commonly used on gas transmission lines, compressor stations, and unfired pressure vessels. This type of valve can achieve full lift without overpressure and will sustain this lift until the reseating pressure is reached, independent of input flow rates. Thus, a valve may open due to a transient pressure surge, depressurize the vessel by the amount of the blowdown, and then reseat itself without operator assistance.

Liquid-relief valves are of similar construction as safety valves and are used on pumps to protect against overload in case the discharge line is blocked or pump demand drops below pump capacity. The liquid relief

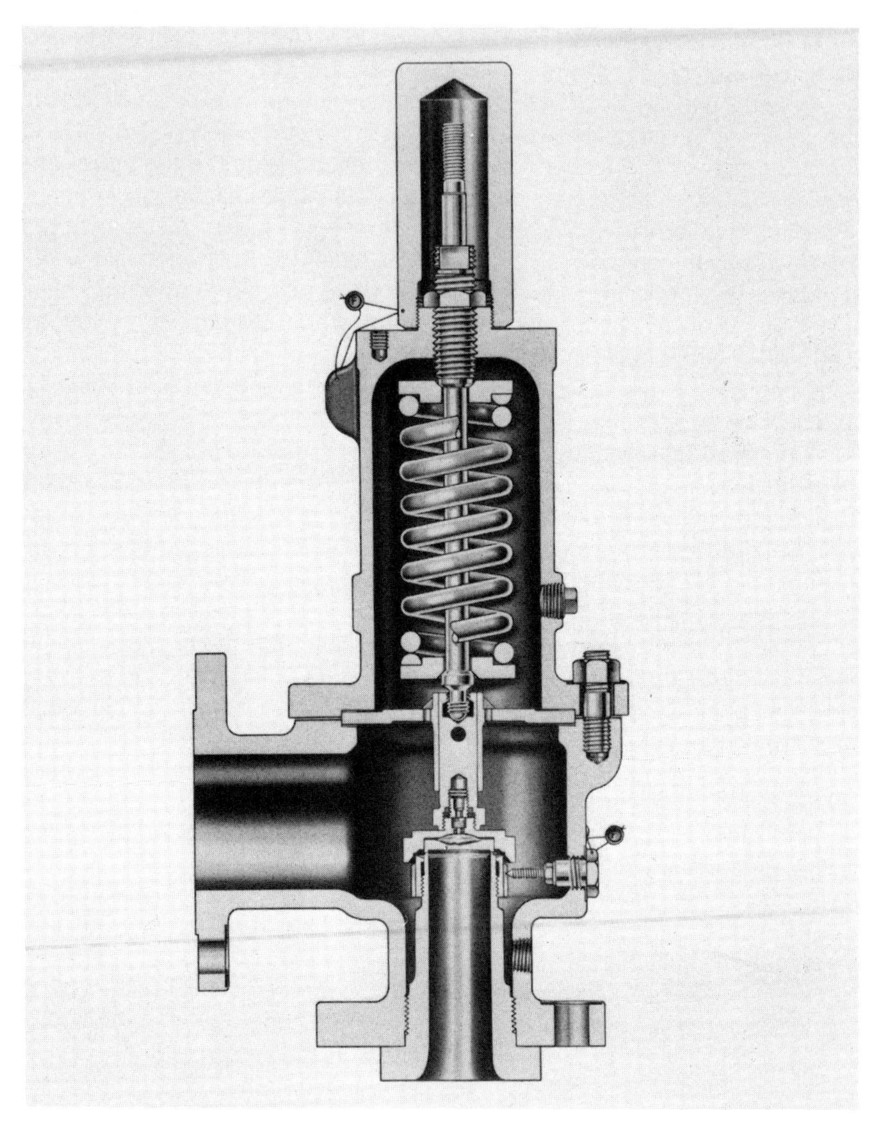

**FIG. 75. Safety relief valve** *(courtesy Teledyne Farris).*

valve is normally mounted on the pump discharge line and relieves overpressure back to the suction line.

A general requirement for all pressure-relief valves is the correct determination of a suitable differential between the normal operating pressure and the set pressure, at which the valve will open. Pressure-relief valves should almost always be installed with the stem in the vertical condition, inlet at the bottom, and outlet on the side.

Before installation, pressure-relief valves should be checked, calibrated, and tested. Such testing should also be performed periodically and should include a check of both opening and reseating performance. Normally, valve name plates will provide the required operating data, which provide the test basis. Because of the infrequent operating need for pressure-relief valves, the importance of periodic testing cannot be emphasized too strongly.

## RUPTURE DISCS

Beside pressure-relief valves, rupture discs present another pressure relief device to comply with the requirements of various codes. A rupture disc is best described as a nonreusable overpressure-relief device that ruptures when it is exposed to the pressure rating for which it has been designed or rated.

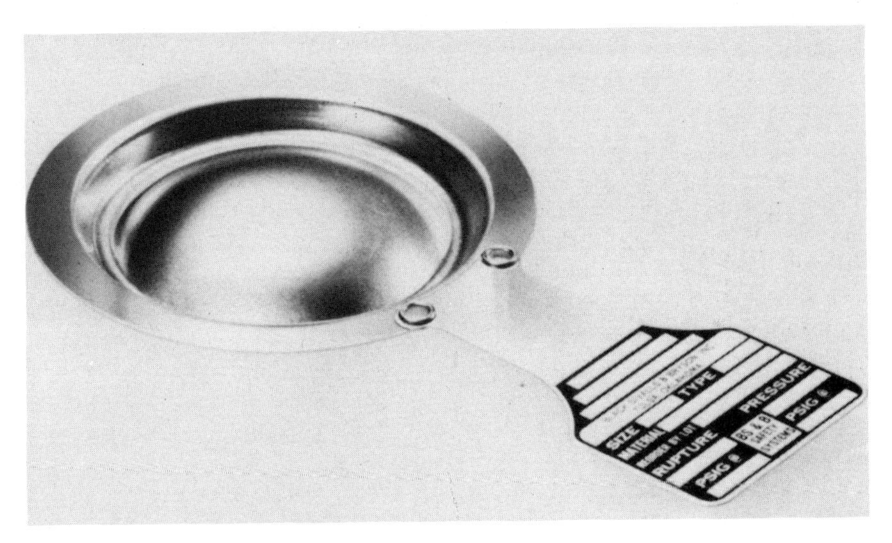


**FIG. 76. Unruptured rupture disc** *(courtesy BS&B).*

A rupture disc may be installed to serve various functions (Fig. 76). It may be installed as the only relief device on a pressure vessel or act side by side with a pressure relief valve to provide an additional margin of safety if the unexpected should occur. Another installation method is in series with a pressure-relief valve, which isolates the valve internals from

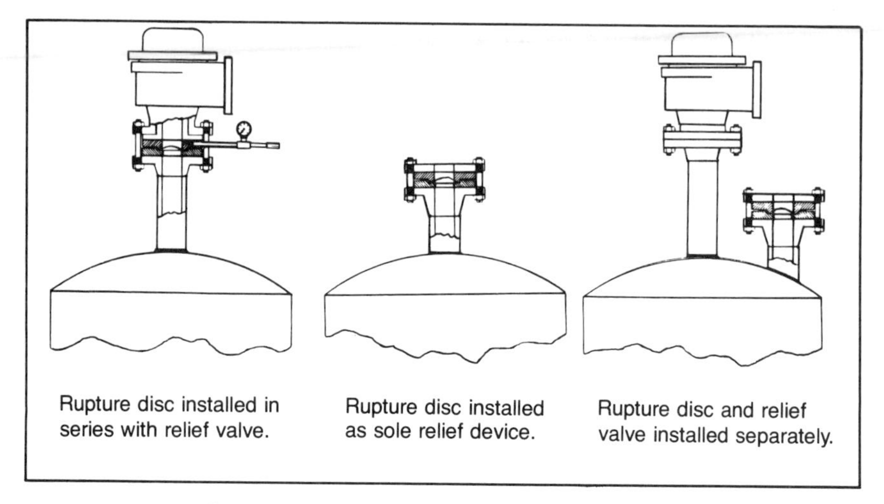


**FIG. 77. Rupture disc installation methods** *(courtesy BS&B Safety Systems).*

possibly corrosive process fluid, thus requiring less costly internals for the pressure-relief valve. In-series installations also reduce downtime during maintenance of the pressure relief valve (Fig. 77).

The rupture disc is a nonmechanical device that has to have a high degree of sensitivity from a functional aspect. As such, it is expected that failure or rupture of the disc will occur at the precise predetermined pressure point for which it has been designed. The sizes and types of metals used in fabricating rupture discs involve a broad spectrum. I.D. sizes range from 1/8 in. to 44 in., and a variety of ductile metals, such as stainless steel, monel, nickel, aluminum, and others are materials of construction for the disc.

Rupture discs in sizes of 1 in. and larger should always have a name plate attached by the manufacturer that provides detailed information as to size, type, material, and rupture pressure rating (Fig. 78).

A major advancement in the design of future discs involved the exposure of metal to compression loading. Previously, the fluid media were only in contact with the concave side of the disc, putting the disc metal under tension. With compression loading, the fluid is in contact with the convex side of the disc. This newer design permits higher operating pressures, and the reverse buckling discs can utilize knife blades, teeth rings, or score lines to open completely after rupture.

When installing a rupture disc, it is important that it is so positioned in the pressurized system that it will have an immediate sensing of an upset from the normal pressure pattern.

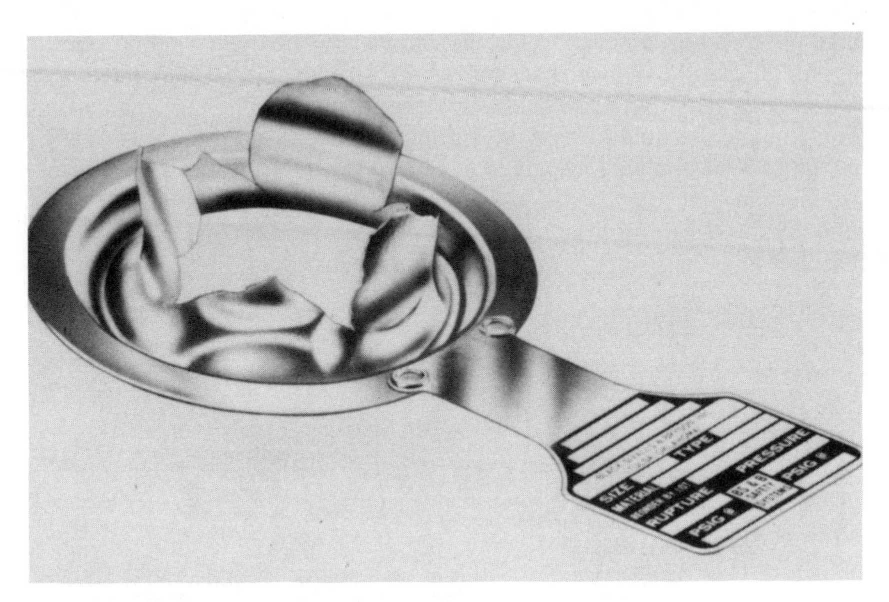


**FIG. 78. Ruptured rupture disc** *(courtesy BS&B Safety Systems).*

## SAFETY-VALVE DISCHARGE ELBOWS

Safety-valve discharge elbows or drip-pan elbows are installed where a safety-valve discharge is vented to atmosphere. A drip pan surrounding the elbow catches rain-water that might enter through the vent line and drains it. The elbow is drilled and tapped so that condensate can be drained from the vent line. Drip-pan elbows are mostly furnished in cast iron or cast steel.

## THERMOWELLS

The thermowell is an adjunct to a thermometer. It is mostly manufactured from metal and can be threaded or welded into a piping system that has no other provision for temperature checking. It transmits the temperature of the transported fluid to a temperature-measuring device that is inserted into the internal of the thermowell. The thermometer is thus protected from contact with the fluid, which otherwise might possibly damage the mercury bulb or other temperature-measuring substance container.

## STRAINERS

A wide variety of strainers are used to filter a fluid, thus preventing contamination and possible mishap. Some are only temporarily installed in a piping system, while others have a permanent function. It is considered good startup practice to install a strainer in any pump suction prior to operation and thus ensure that no debris or sediment that may have been left or may have accumulated during construction or maintenance will contaminate the fluid or damage a pump impeller.

In general applications, there is almost no industry that does not use strainers in their day-by-day operation. Such a use of strainers is limited only by the desire for the protection they afford and by the imagination of the user (Fig. 79).

**FIG. 79. Strainers** *(courtesy Zurn).*

## PLATE OR CONE STRAINERS

The plate strainer is a perforated blind plate that often is additionally covered with wire mesh and which can be inserted between two flanges (Fig. 80). The cone-type strainer is a little more sophisticated. It usually consists of a wire-mesh or perforated-metal cone that is attached to a plate rim that can be held in place wherever a flanged connection presents

**FIG. 80. Plate strainer** *(courtesy Mack Iron Works Co.).*

itself (Fig. 81). These types of strainers are often used in conjunction with a short, flanged section of pipe that can be removed together with the strainer and in a rather cumbersome way allow an intermittent cleanup of sediment accumulation.

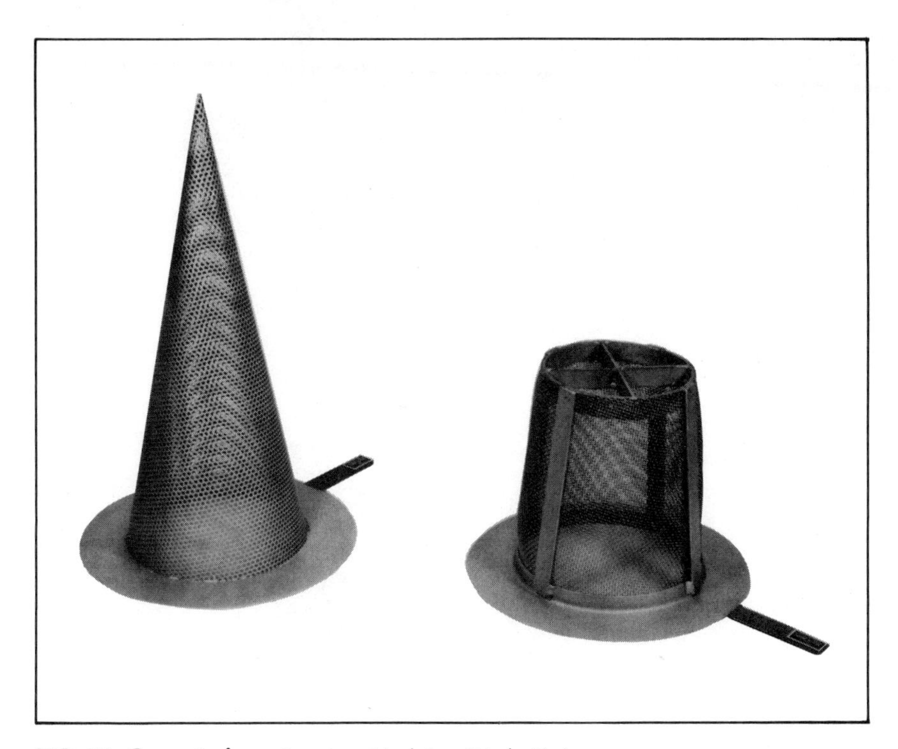

**FIG. 81. Cone strainers** *(courtesy Mack Iron Works Co.).*

Such strainers are often site fabricated; however, operating personnel often frown upon such home-made articles because they suspect their structural integrity and are not quite sure if the open area is sufficient for the required flow. These types of strainers are also commercially available and are manufactured in accordance with specifications that guarantee their performance.

## Y-TYPE STRAINERS

The most frequently used strainer in pipelines of 3-in. NPS (nominal pipe size) or smaller is the Y-type strainer, so called because of its particular configuration (Fig. 82). The flow of liquid is routed through the screen located in the lateral leg and any amount of sediment is trapped. The

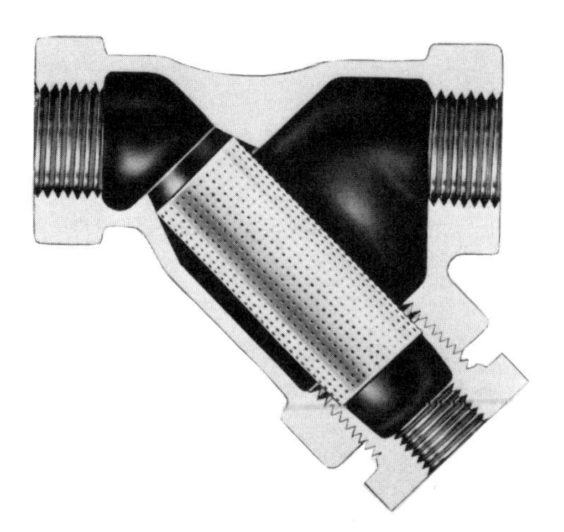

**FIG. 82. Y-type strainer** *(courtesy Sarco).*

screen can be removed and cleaned by opening the lateral leg cover, which is either threaded or flanged for this particular purpose. If such cleaning is considered necessary, a permanent drainage valve should be provided to avoid dumping the liquid trapped behind a shutoff valve on the floor or on the operator, who is supposed to clean the basket. The screen may be in basket or tubular form; when in tubular form, it permits removal of sediment accumulation by simply opening the drain valve, making it unnecessary to remove the basket for cleaning.

Y-type strainers may be installed either in vertical or horizontal pipelines; however, the basket must always be angled downward.

## BASKET STRAINERS

The basket strainer is generally used in larger piping systems, although a good case from an operations point of view can be made for its use in connection with small-diameter piping. Standard construction allows for insertion or removal of the basket-type screen through the top of the

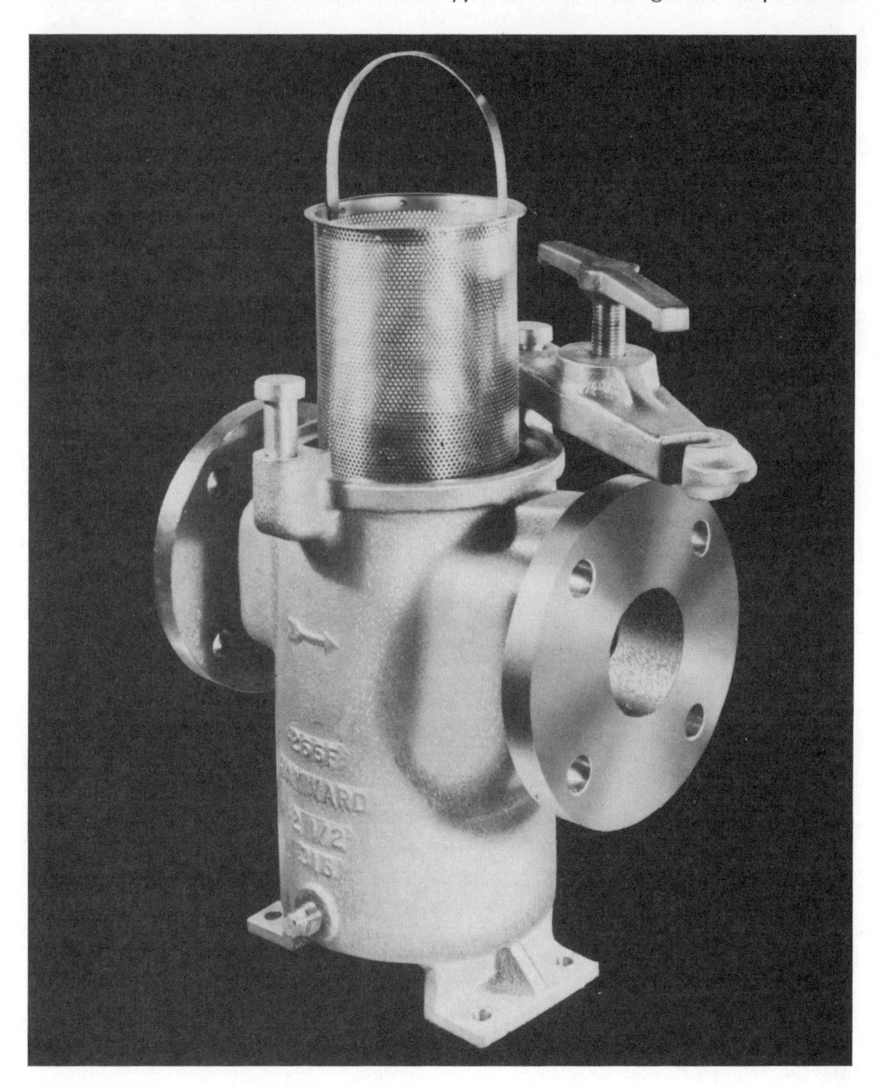

**FIG. 83. Simplex basket strainer** *(courtesy Hayward).*

strainer, which is usually flanged or has a quick-disconnect, yoke-type cover.

Basket strainers are furnished either as simplex or duplex models. The simplex model (Fig. 83) is used where the line can be shut down long enough to allow cleaning of the basket, and the duplex model (Fig. 84) permits continuous operation and does not require system sutdown for cleaning.

An integral diverter valve simply changes the fluid flow from one basket

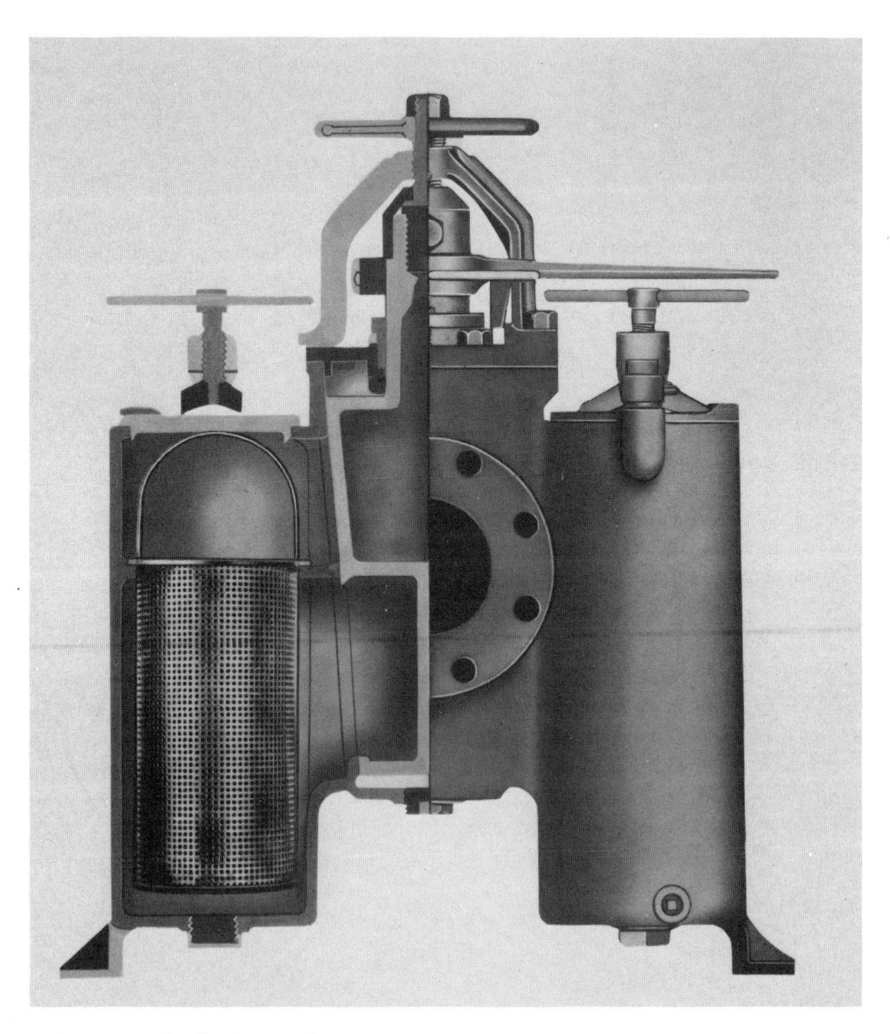

**FIG. 84. Duplex basket strainer** *(courtesy Hayward).*

containment to the other and permits the cleaning of the basket not in use. The main maintenance item for larger-type strainers is the cover, and provisions for an overhead lift should not be overlooked. Some manufacturers also provide a life-cover davit, which assists in easier maintenance. An alternative method to simplify maintenance is the provision of multiple covers and baskets and attachment of the covers to a common yoke.

Basket strainers are available with various accessories, such as a magnet insert to catch microscopic iron or steel particles that otherwise might pass through the straining element. Another accessory is a rotary scraper, which can be operated from outside the strainer body and is usually offered in conjunction with a backflash system. Automatically operated backflush systems are also available and are worth considering when solids loadings are consistent and heavy.

## AUTOMATIC STRAINERS

The development of automatic strainers (Fig. 85) where the throughput, backwash, and drainage are automatically controlled was the logical followup of basket strainers with partial automatic features. The installation of an automatic strainer is a necessity whenever the fluid being handled by a particular piping system requires constant filtration of impurities and where maintenance of predetermined flow characteristics is important. Like any other piece of mechanical equipment, however, engineering data or application sheets are a prerequisite for the design or selection of the correct strainer. It is important to realize at such a time the degree of straining required or desired and to size the strainer and its appurtenance properly.

### Materials of construction

In the manufacture of strainer bodies, a variety of materials are being used with cast iron, cast steel, stainless steel, bronze, and PVC. Straining screens in metallic strainers either are perforated metal or wire mesh made mostly from brass, monel, or stainless steel. Strainer body materials are chosen for compatibility with the inherent piping system, while the choice of the straining element material relates to the possibility of corrosive attacks. End connections are usually the same as otherwise used in the piping system.

### General

Because strainers are a relatively low-priority item, it would seem to be a valid observation that the problems often associated with permanent strainers are a direct result of the use of standard specifications; or of specifications used on previous but nonanalogous systems or plants; or by the

**FIG. 85. Automatic strainer** *(courtesy Zurn).*

use of improper fluid samples, in which case the worst possible conditions may have been removed from consideration.

## STEAM TRAPS

A steam trap is actually an automatic valve that prevents the loss of live steam but permits the release of water (condensate) and air. No drop in line pressure may be registered as a result of a steam trap operation. Steam traps have been in industrial use a long time, but it is sometimes difficult to pick the right type for a specific application from among the many variations that are available. Besides having to make a choice of the type of trap, it is also necessary to select the correct size since under- or oversizing will only provide a piece of inoperative equipment.

Reviewing the various operating principles for steam traps, there are basically five different types. There is no universal steam trap. With the thousands of different applications in the field today, it is virtually impossible to make up a complete steam-trap selection chart. Thus, it is important to understand the operating characteristics of the five main types of traps.

*Balanced-pressure thermostatic*—Responds to changes in temperature between steam and condensate. These changes vary the vapor pressure in the bellows.

*Liquid expansion*—Responds to the changes in temperature through the uniform expansion of a hydrocarbon oil.

*Float and thermostatic*—Responds to difference in density between steam and condensate and difference in temperature between steam and air or an air steam mixture.

*Bucket*—Responds to changes in density between steam and condensate.

*Thermodynamic*—Responds to difference in kinetic energy between steam and condensate.

## BALANCED-PRESSURE THERMOSTATIC STEAM TRAP (Fig. 86)

The heart of a balanced-pressure thermostatic trap is the flexible bellows that moves the valve head to and from its seat. The bellows is partially filled with a volatile fluid and hermetically sealed. This fluid has a pressure-temperature relationship that closely parallels but is approximately 10°F below that of steam. For example, its boiling point at 0 psig is 202°F; water is 212°F.

When cold, the trap is wide open, freely discharging air, noncondensables, and cool condensate. As the condensate temperature reaches approximately 10°F below saturated steam temperature, the liquid filling creates an internal pressure in the bellows. When the condensate temperature approaches that of steam, the internal pressure exceeds the external pressure and the bellows expands, driving the valve head to its seat and closing the trap. As the condensate surrounding the bellows cools, the vaporized filling within the bellows condenses, reducing the internal pressure; and the bellows contracts, opening the trap for discharge.

The thermostatic trap is a blast-type device cycling from a full open to a full-closed position. It requires a temperature depression of the condensate before it will open, which will cause a slight backup of condensate in

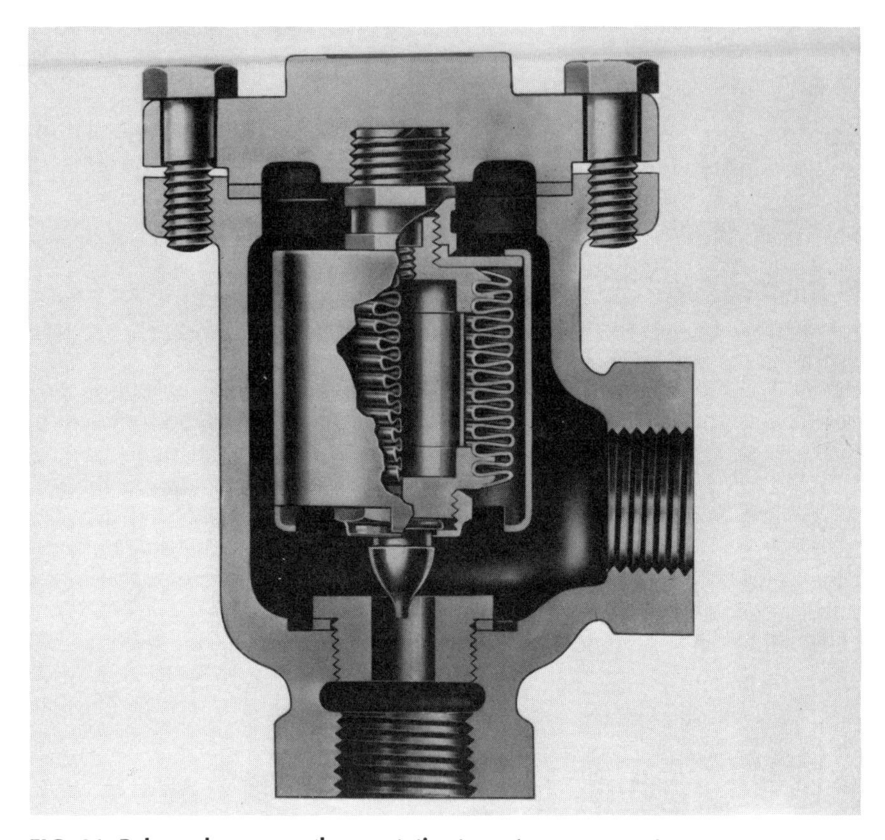

**FIG. 86. Balanced-pressure thermostatic steam trap** *(courtesy Sarco).*

the system. The thermostatic trap has relatively large capacities, has high air-venting capabilities, is freeze-proof when installed with an open discharge, and is completely self-adjusting within its pressure range. It has limited resistance to water hammer and corrosive condensates unless fitted with anticorrosive internals.

## LIQUID-EXPANSION THERMOSTATIC STEAM TRAP (Fig. 87)

This type trap has, for its operating element, a liquid-filled cartridge. Within this cartridge is a hermetically sealed bellows to which is attached the valve head and plunger.

Upon startup, the trap is wide open and freely discharges air and cool condensate until the condensate reaches a predetermined temperature below 212°F. As the warmer condensate flows over the operating element, the liquid filling expands, pushing the plunger forward and moving the

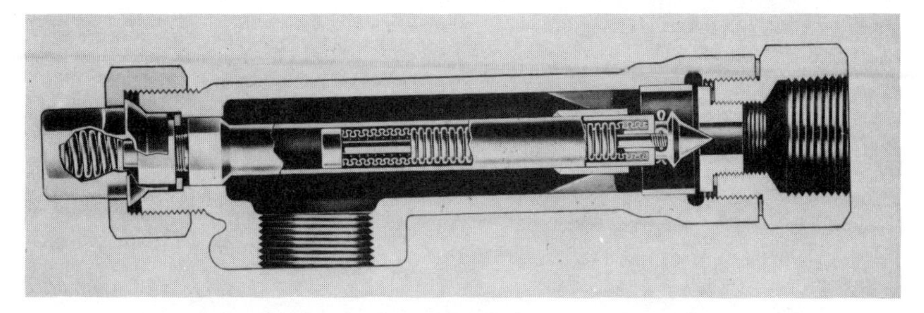

**FIG. 87. Liquid expansion thermostatic steam trap** *(courtesy Sarco).*

valve head toward its seat. This action throttles the flow and, at the set, temperature closes the unit. The maximum discharge temperature is 212°F because the pressure surrounding the element is approximately atmospheric. Any attempt to operate at a temperature above 212°F would result in the trap blowing steam almost continuously. The trap is fitted with a relief spring that prevents damage from over expansion of the element or from water hammer.

Liquid expansion traps permit the use of the sensible heat in a steam system when advantageous. As it is designed to discharge at a temperature below 212°F, there is no flash steam. With an open discharge, it is freezeproof. This trap is limited to only those applications where flooding the equipment can be tolerated. The operating temperature can be field adjusted by turning an adjustment nut.

## FLOAT AND THERMOSTATIC STEAM TRAPS (Fig. 88)

Float and thermostatic steam traps provide immediate and continuous discharge of condensate, air, and noncondensibles from a steam system as soon as they reach the trap. It is actually a steam trap and balanced-pressure air vent combined in one body.

The trap portion consists of a ball float connected by a lever assembly to the main valve head. As condensate reaches the trap, the ball float rises, positioning the valve to discharge the condensate at the same rate as it reaches the trap. The response is immediate, and the discharge is fully modulating and continuous. The condensate level is always maintained above the main valve, providing a positive water seal that prevents any steam leakage.

The integral balanced-pressure air vent unit immediately discharges all air and noncondensible gases that reach the trap. This assures maximum condensate capacity through the main valves at all times. The air vent is self-adjusting to all pressures up to the maximum operating pressure of the trap (see section on Thermostatic Traps for operating method).

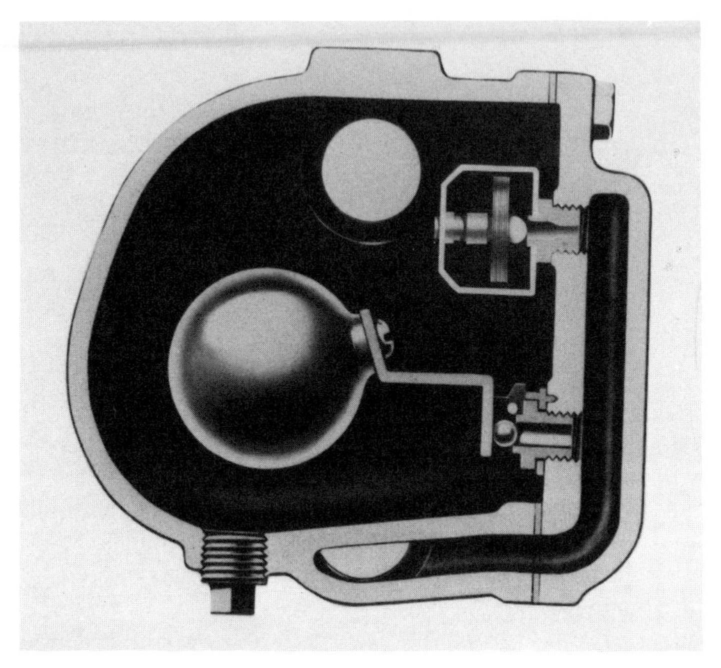

**FIG. 88. Float and thermostatic steam trap** *(courtesy Sarco).*

Float and thermostatic traps respond immediately to all changes in condensate loads reaching the trap and have excellent air-venting capabilities. Their continuous and modulated discharge ensures a minimum of pressure disturbances in both the supply and return system. When installed outdoors, the water seal is subject to freezing and necessary precautions must be taken. It is vulnerable to water hammer and corrosive condensate unless fitted with appropriate anticorrosive internals.

## BUCKET STEAM TRAPS (Fig. 89)

Bucket traps respond to the difference in density between steam and condensate.

Bucket traps have a fair resistance to water hammer and when manufactured with anticorrosive internals have good resistance to contaminated condensate. They have very limited air-venting capabilities, and it is recommended that an auxiliary air vent be utilized if any quantity of air is anticipated. During sudden pressure changes, especially if the trap is operating on light loads, bucket traps have a tendency to lose their "water seal" and blow steam continuously.

Bucket traps require little or no maintenance because of the simplicity of their design. They are usually free of clogging problems and generally

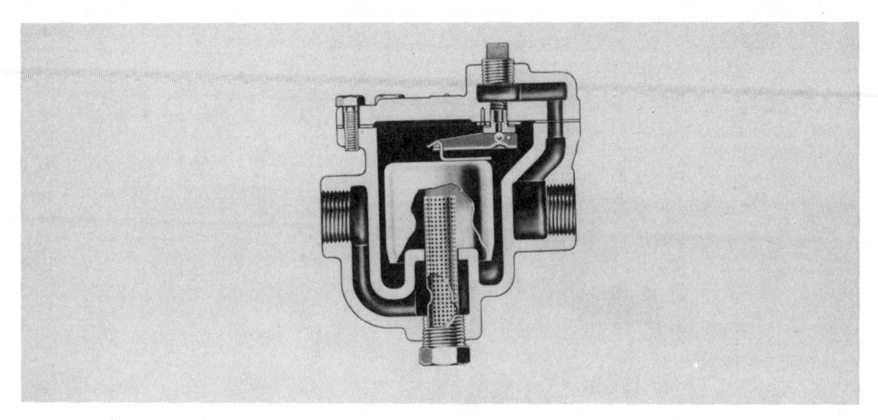

**FIG. 89. Inverted bucket trap** *(courtesy Sarco).*

have a long service record. They are, however, not freeze-proof. They provide mostly service for drip or drainage applications.

The inverted bucket is attached to the valve head by a lever mechanism and opens and closes the trap. On startup, the bucket, by its own weight, rests on the trap bottom and the valve is open, discharging air and non-condensibles. When condensate enters the trap body, it creates a water seal around the bottom of the inverted bucket that, since it is filled with air, becomes bouyant, rises and closes the trap. The trap is now airlocked and would remain so but for a small hole in the top of the bucket. The air leaks through the hole and is replaced by condensate. The bucket slowly loses its buoyancy and sinks to the trap bottom opening the valve. Condensate is discharged until steam enters the trap, filling the bucket that rises and closes the trap. At this point, the steam slowly leaks through the hole in the bucket and is replaced by condensate, causing the bucket to lose its buoyancy, sink, and open the valve, permitting the trap to discharge. This cycle repeats itself, giving the trap an intermittent or blast-type discharge.

The open-bucket trap has a bucket that is connected through linkage to a condensate drainage valve. Condensate entering the trap and bucket causes the bucket to drop and thus activate the linkage that opens the valve. Sufficient steam pressure then lifts the bucket. After the condensate has been removed, the valve closes again. Open bucket traps are usually bulky and subject to air binding.

## THERMODYNAMIC STEAM TRAPS (Fig. 90)

Thermodynamic steam traps have only one moving part: a disc which operates as the valve. Pressure of condensate or air lifts the disc off its seat, allowing free and immediate discharge. Discharge continues until flashing

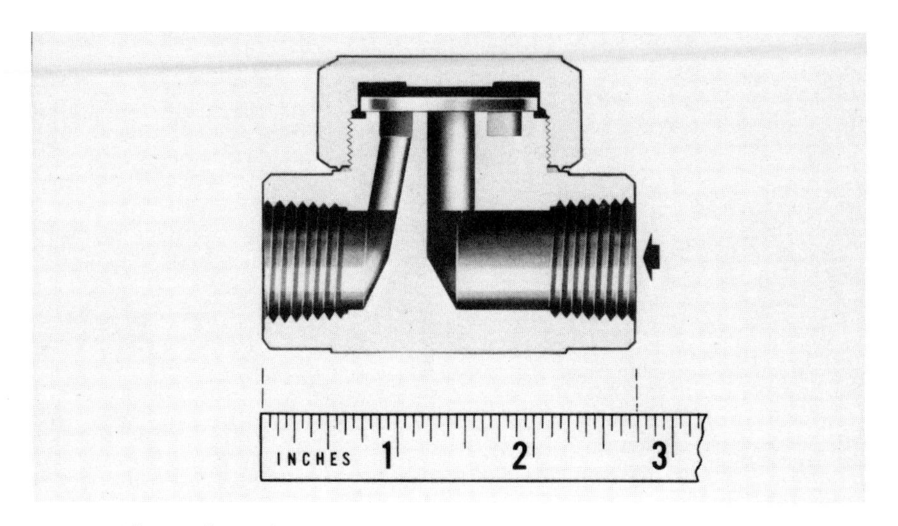


**FIG. 90. Thermodynamic steam trap** *(courtesy Sarco).*

condensate approaches steam temperatures. Then high-velocity jets of flash steam move radially across the underside of the disc, reducing the pressure in this area and, by recompression, build up the pressure in the control chamber above the disc. This combined action drives the disc onto its seat, effecting a tight closure of the trap at saturated steam temperature.

While closed, steam pressure acting over the total upper disc area has a greater force than inlet pressure acting over the smaller inlet area holding the disc tightly on its seat. Heat transfer from the trap inlet maintains control-chamber temperature and pressure, ensuring tight closure during no-load conditions. When condensate reaches the trap inlet, this heat transfer rate is reduced, causing the pressure in the control chamber to fall and the trap immediately opens. This cycle repeats itself.

Since some thermodynamic traps available today can be closed by flashing condensate below steam temperatures, it is important to know the number of degrees below saturated steam temperature at which the trap will close. This is necessary as this temperature depression will cause flooding in the system, reducing the efficiency of the equipment on which the trap is installed.

Most thermodynamic-type traps are of all stainless-steel construction, providing high resistance to corrosion. They are small, lightweight, and unaffected by water hammer. As they can be designed to discharge condensate at saturated temperature, these designs offer high thermal efficiency. They will operate in any position and, when installed vertically discharging down to an open discharge, they are freeze-proof. The minimum operating pressure for some models is 3.5 psi; others cannot operate below 10 psi. Maximum back-pressure toleration is 80% of inlet pressure for some models; others cannot tolerate more than 40 to 50% back pressure.

## THERMOSTATIC AIR (OR GAS) VENTS (Fig. 91)

When air or any other noncondensible gas is present in a steam space, the steam cannot be maintained at its saturated temperature. It can be very serious if air is in the steam system.

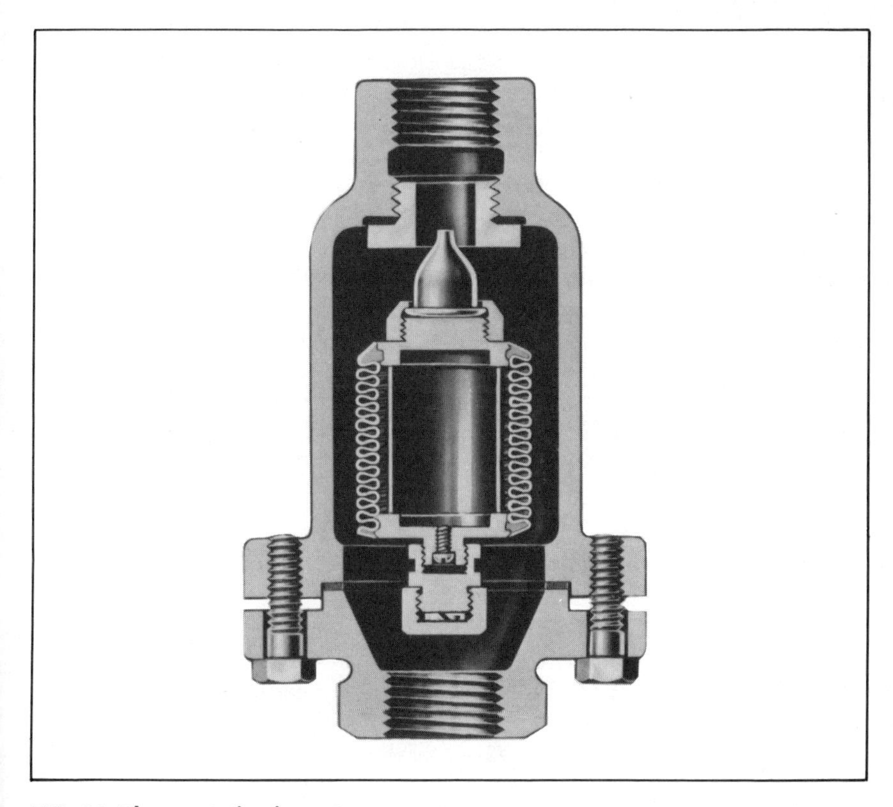

**FIG. 91. Thermostatic air vent** *(courtesy Sarco).*

Thermostatic air vents do an excellent job of removing air from a steam system. The vent is equipped with a bellows to which is attached the valve head. Being sensitive to a change in temperature in the system as soon as air is present, the steam temperature drops and the vent opens, discharging the air present. When all the air has been discharged, the temperature rises and the vent closes tightly.

This type of vent is small in size but has relatively large capacities. As in all equipment of this type, location is important since the air must flow easily to the vent to obtain maximum efficiency in the system.

## FIRE PROTECTION

Numerous piping specialties are in use in fire protection systems. One of the first that comes to mind is the fire hydrant.

This is a specially constructed valve that permits quick opening with either water pressure or mechanical means assisting in full opening. Most hydrants are of the globe-valve-type design with the seat for the plug being either in the inlet section or in the vertical part of the hydrant. Another special feature of most hydrants is a provision for automatic drainage after use. This permits water that has remained in the hydrant body after use to be removed immediately so that the danger of freezing is eliminated.

When a hydrant does not have a provision for automatic drainage or in piping systems where the need may arise, an automatic ball-drip valve can be installed. This valve permits any leakage to flow by the ball, but the ball closes with pressure. Hose connections, hose stations and cabinets, sprinklers, and appurtenances are just some more of the better-known accessories.

## SERVICE CLAMPS

Steel service clamps are being used to make branch connections to a piping system without installing a tee or lateral. This is normally done on piping systems only where the pressure is not too high and can be done with all kinds of piping material. However, most applications of service clamps are with cast-iron or asbestos-cement pipe. The clamp part may be of single or double-strap design with threaded ends fitted through the saddle part and then bolted. The saddle part is drilled and tapped to the size required for the branch connection. Drilling the pipe main can be accomplished before or after installation of the saddle. A gasket of a suitable material is inserted between pipe and saddle to provide the necessary leak tightness.

## REPAIR CLAMPS

Most repair clamps are used only in conjunction with metallic piping. The simplest repair clamp consists of two halves that are tightened over a pipe after putting some gasketing material over the leaky area. However, there are more sophisticated repair clamps or sleeves either in single or two-piece units that can be welded or have better gasketing areas available so that they can be used even for more severe pressure applications.

## NOISE SUPPRESSORS

The most common noise suppressor is a muffler that is used in steam or other hot gas exhaust lines like a car. The muffler is a very simple piece of pipe, manufactured from the same material as the piping system into which it is incorporated. However, the cross section of the muffler is much larger than the pipe to which it is connected. Its function is based on the fact that, by permitting expansion of the discharges gases and thus reducing their nozzle velocity at the point of discharge, considerable reduction of noise is being achieved.

Another noise suppressor for low-pressure steam piping is an exhaust head, which permits expansion of steam discharge at the point of discharge and dissipates the steam discharge.

# 13

# Pipe supports and restraints

There is a great variety of pipe hangers and pipe supports available to ensure that the piping system, which has been designed for a specific purpose, can actually serve its function without being dislocated. Some of the pipe supports can be very simple and are best fabricated at the time of installation by either using standard components or structural steel shapes. Other supports are very sophisticated and go through a lengthy design, layout, and manufacturing process.

The design of a good pipe supporting system is particularly important where stresses induced by high pressure, high temperature, low temperature, or any combination thereof can play havoc with a piping system if not properly restrained or guided. Other conditions, such as possible seismic disturbances or other possible shock effect, may also have to be considered in the pipe-support design. The design of the support system for the Trans-Alaska Pipeline is an example of how intricate such a design might be. Not only was it necessary to design pipe supports to carry the anticipated load of a filled pipeline and review the soil-bearing capabilities of the traversed area, but the existing permafrost in that particular region had to be considered. The unusual conditions encountered, such as permafrost and extreme climatic variables, posed a real challenge to the support-system designers. At the other end of the spectrum, extreme temperature, pressure, and seismic conditions have been encountered by the designers of support systems in nuclear power plants.

The Manufacturers' Standardization Society (MSS) has published various standards dealing with pipe hangers and supports.

| | |
|---|---|
| MSS SP-58 | Pipe Hangers and Supports—Materials, Design and Manufacture. |
| MSS SP-69 | Pipe Hangers and Supports, Selection and Application. |
| MSS SP-89 | Pipe Hangers and Supports, Fabrication and Installation Practice. |
| MSS SP-90 | Guidelines on Terminology for Pipe—Hangers and Supports. |

In Standards SP-58 and SP-69, the various hanger and support types are pictorially shown, and tables are published that deal with hanger and support selection as well as spring-support selection. Other tables in these standards show recommended installation of protection shields for insulated piping, as well as minimum hanger rod requirements (Appendix 10).

All these tables refer in footnotes to one another or to the pictorial index of the various types of pipe hangers. Maximum spacing of pipe hangers and supports, another important aspect of support system design is also covered. At this time, only Standard SP-58 has been approved by ANSI, and incorporation in other codes is still lagging.

As indicated in the various standards tables, there are differences in hanger design depending on pipe size and material of construction. Hangers and supports must always be designed to accommodate insulation and anticipated pipe movements. If a piping system is manufactured from materials that require stress relieving, it is important that pipe-support accessories such as hanger lugs or insulation-protection saddles be attached by the piping manufacturer so that stress relieving away from shop facilities can be held to a minimum. Anchors are used not only to control pipe expansion but are also often needed to prevent disengagement of slip-type or bell-and-spigot-type pipe connections or mechanically connected piping.

Prefabricated pipe supports for any piping system must be identified through a numbering system, and individual sketches of such preengineered supports must be made available to the construction forces. If such supports consist of several pieces, each individual piece must be marked in order to facilitate the construction process.

### Beam clamp

The beam clamp is an easy method to hang piping from overhead beams in a building without resorting to welding. The three basic different designs are top, bottom, or side beam clamp. Depending on the load requirement, beam clamps are manufactured from cast steel or malleable iron components or are fabricated from flat steel stock.

### Wall bracket

This is one of the simplest forms of pipe support and is shaped mostly at a 90-degree angle. It can be manufactured from flat steel or angle iron and the pipe may rest on the cross arm, which can also be cradle-shaped or may be suspended from it. A wall bracket can be bolted or welded or otherwise affixed to a load-bearing wall.

### Clevis hanger

The clevis hanger consists of three parts and is manufactured from flat steel. One part is the pipe cradle, which is shaped half round with extending arms which envelop the pipe that is being supported. The second part is the top bracket, which is suspended from supporting steel by means of a hanger rod. The third part of the clevis hanger is the bolt and nuts assembly, which connects cradle with top bracket.

### Pipe clamp

The pipe clamp can be used in conjunction with some hanger appurtenances for holding suspended pipe in place or it may be used as a riser clamp when vertical piping passes through a floor and hold it thus in place.

Most clamps consist of two halves that are fabricated from flat steel and are known as either 2-bolt, 3-bolt, or 4-bolt pipe clamps. In an offset pipe clamp, one of the halves has its extensions offset so that the pipe cradle will be several inches above the supporting bracket. Another type of pipe clamp has two halves preformed to fit tightly around the pipe and has provisions to hang this assembly or to support it from below. This clamp is manufactured from either forged steel, flat steel stock, or malleable iron.

### Pipe rings

These are simple rings made in one piece or with a joint as a split ring. They have provisions to be hung either from a swivel joint or through a direct connection.

### U-bolt

One of the simplest pipe fasteners, it is fabricated from round steel. It encompasses the pipe to be fastened or supported, and its two threaded legs pass through the supporting steel and are then fastened.

### Pipe saddle support

There are several variations of this type of support, which is also used often for nonmetallic pipe. The saddle is mostly shaped so it fits snugly to more than one-quarter of the circumference of the pipe that it supports and is several inches long. In its simple form, it has a pipe appendage centrally welded on the bottom that fits into a larger pipe stanchion. The adjustable saddle has threaded pipe or bolt stock welded. With the assistance of two locknuts, the saddle can be raised or lowered. Another stanchion type has a U-bolt added so that the saddle can be fastened to the pipe.

## WELDING SUPPORTS

There are many different attachments welded to a pipe that can be part of its supporting system.

A. *Hanger lug*—which is some shaped steel plate with a hole through which a bolting arrangement can be attached.

B. *Base anchor*—a chair-like arrangement that may or may not have bolt holes for anchoring to the floor. It may be fabricated from steel plate or pipe with a baseplate.
C. *Sliding support*—a tee-shaped like structural base that permits movement of a pipe without damaging the pipe that is welded lengthwise to the pipe.
D. *Steel chair*—a tee-shaped base with cutout for the pipe shape into which the pipe is cradled. It may or may not be welded to the pipe.

## ROLLER SUPPORTS

Whenever longitudinal movement of a supported pipe is anticipated and the pipe-support design calls for permitting and facilitating such movement, the use of pipe rolls or another sliding guide is called for. Pipe rolls have a slightly cradle-like shape, and they can be easily accommodated in adjustable or nonadjustable pipe stands. They can also be used in conjunction with a trapeze-like arrangement where the pipe roll is the base of the trapeze or a pipe roll may be the pipe supporting base of a clevis-type hanger assembly.

Since roller supports are mostly used in conjunction with insulated pipe, a method of insulation protection is mostly applied. This may be an insulation shield, which consists of some shaped sheetmetal that covers one-third to one-half of the insulation circumference and is inserted between the pipe roll and the insulation. Another method is to weld a pipe-covering protection saddle to the pipe that is equidistant from the pipe as the insulation and then insulate the rest of the pipe and fill in the saddle area with insulation. Then, only the bare saddle exterior that covers one-quarter of the pipe rides on the pipe roll.

## TRAPEZE HANGERS

These can be fabricated to carry several pipes at the same time on the lower bar and may consist of heavy structural members, depending on the load to be carried.

## SPRING HANGERS

There are two types of spring hangers in use: namely, variable spring hangers or constant supports. Both rely on heavy-duty coiled springs to perform their function.

### Variable-spring hangers

These are furnished as an open spring without any calibration or as a canned spring. The canned spring is calibrated and tested to support any given load for which it has been selected. The spring is contained in a housing or can and is connected through hanger rods or other means directly to the piping. The spring is installed in the compressed or cold position. Once the total piping system is in operation, the limit stops that hold the spring in compression have to be removed so the hot or operating condition is achieved. Most canned-spring hangers have a scale plate fastened to the can, which indicates the required setting.

### Constant supports

Constant-support hangers also make use of canned springs and ensure constant support through the entire deflection of the pipe load. Counterbalancing the load and spring moments about the main pivot is obtained by the use of compression-type load springs, lever, and spring-tension rods. As the lever moves from the high to the low position, the load spring is compressed and the resulting increasing force acting on the decreasing spring-moment arm creates a turning moment about the main pivot that is exactly equal and opposite to the turning moment of the load and load-moment arm. The constant support is also installed with the spring in the cold position. After the whole system is operating, the hot setting must be checked and adjusted.

## PIPELINE SUPPORTS

Pipeline supports or pipe bends are classified separately because their application and materials of construction can differ considerably. In a simple application, wooden railroad ties or precast concrete sleepers may be put down on grade level to support a pipeline without giving it much guidance or restraint. When a pipeline is installed at grade level and requires more than just basic support, concrete sleepers including foundations poured monolithically can provide necessary restraint or guidance through insertion of anchor bolts into the concrete that in turn serve as the anchor point for pipe clamps or other pipe anchors. In order to avoid direct contact between pipe and concrete, some form of structural steel, such as a T-bar or railroad track, is often inserted into the concrete crown to serve as a contact area.

When single or multiple pipelines are located at a higher elevation from ground level, H-frames or T-frames can support such piping. In these instances, the supporting structure must be anchored firmly in the ground. The structure can use concrete, steel, or wood as a supporting base, while the load-carrying beam is mostly structural steel. An H-frame consists of two supporting columns with load-carrying beam fastened across similar to the letter *H*. This type of support was extensively used during the construction of the Alaska Pipeline. The two supporting columns in that case were mostly pipe, which was also used at the same time as a heat-exchanger media. This was filled with a coolant to prevent any disturbance of the underlying permafrost. The T-frame consists of a single load-bearing column with a crossarm attached upon which one or more pipelines can be carried.

In oil refineries or other process plants, pipe racks are used extensively to carry numerous pipelines to interconnect the various plant areas. Pipe racks are mostly fabricated from structural steel and consist of H-frames, which may have several load-carrying beams connected to the supporting columns one over the other at various elevations. Thus, these provide ample supporting surface for the various pipelines. Pipe racks mostly provide sufficient head room to walk underneath. That means that the lowest beam is at least 8–10 feet above grade level. However, pipe racks often provide sufficient head room for automotive traffic to pass underneath. While the average pipe rack consists of a series of H-frame structures, these are mostly interspersed at various distances with a tower-like structure, especially where a change in direction occurs, and thus provide more stability for the entire system.

Pipeline movement is usually provided for through various means. A section of T-bar may be welded to the bottom of the pipe and thus enable pipe movement without adverse effects on the pipe, insofar that the T-bar moves back and forth over the point of contact with the supporting steel. Pipe rollers are discussed elsewhere, but as an alternate, a sliding support, consisting of a plastic or graphite material with the appropriate properties, is often inserted at the point of contact between pipe and support to facilitate movement. Fluorocarbon slide bearings have proven themselves as particularly effective in providing low-friction movement during pipe expansion or contraction. Guidance for such moving pipe is often provided through oversized pipe clamps that restrict lateral movement, while longitudinal expansion may be compensated for through one of the various pipe-expansion devices.

Such longitudinal expansion can additionally be restricted to various segments of the pipeline through the use of pipe anchors, which maintain fixed points at selected intervals. Pipe anchors may be pipe clamps or the attachment of welded or bolted structural shapes.

## ATTACHMENTS

Some of the attachments that are most often called for in conjunction with pipe hangers are:

Threaded rod
Eye rod
Beam attachment

Linked-eye rod
Rod coupling
Eye socket
Forged-steel clevis
Forged-steel turnbuckle

While not actually a hanger attachment, concrete inserts often support whole piping systems and therefore should not be overlooked when reviewing pipe-supporting system components.

## SHOCK SUPPRESSORS

Shock suppressors or snubbers are very much in use in nuclear power plants. They stabilize a piping system and are an additional restraint, so that they can be considered part of the pipe supporting or restraining system. There are two basically different types available: hydraulical and mechanically activated shock absorbers.

The restraining force of shock suppressors varies as a function of load imposed by attached piping and the rate of displacement varies as a function of applied load. The critical nature of operational requirements dictate efficient quality control and preinstallation testing.

### Hydraulic shock suppressors

These consist of a piston-rod assembly encased in a cylinder with the piston-rod end connected through a clevis pin to a pipe clamp or extension at one end and the cylinder connected to a bracket through a pivotal joint at the other end. Grease fittings installed at the pivotal joints permit continued ease of movement. A reservoir for cylinder fluid is mounted directly on top of the cylinder and connected to it. Piston settings for hot and cold positions are precalibrated and should need little adjustment: however, hydraulic fluid loss, due to normal operational function of these snubbers, has been experienced in sensitive areas of nuclear plants.

### Mechanical shock suppressors

Mechanical snubbers contain an inertia mass mounted on a torque transfer drum above a capstan spring in a telescoping cylinder. A ball nut and screw shaft attached to the capstan spring convert piping motion into rotation of the torque transfer drum. Whenever piping motion acceleration exceeds a certain G force, the inertia mass causes the capstan spring to brake the torque transfer drum rotation and limit acceleration to the threshold value. All operating parts in mechanical snubbers are metallic and positive acting.

## VIBRATION CONTROL AND SWAY BRACE

This can be achieved very easily through installing a precalibrated adjustable canned-spring assembly. This sway brace should be adjusted so it is in the neutral position when the piping system is hot and operating. Vibration is then opposed through an instantaneous counterforce, bringing the pipe back to normal position.

# 14

## Insulation for piping

Insulation is normally applied to piping systems to prevent any heat exchange to take place between the fluid carried in the pipe and the exterior surroundings. It is thus applicable to prevent heat loss from a pipe carrying a warm or hot fluid. Alternately, insulation may be designed to keep a low-temperature fluid from increasing its temperature. Heat loss can be expressed in British thermal units (BTU), and the effectiveness of insulation is related to the thermal conductivity of the varying insulation materials. Thermal conductivity is by definition the rate of heat transfer expressed in BTUs per one-inch thickness per square foot per degree Fahrenheit temperature difference per hour (BTU/ft$^2$/°F/hr).

Another use of piping insulation that is not related to the prevention of heat loss is for personnel protection. Piping systems that carry hot fluids and where heat loss is of no importance or even desirable are often insulated to prevent the possibility of accidental flesh burns where such piping is within easy touching distance of personnel.

Until a few years ago, asbestos-derived materials, which are known for their fire and heat-resistant properties, played a major role in general insulation practice. However, with the advent of the Occupational Safety and Health Act of 1970 and the discovery of the cancer-inducing characteristics of asbestos, this material is now almost totally banned from such applications here in the United States.

Major thermal insulating materials today are calcium silicate, diatomaceous earth, polyurethane foam, perlite, and an assortment of various felts, glass fibers and plastics, and cellular glass for cold insulation. A special reflective insulation made from aluminum or stainless steel or a combination of both that is free of dusting and mechanical disintegration has lately been developed for use within the containment vessel of nuclear power plants. A variety of ANSI/ASTM specifications have been issued that deal with materials of construction and their properties and also set dimensional and fabricating standards for miscellaneous insulating materials.

The introduction of OSHA also required in many instances a reduction of prevailing noise levels. This legislation thus assisted the arrival of pipe insulation that is mainly devoted to the problems of noise control. One such product consists of either fiberglass or polyurethane foam laminated to a protective jacket of barium-loaded vinyl or PVC.

With the demise of the asbestos-derived insulation, new products had also to be developed that not only would be preferrable because of low conductivity but also possessed fire or flame-retardant qualities. These qualities are of particular importance since many insurance ratings are based on meeting standards set by the National Fire Protection Association (NFPA). For fire-hazard classification ratings, pipe insulation, insulation jacket, mastics, and adhesives used to secure the jacket are all tested together to check on flame spread, smoke development, and fuel contribution during a fire.

Piping insulation is mostly available as a preshaped product where two halves are slipped over the straight lengths of pipe while fittings, flanged joints, and valves require a craftman's skill to cover their divergent contours unless preshaped materials are available. Pipe insulation in industrial applications is often covered with metallic (aluminum or stainless-steel) covering that provides additional protection and beauty while other types of insulation might make use of a canvas jacket. However, polyurethane-foam application differs markedly, insofar that it is either poured, blown, or sprayed into a partially enclosed area before being covered with a protective covering.

Insulation must be applied on clean piping only and should be continuous through wall or floor openings. Where insulated piping comes in contact with hangers or supports, insulation should be protected from damage through the installation of sheet-metal shields.

## INSULATION IMPACT ON PIPING AND SUPPORT DESIGN

With increasing energy costs, increased insulation can give a greater return on investment than increased plant capacity. To achieve this goal, a number of specific steps can be taken which will implement the economic calculations and rely on more engineering ingenuity during the design phase.

It will be valuable to create and use computerized methods for the calculation of insulation thickness. Review and upgrade insulation materials specified. Give more thought to jacketing to provide protection against weather, mechanical damage, or fire. Provide better values for the heat-transfer properties of insulation. Use a system approach and try to achieve more accuracy in the calculation of total heat loss of a complete system. Recommend the use of engineered insulation, designed to give specific combinations of insulating value, size, weight, and damage resistance. Use combinations—layers of different insulating materials and jackets. All this is the opposite of standard insulation specifications. Make use of newly fabricated forms of insulation designed to make easier the more difficult insulation installation around valves, flanges, and fittings.

There is no difference in pipe and fittings whether they are insulated or not. However, valves that require insulation and particularly in cryogenic service, which requires mostly heavier insulation than any other service, should be specified with extra-long stuffing boxes. Where inspection or maintenance access is provided in the piping design, this should be pointed out in the applicable drawings so that removable insulation can be specified to make such openings accessible when required.

During the design phase of insulated piping, special attention must be directed to the clearance necessary for the additional insulation thickness, which must be added to the piping outside diameter wherever pipe sleeves are foreseen, or otherwise provisions for pipe passage are provided. Since most insulated piping systems can also be expected to have some movement due to temperature differentials, this anticipated movement must also be taken into consideration when checking piping drawings for interference with other plant components.

Where several pipelines are accommodated on the same pipe supports, the distance between the centerlines of the various lines must always provide sufficient room for insulation that may be applied to several lines. Whenever one or more pipelines on a pipe rack or trestle are so designed that pipe loops take care of expansion problems, the designer must show his ingenuity. Such piping loops may be on the two outside pipes only. They may be elevated over other piping systems. Some loops may be nestled one into the other. Loops may be installed vertically or at an angle. It is up to the designer to find the best and most practical solution.

Pipe-support design for insulated piping requires special consideration for several reasons. Not only has the support design to take into account the special clearances required for such piping, but it is of equal importance to make the necessary provisions that preclude damage to the insulation either at the time of installation or during the operation of the piping system.

## PREINSULATED PIPING

Underground steam or hot-water distribution systems or other piping systems that are buried underground and require thermal insulation for heat-loss protection are often prefabricated in sections where the insulation is already factory applied. Lengths of such prefabricated piping are only restricted by the limitations of transportation. Preinsulation calls for insertion of the insulated pipe into a larger jacket or conduit, and often two or more insulated pipes are installed in a single conduit at point of manufacture. This method thus provides for simplified installation of supply and return lines.

At the point of connection between two factory-assembled sections, the piping protrudes several inches from the insulation and jacket so that piping assembly through welding or other methods can be accomplished. After the piping system has been tested for leaks, connecting points are insulated and then continuity of the jacket can be achieved by providing some type of connecting conduit that is welded or otherwise mechanically fastened to both conduit ends.

Materials of construction for pipe and jacket are not restricted, but a preferred type of jacket is made from corrugated steel, which may be galvanized or bitumastic coated in order to enable it to better withstand corrosive elements after embedment. Among other materials, PVC and asbestos cement are recognized as jacketing materials, particularly for applications in the mid-temperature range. These nonmetallic jackets are used mostly in conjunction with polyurethane foam insulation, which is poured or blown—in contrast to most other insulation materials that are hand applied.

Heat-traced piping is also installed in preinsulated shape. For such installation, the use of electrical trace lines and in particular skin effect current traced (SECT)® has proven successful even under arctic conditions (See Fig. 12).

# 15

## Shop fabrication of piping

When we refer to shop-fabricated piping, we mostly refer to such fabrication of piping that can also be performed at the job site but that is done in a fabrication shop for economic reasons. The alternative is prefabricated piping that must be fabricated off the site because construction facilities are mostly inadequate to permit their field fabrication. This refers mostly to lined, coated, or jacketed piping but may also be applicable to large concrete pressure pipe or various types of plastic piping or preinsulated piping.

While the fabrication of carbon steel piping, which will be used later as the pressure piping component of lined pipe, may be performed in a regular pipe fabricating shop, most lining work will be performed in specialty shops that are geared to that particular type of operation. Such specialty fabrication that produces piping as an end product may be lead lining, rubber lining, cement lining, bitumastic enamel lining or coating, plastic lining or spraying, or zinc coating.

Other fabricating plants may manufacture large concrete pressure pipe or large plastic piping assemblies. All these manufacturing facilities provide the necessary services required for pipe lining or coating. In some cases, these fabricators are restricted to only the lining of piping. Others can also perform the fabrication or lining of large pieces of equipment in their shop or in the field. As an example, steel piping that is to be rubber-lined may be fabricated. In this case, that means cut to size, weld on flanges, and remove all burrs in a regular pipe shop. The installation of the rubber lining, which includes covering the flanges, is then performed in the rubber liner's shop, where tanks and other pieces of equipment are also lined. This fabricator is also equipped to handle rubber lining of large tanks in the field but cannot perform any fabrication of metallic piping.

The fiberglass pipe manufacturer, on the other hand, will fabricate sections of any large piping system in such a configuration that will present no difficulties in transportation or installations. Such large fiberglass piping can then either be flanged or fused together at the construction site.

Preinsulated piping is another product whose manufacture precludes on-the-job fabrication. Here we refer to one or several metallic pipes that are insulated and housed inside a metallic or asbestos-cement conduit. This preinsulated piping is furnished normally in either straight length or with fittings or expansion loops fully fabricated. Depending on the manufacturer, there may be air space provided between the insulation and the outer conduit. This air space functions like the air space in a double-pane window and increases the insulation efficiency. This type of preinsulated piping is mostly used for underground installation, and the outer conduit is often protected by a coating of epoxy. Field connections for the internal pipes are mostly done by welding, then covered by insulation, and the conduits are connected through a field closure sleeve or collar that completes the installation. Wherever such preinsulated piping terminates, be it a manhole or inside the walls of a building, the piping is equipped with end seals that permit only the internal piping to protrude from the conduit.

The pipe fabricating shops, most of which are also members of the Pipe Fabricating Institute (PFI), fabricate mostly metallic piping, which may range from carbon steel, copper, nickel, or stainless steel to such exotic materials as titanium or hastelloy. The term *fabrication* is meant to encompass all those operations necessary to produce a finished product: cutting, beveling, welding, bending, heat treating, grinding, threading, cleaning, sand or shotblasting, flushing, pressure or vacuum testing, destructive or nondestructive examinations, and painting.

Offsite pipe fabrication originated with pipe-bending operations of large-sized pipe that could not be accomplished at a construction site. Even today, some of the old established pipe fabrication shops still embody pipe bending in their corporate name. As a result of these early bending operations, there have evolved various bending configurations, which have become an industry-wide standard (Fig. 92).

Today, pipe prefabrication in offsite manufacturing facilities mainly reduces costs and shortens construction time. A pipe fabricating shop normally employs the same craftsmen over a long period of time, so that the teamwork approach is much more engrained than with pipefitters on a construction site who may be there for a year or two before moving to another site with a different crew and other supervisors.

The fabricating shop is also much better equipped with machine tools, automatic welding, and materials-handling equipment. In short, the automation or production that can never be achieved at a construction site is being partially achieved by producing as much prefabrication as possible within the limitations of transportation and final installation parameters. Another asset of pipe fabrication shops is that they normally carry a sizeable inventory of piping and fittings so that any change in drawings requiring different materials can often be accommodated out of existing stocks.

If the pipe fabricator does not furnish the hanger and supporting mate-

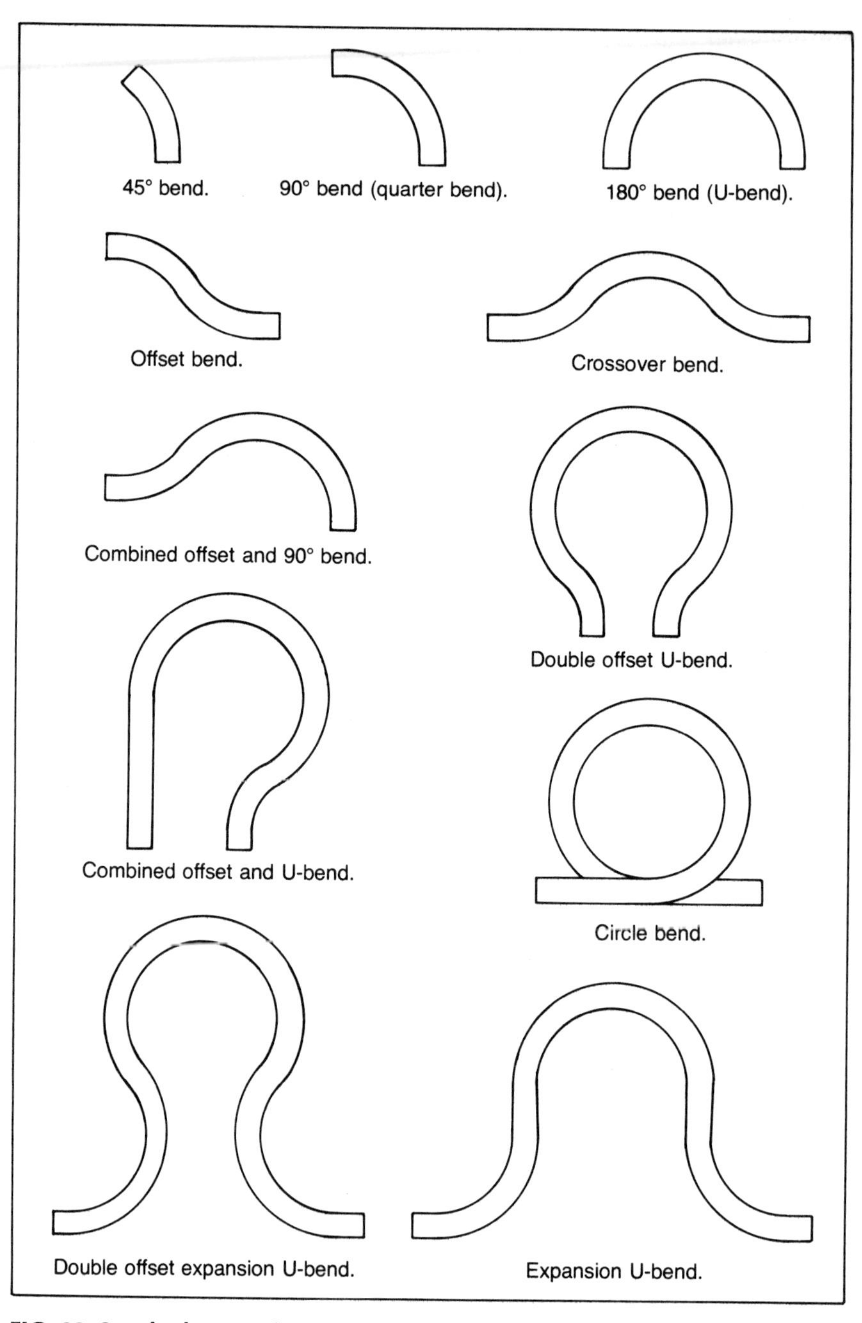


**FIG. 92. Standard types of pipe bends.**

rial, the question often arises, Who installs the hanger lugs or other welded attachments? If the piping material is only carbon steel, this welding can be done in the field as well as in a fabricators shop. But where stress relieving or other special operations are necessary, it is important that all such accessories are available in the fabricators shop so that the necessary welding can be performed prior to shipment of the assembly to the field to avoid additional and much more expensive heat treatment.

In order to prefabricate piping, specifications and drawings must be complete with all dimensions given. The pipe fabricator then prepares pipe-spool drawings. That means that every piping system is broken down into a given number of pipe spools or subassemblies that are the actual fabricated items. A typical pipe-spool drawing has all the information necessary to fabricate this assembly and finally install it. Pipe spools are consecutively numbered by system. To ease installation in the field, it is advisable to have isometric drawings of each piping system and also show the various hangers and supports. The information contained on a pipe-spool drawing is quite detailed and mostly includes piping reference drawing, piping system, piping specification, and applicable codes, working and testing pressure, bill of materials, details of necessary shop operations, examination and testing requirements, packing or closure requirements, and painting, or cleaning requirements.

The pipe fabricator's shop is mostly laid out so an orderly flow of materials, sequence of work, and maximum use of automated welding equipment can be achieved. Most of the fabrication being performed in the fabricator's shop consists of piping only. But lately there has been some tendency to include valves also in piping preassemblies.

In nuclear power plants, the prefabrication of pipe hangers, supports, and restraints is also of major importance. The code requirements are such that more and more restraints must be furnished for critical piping. It is also important to have pipe supports installed before starting pipe erection, and here prefabrication is a must. Most pipe hanger fabricating shops perform the dual role of designing and fabricating. Designing pipe supports or restraints can, of course, only begin after the piping-system design is complete, and this already indicates a major problem. The problem, quite simply, is that the hangers should be onsite before the piping, but their design can only start after piping design is complete.

Pipe-hanger fabricators also prepare separate drawings for each hanger assembly, which contain all the necessary details and code requirements but must also contain a location plan to indicate the exact location of a hanger.

All fabricators who provide prefabrication must possess the code stamps that are required under the design specifications. ASME Section III, which applies to nuclear work, requires authorization of fabrication which is obtainable through implementation surveys performed by ASME and an authorized inspection agency. But even for nonnuclear work, possession of the ASME PP stamp for work related to power piping as per ASME Section I is a precondition for any pipe fabrication.

# 16

# Field installation of piping

Piping erection labor in any industrial facility may vary from between 20–45% of the total erection manhours to be spent. Another statistic of interest is that in instances where large construction projects experience serious time and cost overruns, it is mostly in the piping erection that the worst overrun conditions are apparent. It is often said that a construction project can only meet its cost and scheduling goals if the piping erection is not delayed. In order to meet these goals, it must also be recognized that a good construction performance has its beginnings by being a well-engineered project. It is important to understand how the development of pipe design, procurement, fabrication, delivery and inventory control all play significant parts in their contribution to the achievement of the best possible construction performance.

Initially after a plot plan and set of flow sheets are ready, the piping designer can start to prepare the piping general-arrangement drawings. These drawings will show the plant piping in four different views: plan view, showing a top view of all pipe runs in their respective horizontal position; elevation view of those pipes running in a vertical plane; cross-sectional views of various sections, to show details of where different pipes are relative to each other at more than one elevation; and finally detailed drawings to show connections to equipment or other particular requirements.

Normally, the general-arrangement drawings and flowsheets provide enough information to permit the erection of the piping in the field. From a construction point of view, however, the goal is to obtain this engineering information in such a manner that erection work can proceed in an expeditious format without interruptions caused by inadequate or incorrect information. Most important, the drawings must be presented in such form that they are easily read and understood by the craftsmen performing the actual construction work.

While piping drawings for a project are often prefaced with an introduction to the various symbols used, this is not always the case. The various connecting procedures all have their own particular drafting symbol, and different types of welding are also identified. A legend of significant drafting symbols should always be transmitted to the erection forces to facilitate drawing interpretation (Fig. 93).

Much experimentation has been performed in order to achieve the best engineering information presentation, balancing the erector's requirements with the lowest engineering costs. A difference in philosophy exists between companies that design and construct compared with companies that design only, leaving the erection to someone else. Where the former is forced to provide complete engineering information in order to minimize its field cost, the latter will tend to minimize its engineering manhours so as to build a reputation for an economically priced engineering package. In the overall plant cost, increased engineering costs for detail design are normally more than offset through savings achieved by reducing construction labor costs.

An isometric drawing depicting a single piping system is of utmost value to an erector. It should be furnished for all prefabricated piping because it identifies the location of each individual spool piece, but it should also preferably be furnished for all other piping. The pipe fabricator must point out field welds on the isometric drawings and identify pipe sections as well as valves or specialties, and the hanger designer must indicate hanger type and location.

With all this information incorporated in an isometric drawing, it becomes a valuable tool for the erector. The isometric drawing must then be used in conjunction with the general arrangement drawings, the plot plan, and the applicable specifications, standards, bills of material, and vendor drawings to form a complete erection package.

Often, detailed drawings are not prepared for small-diameter piping of 2 in. and less. In this case, flowsheets are required in addition to the general-arrangement drawings to route the piping. The supervisor in the field must in this case spend much of his time interpreting drawings and preparing field sketches instead of supervising his personnel, causing an adverse effect on productivity.

Ideally, all piping should be shown on isometric drawings. Plant models have often been used to portray a complete project and as such can serve a variety of purposes. An old saying is that one look is worth a thousand words. When anybody looks at a plant model, it certainly means much more than reading a description of all the various components of the plant or looking at design drawings. In the initial stage of plant design, scale models of the various units such as heat exchangers, pumps, and compressors that have already been modelled to scale can be moved around to provide the most attractive design feature. Most plant models are constructed of plastic materials, with pipe and fittings available in a wide range of sizes.

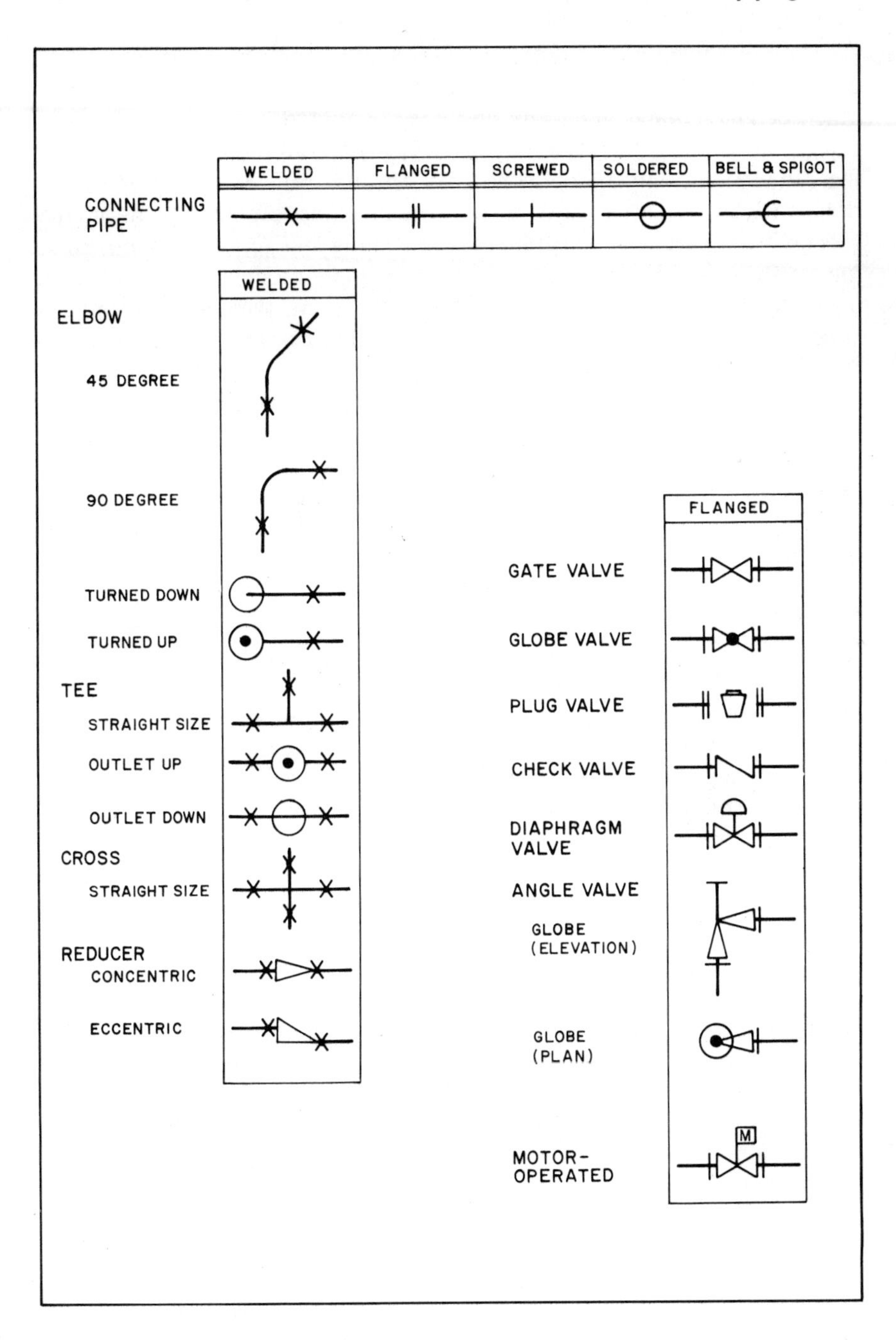


**FIG. 93. Graphical symbols for pipe fittings and valves** *(courtesy ASME).*

The completed model represents a miniature replica of the envisioned project, built to a previously established true scale. By incorporating as much detail as practical in a model, it is possible to reduce the number of design interferences that often play havoc with construction schedules. At the same time, plant models can play an important role in assisting the designer when decisions have to be made as to the routing of plant piping, especially in plants where cable trays or heating and air-conditioning ducts must also be installed in abundance. While plant models can be of significant assistance during the design stage of a plant, their value as a construction aid is limited and is mostly restricted to an easier comprehension of the overall layout of the project.

In order to achieve a reduced construction cost, many innovations have been tried and undoubtedly will continue to be tried. The way to this goal is not furthered through the reduction of engineering manhours. Keep in mind that the ratio of erection manhours to engineering manhours can run as high as seven or eight to one. With this ratio, it seems mandatory from an economical point of view that no shortcuts are taken in engineering that lead to incomplete information for the erector. To obtain the lowest cost in the field, the engineering information must be complete and easy to understand for those who use it. It is quite evident that incomplete, incorrect, or unclear engineering information can have an adverse effect on productivity and unit costs. Thus, a good construction job must start with a good engineering job.

Scheduling construction activities has become more important as the size of construction projects has increased. This can play a significant part in assisting an orderly approach to the work at hand. However, scheduling is only a tool that can be used when it is beneficial to the overall project. Scheduling is a means toward an end that should not be overemphasized as a solution to all the ills of modern construction. While achieving project milestones within their timeframes is certainly a requirement that cannot be ignored without grave consequences, detailed schedules of minor importance may well be ignored without affecting the overall goal.

Actual construction activities start with a review of material, tool, and construction equipment requirements for the project at hand. The erection work itself then proceeds and should follow a standard pattern that sets as a first requirement the preparation of the work area. This involves checking out the equipment installation or setting foundations and anchorbolts if necessary. At the same time, installing pipe hangers, supports, and restraints can also be accomplished.

It is important from an economical point of view to install pipe hangers and supports that form part of the piping system early so the necessity to install temporary hangers will not arise. With pipe supports already in place, it is also much easier to install piping spools; installation and lining up of piping systems can proceed with a minimum of delay. Thus, pipe-hanger installation is followed by the erection of prefabricated large-bore piping and accompanying valves. Since all installation should follow this pattern, it can easily be seen that unavailability or delay in receipt of parts included in this sequence leads to costly delays. Similarly, the storage and marking of piping materials should be such that component parts of any given system are easily recognizable and retrievable and not too much time of personnel and equipment is required to locate required parts.

So far, it has been pointed out that hangers should be installed first and then followed by piping materials. Large piping and even hangers require lifting equipment for the operation for which sufficient clear space must be available. After the pipe spools have been lifted and are aligned, preferably starting from a fixed position such as a pump outlet or vessel nozzle, necessary connections between various pipe spools, valves, and vessels or pumps can be accomplished.

When flanged joints are involved, the necessary gaskets are inserted and the bolts will be lightly tightened. Final tightening of bolts on flanged joints should only proceed when a piping system is completely installed between two fixed-end positions. A similar approach should be taken with welded circumferential joints, which should be tack-welded first or sufficiently guided through alignment clamps, and completion of welded joints should be held in abeyance until a complete pipe run can be welded.

In spite of good planning and scheduling, it is a fact of life that very often a valve or specialty item may be missing in a piping system that is scheduled for early completion. It may be best to install or tack weld a temporary pipe spool in lieu of the missing item so that the piping system can be otherwise completed. When the missing item arrives, it is then an easy matter to remove the temporary insert and replace it.

After the erection of large-bore piping is well advanced, the installation of small-bore piping can begin. Piping of 2-in. diameter and smaller is normally considered in this category and consists mostly of field-run piping; i.e., this piping is not prefabricated and often has to be routed differently from the details shown on drawings due to some unforeseen obstruction. Because of field conditions, hangers for this type piping are also not preinstalled and generally are installed simultaneously with the piping itself.

The subject of interferences warrants considerable attention on the part of engineers and designers. A construction project that encounters a large number of interferences during erection will find difficulty in realizing its schedule and cost goals. There is nothing so time consuming and costly for a project as being unexpectedly confronted with frequent interferences during the normal erection routine.

Since most interferences cannot be resolved without some waiting period, the disruptive effect on work crew efficiency can become quite severe if more than a few interferences are encountered. Work crews must either wait for the project engineering office to resolve the interference or be shifted to another work area, either of which results in inefficient use of manpower, lost productivity, schedule delays, and extra cost. In addition, the material flow to the area with the interference must now be interrupted and other material routed to a new work area. Of course, this introduces more complexity to the situation which results in more lost time.

The project schedule may require completion of an area with an interference as soon as possible to maintain progress on the critical-path activities, in which case a work crew may have to be shifted back to that work area as soon as the interference is resolved.

Most construction interferences result from design errors or from construction sequencing errors. Among design errors, there are those interferences where the design engineers during their checkout have failed to pick up interferences within a particular design discipline or, more commonly, have failed to recognize interferences of one discipline with another. For example, a pipe has been routed in a location where it interferes with a steel channel forming a staircase support. Although it is part of designer's responsibility to eliminate these kinds of interferences, at least some always remain.

Construction sequencing errors include interferences which result from poor scheduling or noncoordination of work groups at the job site.

It is easily seen that avoiding the underlying causes, whatever they may be, and eliminating possible resulting interferences will assist substantially in keeping project costs within an allocated budget.

Finally, after a piping system is completely installed and the required inspection of its component parts (i.e., X-ray of butt welds) has been satisfactorily completed, a hydrostatic test can be performed. A system so tested should then be described on a data sheet, indicating the exact boundaries of the tested area. Such a data sheet will then represent a record that gives all available date, including testing media and applied pressure.

# 17

# Construction tools & equipment

The various pieces of equipment and tools used during piping erection represent a tremendous variety. The different manufacturers of similar items all try to outdo one another by introducing special features they hope will make their equipment and tools more attractive than others.

While a broad assortment of tools are mentioned below, this listing can by no means be considered complete, and different names abound for similar tools. In order to review this large assortment a little more coherently, equipment and tools are broken down into seven categories:

Automotive equipment and cranes
Material handling equipment
Welding equipment
Shop equipment
Safety equipment
Pipefitters' tools
Miscellaneous tools and equipment

## AUTOMOTIVE EQUIPMENT AND CRANES

This represents vehicles used in the hauling, transporting, loading, and unloading of pipe materials. It consists mostly of self-propelled equipment, the most prominent of which are listed below:

Trucks (also with winch hoist and/or hydraulic lifting platform)
Pickup trucks
Low-bed trailers
Pipe trailers
Cherry pickers or hydraulic cranes
Truck cranes
Forklift trucks
Sideboom tractors

## MATERIAL-HANDLING EQUIPMENT

Equipment, tools, and accessories used in the performance of pipe fabrication or pipe erection are included in this category. There is no differentiation made as to the driving mechanism of some of the equipment listed, which may be electric, pneumatic, or by internal combustion engine:

Chain puller, lug-all
Chain hoist, ratchet chain hoist
Wire rope hoist
Grip hoist with pulleys
Floor cranes and gantries
Roller skids
Dolly
Wire rope and blocks
Block and tackle
Wire slings and cable spreaders
Chain slings
Shackles, clips
Eye hooks
Turnbuckles
Wire rope, thimbles
Pipe hooks

## WELDING EQUIPMENT

Only accessories and equipment used in the performance of welding operations are included herein; specialty equipment as required for joining of plastic piping are not included:

Welding and cutting torches (oxyacetylene)
Pressure regulators
Oxyacetylene hose
Cutting and beveling machine
Oxyacetylene cart
Welding machines and rectifiers
Heliarc welding equipment
Welding cable
Electrode holder
Electrode oven
Tempil sticks
Wire brushes
Wire cup brushes
Chipping hammer
Grinders (pencil and wheel)
Stress-relieving equipment

## SHOP EQUIPMENT

Mostly stationary equipment that form part of a pipe fabricating shop, either as a commercial enterprise or as an adjunct to pipe erection at a construction site, are considered in this listing:

Pipe and bolt threading machine
Cutoff machines
Metal cutting band saw
Hack saw
Abrasive cutoff wheel
Bench grinder
Drill press
Lathe
Shaper
Arbor press
Anvil
Parallel vise
Pipe vise
Work bench
Air compressor

## SAFETY EQUIPMENT

The prevention of accidents is of utmost importance and a variety of accessories have been introduced and today are considered standard equipment. Their sole purpose is the protection of the individual performing construction work:

Safety hats
Goggles
Work gloves
Ear protectors
Welding helmets
Welding visor
Welding goggles
Welding apron and sleeves
Welding gloves
Safety shoes

## PIPEFITTERS' TOOLS

Some tools are used only in conjunction with the fabrication or installation of piping and are applicable only to piping. Some of these tools are listed below:

Pipe wrench
Chain wrench
Strap wrench
Chain pipe tongs
Pipe extractors (threads)
Pipe dies and ratchet
Pipe reamer (conical and taper)
Pipe cutter
Pipe bender
Pipe roller
Pipe alignment clamps
Yarning and caulking tools
Pipe taps
Gasket cutter
Flaring tools
Testing equipment

## MISCELLANEOUS TOOLS AND EQUIPMENT

This listing includes a variety of hand tools, which are being utilized at one time or another during the fabrication and installation of piping materials and that should be considered as general tools:

Power drills
Rotary hammers
Impact wrenches
Ladders
Scaffolds (rolling and portable)
Sledge hammers
Ball-peen hammers
Chisels
Levels
Squares
Straightedge
Monkey wrenches
Adjustable (crescent) wrenches
Socket wrench
Open-end wrenches
Drift pins
Tinsnips
Hack saw
Steel tape
Folding ruler
Micrometer
Combination square sets
Calipers and vernier calipers
Twist drill and thread gauges
Plumb bob
Bolt dies and taps
Screw extractors
Files (bastard, etc.)
Hole saw
Jacks (hand and hydraulic)

Box wrench
Construction wrenches (alignment)
Torque-indicating wrenches
Pliers
Crowbars
C-clamps
Precision level (equipment)
Dial indicator
Feeler gauges
Punches (holes and letters)
Wedges

# 18

# Nondestructive testing

Nondestructive testing, mostly referred to as NDT, is of prime importance in pipe manufacturing and construction. By definition, nondestructive testing does not impair the service ability of the piping being tested. Destructive tests can be made only as a sampling or as a percentage of total production. However, for high pressure and high temperature as well as corrosive operating conditions, such sampling is insufficient to ensure the reliability of total piping systems.

Only nondestructive testing methods are capable of detecting material discontinuities, which, at some later time, might lead to failure of an entire piping system while under operating conditions. Nondestructive testing is not a new field but has been developing at a fast pace, where new methods are continually introduced and existing methods are improved. Most nondestructive test indications are qualitative and not quantitative; and interpretation of test results involves judgment based on experience. This reliance on human experience factors is much better expressed through the term nondestructive examination, which provides a better description of the procedures being followed and the interpretation of the test results.

Section V of the ASME Boiler and Pressure Vessel Code is dedicated entirely to nondestructive examinations and deals with the following methods:

1. Liquid penetrant examination
2. Magnetic particle examination
3. Ultrasonic examination
4. Eddy current examination of tubular products
5. Radiography
6. Visual examination
7. Leak testing
   a. Gas and bubble-formation testing
   b. Halogen-diode detector (sniffer)
   c. Helium-mass spectrometer (reverse probe sniffer)
   d. Helium-mass spectrometer (hood method)

## LIQUID-PENETRANT EXAMINATION

Liquid-penetrant examination is a method of detecting and indicating surface discontinuities in relatively nonporous ferrous and nonferrous metals. It is the oldest NDT method, and has been in use for a long time.

This NDT method is being used mostly to detect cracks or porosities and other surface defects in pipe welds. While grinding the weld area is not a prerequisite for this type of examination, it is helpful insofar that it removes weld spatter and surface irregularities and other extraneous matters that might lead to an incorrect interpretation of the examination results.

The actual examination is performed by painting the weld area with a liquid dye or searching liquid first that penetrates any surface openings. After permitting time for penetration, the excess liquid is removed by a special cleaner. Then the area under examination is painted again, this time with a coating or developer. Application of the developer is crucial since either too much or not enough might give poor results. If there are any cracks, they are shown as visible lines through the action of the remaining searching liquid on the developer.

Liquid-penetrant examination is one of the least costly nondestructive testing methods and requires only a short training period for the operator. Application of this method is, however, also limited and can only be performed within strict temperature limitations, which are approximately between 60°F and 130°F.

## MAGNETIC-PARTICLE EXAMINATION

This examination method uses the magnetic properties of the material to be inspected and is therefore limited to the ferritic metals and cannot be applied to austenitic stainless steels and most nonferrous metals. Results of this examination are based on knowledge of the operation of magnetic fields. Wherever magnetic lines of force are interrupted, they cause the formation of poles. Such a discontinuity is created by flaws in welded areas, and miniature poles are established wherever a flaw in welding has thus interrupted the magnetic force lines.

The weld area to be examined must be smooth and clean, and grinding is mostly required for such a preparation when nonautomatic welding processes are involved.

Iron particles, mostly dry magnetic powders, are dusted or sprayed onto the area to be examined. The flaws, or rather the resultant magnetic poles even if below the surface, will exert sufficient pull on the surface-applied powder to align it in such a way that cracks are clearly indicated.

Wet powders contained in fluorescent liquids or colored magnetic particles are also being utilized in this type of examination and are often preferred because of the easier identification of the appearing flaw lines.

Magnetization of the area to be examined may not be sufficient at times and can be assisted through the application of high-amperage current. Such additional magnetization requirements depend also on the thickness of the weld to be examined.

In order to achieve a good result, two examinations at right angles should be performed. This will thus ensure the most effective detection of weld flaws. Like liquid-penetrant examinations, operators for magnetic-particle inspection can be trained in a relatively short time, and the interpretation of examination results is relatively simple, although sometimes nonrelevant indications of magnetic permeability may interfere with the interpretation of accepted criteria.

## ULTRASONIC EXAMINATIONS

Ultrasonic examination of piping or tubing is standard procedure in many manufacturing processes and is often a requirement of the various applicable standards. There are two different basic methods:

1. A beam of ultrasonic energy is directed into the specimen, and the energy transmitted through it is indicated.
2. The energy of the ultrasonic beam as reflected from areas within the specimen is indicated.

Ultrasonic inspection utilizes high-frequency mechanical vibration and sound frequencies in the range between 200 kilohertz and 25 megahertz. Today there are a number of examination techniques available, and various special instruments have been developed to furnish examination data and assist in their interpretation. Ultrasonic examinations employ only low-amplitude stresses and do not affect the material undergoing such tests.

Among major advantages of this type of examination are the following:

1. Readout and interpretation of test results can be accomplished almost instantly.
2. The accuracy of fault detection is not hampered by the direction a particular crack may be following.
3. Even minute cracks and flaws that might defy other examination methods can be detected.

As opposed to advantages, there are also disadvantages that cannot be overlooked:

1. This method does not normally provide a permanent record of the inspection, although tape recordings of a test can be made.
2. Only trained and experienced operators who are well versed in the use of the special equipment can be trusted to use it effectively and provide a true analysis of test results.
3. Good contact between test material and the search unit is essential, and air is normally excluded from such contact areas through the application of a suitable couplant.

Finally, ultrasonic examination is usually restricted to small areas, such as a longitudinal-weld seam of pipe, and should not be confused with low-frequency resonance methods, in which a large test specimen is vibrated at sonic frequency.

Since ultrasonic testing equipment is readily portable, it is well suited to field examination of circumferential pipe joints (butt welds) and is being used extensively on large-diameter overland transmission lines in the oil and gas industry.

Since almost instantaneous readout of the results is one of the strong points of this examination, it is being put to good use already before completion of the final weld. Preceding the testing, the equipment should be calibrated to provide a reference standard for the examination. The weld surface to be examined must be reasonably smooth to permit a valid interpretation of the test results. Since good mechanical bonding between the weld area that is being examined and the testing unit is a prerequisite, a coating of a suitable liquid or paste coupler is applied to that area before contact with the testing apparatus is established. Some sonar emission patterns are particularly difficult to interpret, and even an experienced operator may not be able to differentiate between defective welding and weld buildup inside the pipe.

## EDDY-CURRECT EXAMINATION

Electric impulses generated in close proximity of the area to be examined are affected by the properties of the material being examined. Thus, this examination method uses the application of electricity to investigate electrical and physical characteristics of the test object.

Eddy-current methodology is used mainly to detect surface cracks but can also be used to detect variations in material buildup and thickness. Here, also, well-trained operators are required—especially since this testing method can also be very easily influenced by a variety of material conditions that must be recognized to avoid distorting the examination results.

## RADIOGRAPHY

Radiography is extensively used in the examination of both shop as well as field-fabricated pipe welds. Sections I and III of the ASME Boiler and Pressure Vessel Code refer extensively to radiographic examinations, and this is one type of examination that also produces a permanent record of its findings. Additional interpretations of any examination can be undertaken should the necessity arise.

In essence, a radiograph is a photographic record of a test specimen that is exposed to the passage of X-rays or gamma rays, with the result then captured on film. The nomenclature X-rays and gamma rays refer mostly to the methodology or technique that is being used to inspect certain weldments; otherwise, X-rays and gamma rays of the same wave length are physically identical. X-ray inspection is more often used in manufacturing facilities where an unwieldy machine with an X-ray tube may be permanently installed. For inspection of field welds, however, portable radioactive isotopes such as Cobalt-60 or Iridium-192 provide an ideal source of radiation.

In either case, radiation from either source proceeds in straight lines to the test object. Some rays are absorbed and others pass through the material being examined. Less density of the test object, which may result from slag inclusion, gas bubbles, porosity, or other internal weld defects, absorbs less radiation and thus results in a darker spot on the film, indicating the size and the exact location of any inclusion.

The ASME Boiler and Pressure Vessel Code provides instructions as to the application of a penetrameter, a standard test piece that is placed on the source side of the test specimen. Later, it shows in the developed film and indicates whether the radiographic procedure is adequate. The penetrameter may also indicate permissible sizes of weld inclusions under code requirements. Although a standard for acceptability of radiographic techniques and sizes of inclusions has been provided, the operator's judgment is still required in the interpretation of any radiographic film.

Radiography requires extensive training of the operator for both operating of the equipment and interpretation of the resultant films. Not only is the operator required to be skilled in the various techniques of radiography, but he or she must also follow strict federal regulations that require time records of the exposure to radiation sources and limit personal exposure during prescribed periods.

Weld areas to be radiographed should be smooth, and any weld ripples must be removed through grinding so that a smooth weld crown can be examined. The actual performance of one type of radiography of a pipe butt weld requires attaching or placing the film on the outside of the pipe and having the radiation source at the opposite side of the pipe. The X-rays or gamma rays then pass through both pipe walls to provide film exposure that only encompasses a segment of the weld. Depending on pipe size and wall thickness, an average of three exposures is required, each encompassing one-third or 120 degrees of a pipe circumference.

An alternative method provides placing the radiation source inside the pipe, permitting exposure of the total weld at one time. This mostly requires an access hole through which the radiation source is entered into the pipe. With the help of a mechanical device, it is placed in the exact center of the pipe and butt weld. After satisfactory completion of such radiographs, the access hole is then closed, which most often does not lend itself to radiography. Radiography on construction sites is mostly done after working hours or on weekends to prevent exposure of nonoperators.

## LEAK TESTING

*Gas and bubble formation testing* is basically a pneumatic test where a pressured piping systems is examined to detect escaping test media. This test medium is normally air, but nitrogen or helium may also be used. While observing a pressure-indicating device for a given period of time gives an indication of leakproof construction, temperature differentials can interfere in such observation. A bubble-forming solution (i.e., soap bubbles) applied to the areas to be examined will form bubbles when gas passes through it and thus indicate if and where a leak exists.

*Halogen diode detection* uses the general principle that halogen vapor is ionized by a heated platinum element (anode), and the resulting ions are then collected by an ion collector plate (cathode). Various refrigerants are being used as test media and serve to search out leaks and pass through them. A meter indicates a current that is proportional to the rate of ion formation and thus provides some measurement. The relative concentration of halogen present can be measured by comparing the meter reading for the gas leakage of a component with that for a standard gas leakage.

*Helium mass spectrometer examination (reverse probe sniffer)* makes use of a portable mass spectrometer and its sensitivity to traces of helium. The system under examination should be pressurized at not more than 60 psi and should remain thus for at least one hour before testing begins. The leak detector and its strong sensitivity when applied to possible leakage points permit easy detection of any flow of helium even through minute leaks and determine the presence of helium in any gaseous mixture.

*Helium mass spectrometer examination (hood test)* uses similar equipment as in the aforementioned examination, but the components to be examined are placed in a suitable hood container. This method is by its very nature applicable only to small piping components.

While bubble testing is a procedure that can easily be applied even by the itinerant, halogen detection and mass spectrometer examinations require operators well versed in the application and examination procedures.

## HYDROSTATIC-PRESSURE TESTS

While the ASME Boiler Code Section V makes reference to a number of nondestructive examination methods, no mention is made of hydrostatic pressure tests. However, the Code for Pressure Piping ANSI B31.1 gives specifics for such tests, while also relying on the ASME Boiler Code for other nondestructive examinations.

Besides visual examination, the simplest and easiest test procedure is a hydrostatic pressure test. For hydrostatic testing, the piping or vessel is filled with the testing medium, which is normally water, and all openings are closed with the exception of a connection to a test pump and other openings that may be required to vent all air out so that no air pockets impede the pressure test.

It is standard practice that during hydrostatic testing all expansion joints are being restrained or isolated and all safety or relief valves are gagged. Packing boxes of all valves in the system should be given special attention since they are an obvious source of possible leakage. After completing the filling operation, the test pump will start operating until the required test pressure has been reached. The actual test itself consists then of two basically distinct parts. One is observing the pressure-measuring instrument to see that pressure remains constant. Two, if a pressure drop is observed, one must visually check if and where a leak can be detected.

Test pressure for hydrostatic tests as prescribed in the ANSI Code for Pressure Piping B31.1 should be at least $1\frac{1}{2}$ times design pressure subject to other code-prescribed conditions. The design pressure rating is used as a base because operating pressure may be applicable to higher or lower than ambient temperature ratings. Thus, design pressure rating is the relative comparative vehicle at ambient temperature. A minimum time limit of 10 minutes pressure holding time is required in the Pressure Piping Code plus additional time to look for leaks.

If a pressure test is maintained for a period of time, during which the test medium is subject to thermal expansion, precautions must be taken to avoid excessive pressure. A pressure-relief valve set above the test pressure is often used in such a case if such higher pressure meets all other pressure piping code requirements. These requirements are that the test pressure shall not exceed the maximum allowable test pressure of any nonisolated component part of that particular system and that no part of the system shall be subjected to a stress greater than 90 percent of its yield strength (0.2 percent offset) at test temperature.

In order to facilitate field testing and eliminate as many possible sources of leaks before field installation, shop-fabricated welded piping are often subjected to a hydrostatic test before shipment to a construction site. Thus, when a hydrostatic test is applied at a construction site and pressure drop at the pressure gauge indicates a leak, the search for the source of the leak can be restricted to pipe connections made in the field and possible intermediate valves or piping specialties.

Hydrostatic testing of pipelines, tanks or other equipment that involves large amounts of test fluid run into the additional problems of supply and disposal of the test media after successful completion of the test. The removal of air pockets in long pipelines poses also a special problem, and vent-valve attendance over a long distance makes considerable requirements on manpower.

## DESTRUCTIVE TESTS

One of the most widely used destructive tests is performed to evaluate the qualifications of welders. In order to show his proficiency, the welder is required to weld two short pieces of pipe together for a weld coupon. After completion, the weld coupon is cut into a number of segments that serve as test vehicles for the various tests. Section IX of the ASME Boiler Code provides full details of these requirements.

Another application of destructive testing of pipe or vessel welding that has by now become nearly obsolete is removing a test coupon by circular-hole saw from the actual weld area, for examination of the coupon by etching and other means. If the examination of the coupon does not disclose any weld discontinuity, a plug is inserted into the previously drilled hole and rewelded.

# 19

## Pipe cleaning

Pipe cleanliness begins when the piping reaches the construction site. Prefabricated piping or random-length material should never be permitted to be stored on the ground in the field. It should be located on dunnage at least 4 to 6 in. above the ground or on pipe racks in order to prevent the accumulation of dirt, mud, sand, stones, or even dead animals inside the pipe. Before straight-run pipe is erected, it must be checked visually if it is clean internally and sometimes swabbed out to remove dirt or other foreign matter.

All fabricated pipe spools when finished should be visually inspected, possibly flushed out with water, swabbed out with rags, and have the ends closed with sheet metal or plastic caps sealed with a good tape. When flanged, they should be covered with temporary flange covers.

Piping can be cleaned by flushing warm or cold water through it. After the flushing looks complete, the water flow is cut down and run through cheesecloth placed about 2 feet from the end of the pipe to check for dirt particles.

Piping can also be cleaned by steam under pressure, which is standard procedure in fossil-fuel power plants. Steam should be blown until the line gets hot then allowed to cool down about one-half hour. Then steam is blown again. After the third or fourth steam blowing, a polished stainless-steel plate about $^1/_8$-in. thick should be installed and tack welded to a heavier plate at the end of the steam main being blown out. If dotted spots appear on the polished plate, a new polished plate is installed. Generally, one needs three of these polished plates before piping is blown clean. After cleaning, the end of the pipe should be sealed again with temporary caps securely taped.

In oil and gas transmission lines, the use of pigs for pipeline cleaning prior to service or at periodic intervals has long been established. These are oblong or spherical scrapers that rotate and clean pipe walls while they are being propelled forward inside the pipeline with hydraulic or pneumatic pressure. Modern instrumentation technology has been incorporated in the design of some of those pipe-cleaning apparatus to help locate a pig if it gets stuck in a pipeline. A special installation called a pig trap is provided to allow insertion or removal of the pig. In addition to cleaning, pigs can also dewater pipelines or provide product batch separation.

On certain systems, such as compressor piping, oil systems, etc., the piping may have to be chemically cleaned. Various piping systems require specialty cleaning, which is applicable to the service for which the system is intended. For instance, the lube-oil piping system for turbogenerators in power-generating stations, even after chemical cleaning, must be thoroughly flushed with oil to ensure utmost cleanliness. Such flushing may often take more than 1 week of uninterrupted oil recycling before satisfactory results are achieved.

Other special cleaning includes the use of solvents or other chemicals. This is mostly done by specialty contractors who have gained wide experience under a variety of field conditions and who are well versed to handle possible adverse conditions.

Various methods are being used to clean piping-systems components before field installation. Pickling is achieved by submerging the pipe in an acid bath. Two open-top tanks must be used, one for acid solution and one for oil. The tanks must be long and deep enough to submerge the piping being chemically cleaned. A solution of various types of acids, such as muratic, sulphuric, caustic soda, etc., mixed with water will chemically clean and remove rust, scale, etc.

The piping must be submerged in the tank, giving enough time to clean the piping. The time piping must be submerged in an acid tank will depend on the acid solution mix. After approximately 2 hours in the solution mix, the piping must be removed from the acid tank and washed down with water. Great care must be taken by the workers for safety. Rubber boots, pants, jackets, gloves, and face masks must be worn, since contact with any acid may easily result in painful injuries. After washdown, piping should be allowed to dry to remove any remaining water. Piping can then be submerged in oil that protects the cleaning from rust and can then be stored until installed. If piping is installed immediately after chemical cleaning, it will not require oil dip.

Other methods of cleaning involve sandblasting or shotblasting both the interior and exterior of piping or manually cleaning the interior of large O.D. piping.

But regardless of service, it is very important that any and every piping system be completely clean, whether such cleaning be effected before or after installation.

# 20

# Cost comparison of installed cost of piping materials

Cost comparison data are for general information only. Size and configuration of a piping system, as well as project location and the cost of construction labor involved are all causal contributors, which affect a change in the cost interrelationship of the various piping materials.

| | Pipe Schedule | Cost Ratio |
|---|---|---|
| Polypropylene | 40 | 0.82 |
| Polyvinylchloride (PVC) | 40 | 0.86 |
| Carbon steel, seamless | 40 | 1.00 |
| Epoxy, fiberglass reinforced | 40 | 1.16 |
| Saran-lined carbon steel | 40 | 1.48 |
| Rubber-lined carbon steel | 40 | 1.64 |
| 85% red brass | 40 | 1.66 |
| Stainless steel, type 304, welded | 10 | 1.73 |
| Aluminum | 40 | 2.04 |
| Stainless steel, type 304, seamless | 10 | 2.06 |
| Chrome-moly 1¼ Cr - ½ moly | 40 | 2.13 |
| Chrome-moly 2¼ Cr - 1 moly | 40 | 2.42 |
| Stainless steel, type 304, welded | 40 | 2.42 |
| Chrome-moly 5 cr -½ moly | 40 | 2.53 |
| Stainless steel, type 304, seamless | 40 | 2.90 |
| Teflon-lined carbon steel (FEP) | 40 | 3.05 |
| Teflon-lined carbon steel (TFE) | 40 | 3.45 |
| Stainless steel, type 347, seamless | 40 | 3.47 |
| Stainless steel, type 316, welded | 40 | 3.77 |
| Stainless steel, type 316, seamless | 40 | 4.04 |
| Monel | 40 | 4.05 |
| Nickel | 40 | 4.53 |
| Inconel | 40 | 4.99 |

# APPENDIX 1

## Table of properties of pipe

*(Courtesy Taylor Forge Group, Gulf & Western Mfg. Co.)*

Tabulated below are the most generally required data used in piping design. This table is believed to be the most comprehensive published up to this time. Many thicknesses traditionally included in such tables have been omitted because of their having become obsolete through disuse and lack of coverage by any standard.

Sizes and thicknesses listed herein are covered by the following standards:—

1) American National Standard Institute B36.10
2) American National Standard Institute B36.19
3) American Petroleum Institute Standard API 5L
4) American Petroleum Institute Standard API 5LX
5) New United States Legal Standard for Steel Plate Gauges.

Sizes and thicknesses to which no standard designation applies are largely the more commonly used dimensions produced for a wide variety of applications including river crossings, penstocks, power plant and other piping.

All data is computed from the *nominal* dimensions listed and the effect of tolerances is not taken into account. Values are computed by application of the following formulas:

Radius of Gyration: $R = .25\sqrt{D^2 + d^2}$

Moment of Inertia: $I = R^2 A$

Section Modulus: $Z = \dfrac{I}{0.5D}$

ANSI American National Standards Institute

| Nominal | | Designation | Wall Thickness | Inside Diam. | Weight per Foot | Wt. of Water per Ft. of Pipe | Sq. Ft. Outside Surface per Ft. | Sq. Ft. Inside Surface per Ft. | Transverse Area in.² | Area of Metal in.² | Moment of Inertia in.⁴ | Section Modulus in.³ | Radius of Gyration in. |
|---|---|---|---|---|---|---|---|---|---|---|---|---|---|
| Pipe Size | Outside Diam. D | | | d | | | | | A | A | I | Z | R |
| 1/8 | .405 | 10S | .049 | .307 | .186 | .0320 | .106 | .0804 | .0740 | .0548 | .00090 | .00440 | .1270 |
| | | Std. | .068 | .269 | .244 | .0246 | .106 | .0705 | .0568 | .0720 | .00106 | .00530 | .1215 |
| | | X-Stg | .095 | .215 | .314 | .0157 | .106 | .0563 | .0364 | .0925 | .00122 | .00600 | .1146 |

| | | | | | | | | | | | | |
|---|---|---|---|---|---|---|---|---|---|---|---|---|
| 1/4 | .540 | 10S | .065 | .410 | .330 | .0570 | .141 | .1073 | .1320 | .0970 | .00280 | .01030 |
| | | Std. | .088 | .364 | .424 | .0451 | .141 | .0955 | .1041 | .1250 | .00331 | .01230 |
| | | X-Stg. | .119 | .302 | .535 | .0310 | .141 | .0794 | .0716 | .1574 | .00378 | .01395 |
| 3/8 | .675 | 10S | .065 | .545 | .423 | .1010 | .177 | .1427 | .2333 | .1245 | .00590 | .01740 |
| | | Std. | .091 | .493 | .567 | .0827 | .177 | .1295 | .1910 | .1670 | .00730 | .02160 |
| | | X-Stg. | .126 | .423 | .738 | .0609 | .177 | .1106 | .1405 | .2173 | .00862 | .02554 |
| 1/2 | .840 | 10S | .083 | .670 | .671 | .1550 | .220 | .1764 | .3568 | .1974 | .01430 | .03410 |
| | | Std. | .109 | .622 | .850 | .1316 | .220 | .1637 | .3040 | .2503 | .01710 | .04070 |
| | | X-Stg. | .147 | .546 | 1.087 | .1013 | .220 | .1433 | .2340 | .3200 | .02010 | .04780 |
| | | 160 | .138 | .464 | 1.310 | .0740 | .220 | .1220 | .1706 | .3836 | .02213 | .05269 |
| | | XX-Stg. | .294 | .252 | 1.714 | .0216 | .220 | .0660 | .0499 | .5043 | .02424 | .05772 |
| 3/4 | 1.050 | 10S | .083 | .884 | .857 | .2660 | .275 | .2314 | .6138 | .2522 | .02970 | .05660 |
| | | Std. | .113 | .824 | 1.130 | .2301 | .275 | .2168 | .5330 | .3326 | .03704 | .07055 |
| | | X-Stg. | .154 | .742 | 1.473 | .1875 | .275 | .1948 | .4330 | .4335 | .04479 | .08531 |
| | | 160 | .219 | .612 | 1.940 | .1280 | .275 | .1607 | .2961 | .5698 | .05270 | .10038 |
| | | XX-Stg. | .308 | .434 | 2.440 | .0633 | .275 | .1137 | .1479 | .7180 | .05792 | .11030 |
| 1 | 1.315 | 10S | .109 | 1.097 | 1.404 | .4090 | .344 | .2872 | .9448 | .4129 | .07560 | .1150 |
| | | Std. | .133 | 1.049 | 1.678 | .3740 | .344 | .2740 | .8640 | .4939 | .08734 | .1328 |
| | | X-Stg. | .179 | .957 | 2.171 | .3112 | .344 | .2520 | .7190 | .6388 | .10560 | .1606 |
| | | 160 | .250 | .815 | 2.850 | .2261 | .344 | .2134 | .5217 | .8364 | .12516 | .1903 |
| | | XX-Stg. | .358 | .599 | 3.659 | .1221 | .344 | .1570 | .2818 | 1.0760 | .14050 | .2136 |
| 1 1/4 | 1.660 | 10S | .109 | 1.442 | 1.806 | .7080 | .434 | .3775 | 1.633 | .5314 | .1606 | .1934 |
| | | Std. | .140 | 1.380 | 2.272 | .6471 | .434 | .3620 | 1.495 | .6685 | .1947 | .2346 |
| | | | .191 | 1.278 | 2.996 | .5553 | .434 | .3356 | 1.283 | .8815 | .2418 | .2913 |
| | | | .250 | 1.160 | 3.764 | .4575 | .434 | .3029 | 1.057 | 1.1070 | .2833 | .3421 |
| | | | .382 | .896 | 5.214 | .2732 | .434 | .2331 | .6305 | 1.5340 | .3411 | .4110 |

| | | |
|---|---|---|
| 1/4 | .540 | .1695 |
| | | .1628 |
| | | .1547 |
| 3/8 | .675 | .2160 |
| | | .2090 |
| | | .1991 |
| 1/2 | .840 | .2692 |
| | | .2613 |
| | | .2505 |
| | | .2402 |
| | | .2192 |
| 3/4 | 1.050 | .3430 |
| | | .3337 |
| | | .3214 |
| | | .3041 |
| | | .2840 |
| 1 | 1.315 | .4282 |
| | | .4205 |
| | | .4066 |
| | | .3868 |
| | | .3613 |
| 1 1/4 | 1.660 | .5499 |
| | | .5397 |
| | | .5237 |
| | | .5063 |
| | | .4716 |

APPENDIX 1 *continued*

| Nominal Pipe Size | Nominal Outside Diam. D | Designation | Wall Thickness | Inside Diam. d | Weight per Foot | Wt. of Water per Ft. of Pipe | Sq. Ft. Outside Surface per Ft. | Sq. Ft. Inside Surface per Ft. | Transverse Area in.² A | Area of Metal in.² A | Moment of Inertia in.⁴ I | Section Modulus in.³ Z | Radius of Gyration in. R |
|---|---|---|---|---|---|---|---|---|---|---|---|---|---|
| 1½ | 1.900 | 10S | .109 | 1.682 | 2.085 | .9630 | .497 | .4403 | 2.221 | .613 | .2469 | .2599 | .6344 |
| | | Std. | .145 | 1.610 | 2.717 | .8820 | .497 | .4213 | 2.036 | .800 | .3099 | .3262 | .6226 |
| | | X-Stg. | .200 | 1.500 | 3.631 | .7648 | .497 | .3927 | 1.767 | 1.068 | .3912 | .4118 | .6052 |
| | | 160 | .281 | 1.337 | 4.862 | .6082 | .497 | .3519 | 1.405 | 1.430 | .4826 | .5080 | .5809 |
| | | XX-Stg. | .400 | 1.100 | 6.408 | .4117 | .497 | .2903 | .950 | 1.885 | .5678 | .5977 | .5489 |
| 2 | 2.375 | 10S | .109 | 2.157 | 2.638 | 1.583 | .622 | .5647 | 3.654 | .775 | .5003 | .4213 | .8034 |
| | | Std. | .154 | 2.067 | 3.652 | 1.452 | .622 | .5401 | 3.355 | 1.075 | .6657 | .5606 | .7871 |
| | | X-Stg. | .218 | 1.939 | 5.022 | 1.279 | .622 | .5074 | 2.953 | 1.477 | .8679 | .7309 | .7665 |
| | | - - | .250 | 1.875 | 5.673 | 1.196 | .622 | .4920 | 2.761 | 1.669 | .9555 | .8046 | .7565 |
| | | 160 | .344 | 1.687 | 7.450 | .970 | .622 | .4422 | 2.240 | 2.190 | 1.162 | .9790 | .7286 |
| | | XX-Stg. | .436 | 1.503 | 9.029 | .769 | .622 | .3929 | 1.774 | 2.656 | 1.311 | 1.1040 | .7027 |
| 2½ | 2.875 | 10S | .120 | 2.635 | 3.53 | 2.360 | .753 | .6900 | 5.453 | 1.038 | .9878 | .6872 | .9755 |
| | | Std. | .203 | 2.469 | 5.79 | 2.072 | .753 | .6462 | 4.788 | 1.704 | 1.530 | 1.064 | .9474 |
| | | X-Stg. | .276 | 2.323 | 7.66 | 1.834 | .753 | .6095 | 4.238 | 2.254 | 1.924 | 1.339 | .9241 |
| | | 160 | .375 | 2.125 | 10.01 | 1.535 | .753 | .5564 | 3.547 | 2.945 | 2.353 | 1.638 | .8938 |
| | | XX-Stg. | .552 | 1.771 | 13.69 | 1.067 | .753 | .4627 | 2.464 | 4.028 | 2.871 | 1.997 | .8442 |
| 3 | 3.500 | 10S | .120 | 3.260 | 4.33 | 3.62 | .916 | .853 | 8.346 | 1.272 | 1.821 | 1.041 | 1.196 |
| | | API | .125 | 3.250 | 4.52 | 3.60 | .916 | .851 | 8.300 | 1.329 | 1.900 | 1.086 | 1.195 |
| | | API | .156 | 3.188 | 5.58 | 3.46 | .916 | .835 | 7.982 | 1.639 | 2.298 | 1.313 | 1.184 |
| | | API | .188 | 3.125 | 6.65 | 3.34 | .916 | .819 | 7.700 | 1.958 | 2.700 | 1.545 | 1.175 |
| | | Std. | .216 | 3.068 | 7.57 | 3.20 | .916 | .802 | 7.393 | 2.228 | 3.017 | 1.724 | 1.164 |
| | | API | .250 | 3.000 | 8.68 | 3.06 | .916 | .785 | 7.184 | 2.553 | 3.388 | 1.936 | 1.152 |

| | | | | | | | | | | | | |
|---|---|---|---|---|---|---|---|---|---|---|---|---|
| | | API | .281 | 2.938 | 9.65 | 2.94 | .916 | .769 | 6.780 | 2.842 | 3.819 | 2.182 | 1.142 |
| | | X-Stg. | .300 | 2.900 | 10.25 | 2.86 | .916 | .761 | 6.605 | 3.016 | 3.892 | 2.225 | 1.136 |
| | | 160 | .438 | 2.624 | 14.32 | 2.34 | .916 | .687 | 5.407 | 4.214 | 5.044 | 2.882 | 1.094 |
| | | XX-Stg. | .600 | 2.300 | 18.58 | 1.80 | .916 | .601 | 4.155 | 5.466 | 5.993 | 3.424 | 1.047 |
| 3½ | 4.000 | 10S | .120 | 3.760 | 4.97 | 4.81 | 1.047 | .984 | 11.10 | 1.46 | 2.754 | 1.377 | 1.372 |
| | | API | .125 | 3.750 | 5.18 | 4.79 | 1.047 | .982 | 11.04 | 1.52 | 2.859 | 1.430 | 1.371 |
| | | API | .156 | 3.688 | 6.41 | 4.63 | 1.047 | .966 | 10.68 | 1.88 | 3.485 | 1.743 | 1.360 |
| | | API | .188 | 3.624 | 7.71 | 4.48 | 1.047 | .950 | 10.32 | 2.27 | 4.130 | 2.065 | 1.350 |
| | | Std. | .226 | 3.548 | 9.11 | 4.28 | 1.047 | .929 | 9.89 | 2.68 | 4.788 | 2.394 | 1.337 |
| | | API | .250 | 3.500 | 10.02 | 4.17 | 1.047 | .916 | 9.62 | 2.94 | 5.201 | 2.601 | 1.329 |
| | | API | .281 | 3.438 | 11.17 | 4.02 | 1.047 | .900 | 9.28 | 3.29 | 5.715 | 2.858 | 1.319 |
| | | X-Stg. | .318 | 3.364 | 12.51 | 3.85 | 1.047 | .880 | 8.89 | 3.68 | 6.280 | 3.140 | 1.307 |
| | | XX-Stg. | .636 | 2.728 | 22.85 | 2.53 | 1.047 | .716 | 5.84 | 6.72 | 9.848 | 4.924 | 1.210 |
| 4 | 4.500 | 10S | .120 | 4.260 | 5.61 | 6.18 | 1.178 | 1.115 | 14.25 | 1.65 | 3.97 | 1.761 | 1.550 |
| | | API | .125 | 4.250 | 5.84 | 6.15 | 1.178 | 1.113 | 14.19 | 1.72 | 4.12 | 1.829 | 1.548 |
| | | API | .156 | 4.188 | 7.24 | 5.97 | 1.178 | 1.096 | 13.77 | 2.13 | 5.03 | 2.235 | 1.537 |
| | | API | .188 | 4.124 | 8.56 | 5.80 | 1.178 | 1.082 | 13.39 | 2.52 | 5.86 | 2.600 | 1.525 |
| | | API | .219 | 4.062 | 10.02 | 5.62 | 1.178 | 1.063 | 12.96 | 2.94 | 6.77 | 3.867 | 1.516 |
| | | Std. | .237 | 4.026 | 10.79 | 5.51 | 1.178 | 1.055 | 12.73 | 3.17 | 7.23 | 3.214 | 1.510 |
| | | API | .250 | 4.000 | 11.35 | 5.45 | 1.178 | 1.049 | 12.57 | 3.34 | 7.56 | 3.360 | 1.505 |
| | | API | .281 | 3.938 | 12.67 | 5.27 | 1.178 | 1.031 | 12.17 | 3.73 | 8.33 | 3.703 | 1.495 |
| | | API | .312 | 3.876 | 14.00 | 5.12 | 1.178 | 1.013 | 11.80 | 4.11 | 9.05 | 4.020 | 1.482 |
| | | X-Stg. | .337 | 3.826 | 14.98 | 4.98 | 1.178 | 1.002 | 11.50 | 4.41 | 9.61 | 4.271 | 1.477 |
| | | 120 | .438 | 3.624 | 19.00 | 4.47 | 1.178 | .949 | 10.32 | 5.59 | 11.65 | 5.177 | 1.444 |
| | | - - | .500 | 3.500 | 21.36 | 4.16 | 1.178 | .916 | 9.62 | 6.28 | 12.77 | 5.676 | 1.425 |
| | | 160 | .531 | 3.438 | 22.60 | 4.02 | 1.178 | .900 | 9.28 | 6.62 | 13.27 | 5.900 | 1.416 |
| | | XX-Stg. | .674 | 3.152 | 27.54 | 3.38 | 1.178 | .826 | 7.80 | 8.10 | 15.28 | 6.793 | 1.374 |

APPENDIX 1 *continued*

| Nominal Pipe Size | Nominal Outside Diam. D | Designation | Wall Thickness | Inside Diam. d | Weight per Foot | Wt. of Water per Ft. of Pipe | Sq. Ft. Outside Surface per Ft. | Sq. Ft. Inside Surface per Ft. | Transverse Area in.$^2$ A | Area of Metal in.$^2$ A | Moment of Inertia in.$^4$ I | Section Modulus in.$^3$ Z | Radius of Gyration in. R |
|---|---|---|---|---|---|---|---|---|---|---|---|---|---|
| 5 | 5.563 | 10S | .134 | 5.295 | 7.77 | 9.54 | 1.456 | 1.386 | 22.02 | 2.29 | 8.42 | 3.028 | 1.920 |
| | | API | .156 | 5.251 | 9.02 | 9.39 | 1.456 | 1.375 | 21.66 | 2.65 | 9.70 | 3.487 | 1.913 |
| | | API | .188 | 5.187 | 10.80 | 9.16 | 1.456 | 1.358 | 21.13 | 3.17 | 11.49 | 4.129 | 1.902 |
| | | API | .219 | 5.125 | 12.51 | 8.94 | 1.456 | 1.342 | 20.63 | 3.68 | 13.14 | 4.726 | 1.891 |
| | | Std. | .258 | 5.047 | 14.62 | 8.66 | 1.456 | 1.321 | 20.01 | 4.30 | 15.16 | 5.451 | 1.878 |
| | | API | .281 | 5.001 | 15.86 | 8.52 | 1.456 | 1.309 | 19.64 | 4.66 | 16.31 | 5.862 | 1.870 |
| | | API | .312 | 4.939 | 17.51 | 8.31 | 1.456 | 1.293 | 19.16 | 5.15 | 17.81 | 6.402 | 1.860 |
| | | API | .344 | 4.875 | 19.19 | 8.09 | 1.456 | 1.276 | 18.67 | 5.64 | 19.28 | 6.932 | 1.849 |
| | | X-Stg. | .375 | 4.813 | 20.78 | 7.87 | 1.456 | 1.260 | 18.19 | 6.11 | 20.67 | 7.431 | 1.839 |
| | | 120 | .500 | 4.563 | 27.10 | 7.08 | 1.456 | 1.195 | 16.35 | 7.95 | 25.74 | 9.253 | 1.799 |
| | | 160 | .625 | 4.313 | 32.96 | 6.32 | 1.456 | 1.129 | 14.61 | 9.70 | 30.03 | 10.800 | 1.760 |
| | | XX-Stg. | .750 | 4.063 | 38.55 | 5.62 | 1.456 | 1.064 | 12.97 | 11.34 | 33.63 | 12.090 | 1.722 |
| 6 | 6.625 | 12 Ga. | .104 | 6.417 | 7.25 | 14.02 | 1.734 | 1.680 | 32.34 | 2.13 | 11.33 | 3.42 | 2.31 |
| | | 10S | .134 | 6.357 | 9.29 | 13.70 | 1.734 | 1.660 | 31.75 | 2.73 | 14.38 | 4.34 | 2.29 |
| | | 8 Ga. | .164 | 6.297 | 11.33 | 13.50 | 1.734 | 1.649 | 31.14 | 3.33 | 17.38 | 5.25 | 2.28 |
| | | API | .188 | 6.249 | 12.93 | 13.31 | 1.734 | 1.639 | 30.70 | 3.80 | 19.71 | 5.95 | 2.28 |
| | | 6 Ga. | .194 | 6.237 | 13.34 | 13.25 | 1.734 | 1.633 | 30.55 | 3.92 | 20.29 | 6.12 | 2.27 |
| | | API | .219 | 6.187 | 15.02 | 13.05 | 1.734 | 1.620 | 30.10 | 4.41 | 22.66 | 6.84 | 2.27 |
| | | API | .250 | 6.125 | 17.02 | 12.80 | 1.734 | 1.606 | 29.50 | 5.01 | 25.55 | 7.71 | 2.26 |
| | | API | .277 | 6.071 | 18.86 | 12.55 | 1.734 | 1.591 | 28.95 | 5.54 | 28.00 | 8.46 | 2.25 |
| | | Std. | .280 | 6.065 | 18.97 | 12.51 | 1.734 | 1.587 | 28.90 | 5.58 | 28.14 | 8.50 | 2.24 |
| | | API | .312 | 6.001 | 21.05 | 12.26 | 1.734 | 1.571 | 28.28 | 6.19 | 30.91 | 9.33 | 2.23 |
| | | API | .344 | 5.937 | 23.09 | 12.00 | 1.734 | 1.554 | 27.68 | 6.79 | 33.51 | 10.14 | 2.22 |
| | | API | .375 | 5.875 | 25.10 | 11.75 | 1.734 | 1.540 | 27.10 | 7.37 | 36.20 | 10.90 | 2.21 |

| | | | | | | | | | | | | |
|---|---|---|---|---|---|---|---|---|---|---|---|---|
| | | X-Stg. | .432 | 5.761 | 28.57 | 11.29 | 1.734 | 1.510 | 26.07 | 8.40 | 40.49 | 12.22 | 2.19 |
| | | - - | .500 | 5.625 | 32.79 | 10.85 | 1.734 | 1.475 | 24.85 | 9.63 | 45.60 | 13.78 | 2.16 |
| | | 120 | .562 | 5.501 | 36.40 | 10.30 | 1.734 | 1.470 | 23.77 | 10.74 | 49.91 | 15.07 | 2.15 |
| | | 160 | .719 | 5.187 | 45.30 | 9.16 | 1.734 | 1.359 | 21.15 | 13.36 | 58.99 | 17.81 | 2.10 |
| | | XX-Stg. | .864 | 4.897 | 53.16 | 8.14 | 1.734 | 1.280 | 18.83 | 15.64 | 66.33 | 20.02 | 2.06 |
| 8 | 8.625 | 12 Ga. | .104 | 8.417 | 9.47 | 24.1 | 2.26 | 2.204 | 55.6 | 2.78 | 25.3 | 5.86 | 3.01 |
| | | 10 Ga. | .134 | 8.357 | 12.16 | 23.8 | 2.26 | 2.188 | 54.8 | 3.57 | 32.2 | 7.46 | 3.00 |
| | | 10S | .148 | 8.329 | 13.40 | 23.6 | 2.26 | 2.180 | 54.5 | 3.94 | 35.4 | 8.22 | 3.00 |
| | | 8 Ga. | .164 | 8.297 | 14.83 | 23.4 | 2.26 | 2.172 | 54.1 | 4.36 | 39.1 | 9.06 | 2.99 |
| | | API | .188 | 8.249 | 16.90 | 23.2 | 2.26 | 2.161 | 53.5 | 5.00 | 44.5 | 10.30 | 2.98 |
| | | 6 Ga. | .194 | 8.237 | 17.48 | 23.1 | 2.26 | 2.156 | 53.3 | 5.14 | 45.7 | 10.60 | 2.98 |
| | | API | .203 | 8.219 | 18.30 | 23.1 | 2.26 | 2.152 | 53.1 | 5.38 | 47.7 | 11.05 | 2.98 |
| | | API | .219 | 8.187 | 19.64 | 22.9 | 2.26 | 2.148 | 52.7 | 5.80 | 51.3 | 11.90 | 2.97 |
| | | 3 Ga. | .239 | 8.147 | 21.42 | 22.6 | 2.26 | 2.133 | 52.1 | 6.30 | 55.4 | 12.84 | 2.96 |
| | | 20 | .250 | 8.125 | 22.40 | 22.5 | 2.26 | 2.127 | 51.8 | 6.58 | 57.7 | 13.39 | 2.96 |
| | | 30 | .277 | 8.071 | 24.70 | 22.2 | 2.26 | 2.115 | 51.2 | 7.26 | 63.3 | 14.69 | 2.95 |
| | | API | .312 | 8.001 | 27.72 | 21.8 | 2.26 | 2.095 | 50.3 | 8.15 | 70.6 | 16.37 | 2.94 |
| | | Std. | .322 | 7.981 | 28.55 | 21.6 | 2.26 | 2.090 | 50.0 | 8.40 | 72.5 | 16.81 | 2.94 |
| | | API | .344 | 7.937 | 30.40 | 21.4 | 2.26 | 2.078 | 49.5 | 8.94 | 76.8 | 17.81 | 2.93 |
| | | API | .375 | 7.875 | 33.10 | 21.1 | 2.26 | 2.062 | 48.7 | 9.74 | 83.1 | 19.27 | 2.92 |
| | | 60 | .406 | 7.813 | 35.70 | 20.8 | 2.26 | 2.045 | 47.9 | 10.48 | 88.8 | 20.58 | 2.91 |
| | | API | .438 | 7.749 | 38.33 | 20.4 | 2.26 | 2.029 | 47.2 | 11.27 | 94.7 | 21.97 | 2.90 |
| | | X-Stg. | .500 | 7.625 | 43.39 | 19.8 | 2.26 | 2.006 | 45.6 | 12.76 | 105.7 | 24.51 | 2.88 |
| | | 100 | .594 | 7.437 | 50.90 | 18.8 | 2.26 | 1.947 | 43.5 | 14.96 | 121.4 | 28.14 | 2.85 |
| | | - - | .625 | 7.375 | 53.40 | 18.5 | 2.26 | 1.931 | 42.7 | 15.71 | 126.5 | 29.33 | 2.84 |
| | | 120 | .719 | 7.187 | 60.70 | 17.6 | 2.26 | 1.882 | 40.6 | 17.84 | 140.6 | 32.61 | 2.81 |
| | | 140 | .812 | 7.001 | 67.80 | 16.7 | 2.26 | 1.833 | 38.5 | 19.93 | 153.8 | 35.65 | 2.78 |
| | | XX-Stg. | .875 | 6.875 | 72.42 | 16.1 | 2.26 | 1.800 | 37.1 | 21.30 | 162.0 | 37.56 | 2.76 |
| | | 160 | .906 | 6.813 | 74.70 | 15.8 | 2.26 | 1.784 | 36.4 | 21.97 | 165.9 | 38.48 | 2.76 |

APPENDIX 1 *continued*

| Nominal Pipe Size | Nominal Outside Diam. D | Designation | Wall Thickness | Inside Diam. d | Weight per Foot | Wt. of Water per Ft. of Pipe | Sq. Ft. Outside Surface per Ft. | Sq. Ft. Inside Surface per Ft. | Transverse Area in.$^2$ A | Area of Metal in.$^2$ A | Moment of Inertia in.$^4$ I | Section Modulus in.$^3$ Z | Radius of Gyration in. R |
|---|---|---|---|---|---|---|---|---|---|---|---|---|---|
| 10 | 10.750 | 12 Ga. | .104 | 10.542 | 11.83 | 37.8 | 2.81 | 2.76 | 87.3 | 3.48 | 49.3 | 9.16 | 3.76 |
| | | 10 Ga. | .134 | 10.482 | 15.21 | 37.4 | 2.81 | 2.74 | 86.3 | 4.47 | 63.0 | 11.71 | 3.75 |
| | | 8 Ga. | .164 | 10.422 | 18.56 | 37.0 | 2.81 | 2.73 | 85.3 | 5.45 | 76.4 | 14.22 | 3.74 |
| | | 10S | .165 | 10.420 | 18.65 | 36.9 | 2.81 | 2.73 | 85.3 | 5.50 | 76.8 | 14.29 | 3.74 |
| | | API | .188 | 10.374 | 21.12 | 36.7 | 2.81 | 2.72 | 84.5 | 6.20 | 86.5 | 16.10 | 3.74 |
| | | 6 Ga. | .194 | 10.362 | 21.89 | 36.6 | 2.81 | 2.71 | 84.3 | 6.43 | 89.7 | 16.68 | 3.73 |
| | | API | .203 | 10.344 | 22.86 | 36.5 | 2.81 | 2.71 | 84.0 | 6.71 | 93.3 | 17.35 | 3.73 |
| | | API | .219 | 10.310 | 24.60 | 36.2 | 2.81 | 2.70 | 83.4 | 7.24 | 100.5 | 18.70 | 3.72 |
| | | 3 Ga. | .239 | 10.272 | 28.05 | 35.9 | 2.81 | 2.69 | 82.9 | 7.89 | 109.2 | 20.32 | 3.72 |
| | | 20 | .250 | 10.250 | 28.03 | 35.9 | 2.81 | 2.68 | 82.6 | 8.26 | 113.6 | 21.12 | 3.71 |
| | | API | .279 | 10.192 | 31.20 | 35.3 | 2.81 | 2.66 | 81.6 | 9.18 | 125.9 | 23.42 | 3.70 |
| | | 30 | .307 | 10.136 | 34.24 | 35.0 | 2.81 | 2.65 | 80.7 | 10.07 | 137.4 | 25.57 | 3.69 |
| | | API | .344 | 10.062 | 38.26 | 34.5 | 2.81 | 2.63 | 79.5 | 11.25 | 152.3 | 28.33 | 3.68 |
| | | Std. | .365 | 10.020 | 40.48 | 34.1 | 2.81 | 2.62 | 78.9 | 11.91 | 160.7 | 29.90 | 3.67 |
| | | API | .438 | 9.874 | 48.28 | 33.2 | 2.81 | 2.58 | 76.6 | 14.19 | 188.8 | 35.13 | 3.65 |
| | | X-Stg. | .500 | 9.750 | 54.74 | 32.3 | 2.81 | 2.55 | 74.7 | 16.10 | 212.0 | 39.43 | 3.63 |
| | | 80 | .594 | 9.562 | 64.40 | 31.1 | 2.81 | 2.50 | 71.8 | 18.91 | 244.9 | 45.56 | 3.60 |
| | | 100 | .719 | 9.312 | 77.00 | 29.5 | 2.81 | 2.44 | 68.1 | 22.62 | 286.2 | 53.25 | 3.56 |
| | | - - | .750 | 9.250 | 80.10 | 29.1 | 2.81 | 2.42 | 67.2 | 23.56 | 296.2 | 55.10 | 3.54 |
| | | 120 | .844 | 9.062 | 89.20 | 27.9 | 2.81 | 2.37 | 64.5 | 26.23 | 324.3 | 60.34 | 3.51 |
| | | 140 | 1.000 | 8.750 | 104.20 | 26.1 | 2.81 | 2.29 | 60.1 | 30.63 | 367.8 | 68.43 | 3.46 |
| | | 160 | 1.125 | 8.500 | 116.00 | 24.6 | 2.81 | 2.22 | 56.7 | 34.01 | 399.4 | 74.31 | 3.43 |

| | | | | | | | | | | | | | |
|---|---|---|---|---|---|---|---|---|---|---|---|---|---|
| 12 | 12.750 | 12 Ga. | .104 | 12.542 | 14.1 | 53.6 | 3.34 | 3.28 | 123.5 | 4.13 | 82.6 | 12.9 | 4.47 |
| | | 10 Ga. | .134 | 12.482 | 18.1 | 53.0 | 3.34 | 3.27 | 122.4 | 5.31 | 105.7 | 16.6 | 4.46 |
| | | 8 Ga. | .164 | 12.422 | 22.1 | 52.5 | 3.34 | 3.25 | 121.2 | 6.48 | 128.4 | 20.1 | 4.45 |
| | | 10S | .180 | 12.390 | 24.2 | 52.2 | 3.34 | 3.24 | 120.6 | 7.11 | 140.4 | 22.0 | 4.44 |
| | | 6 Ga. | .194 | 12.362 | 26.0 | 52.0 | 3.34 | 3.23 | 120.0 | 7.65 | 150.9 | 23.7 | 4.44 |
| | | API | .203 | 12.344 | 27.2 | 52.0 | 3.34 | 3.23 | 119.9 | 7.99 | 157.2 | 24.7 | 4.43 |
| | | API | .219 | 12.312 | 29.3 | 51.7 | 3.34 | 3.22 | 119.1 | 8.52 | 167.6 | 26.3 | 4.43 |
| | | 3 Ga. | .239 | 12.272 | 32.0 | 51.3 | 3.34 | 3.21 | 118.3 | 9.39 | 183.8 | 28.8 | 4.42 |
| | | 20 | .250 | 12.250 | 33.4 | 51.3 | 3.34 | 3.12 | 118.0 | 9.84 | 192.3 | 30.2 | 4.42 |
| | | API | .281 | 12.188 | 37.4 | 50.6 | 3.34 | 3.19 | 116.7 | 11.01 | 214.1 | 33.6 | 4.41 |
| | | API | .312 | 12.126 | 41.5 | 50.1 | 3.34 | 3.17 | 115.5 | 12.19 | 236.0 | 37.0 | 4.40 |
| | | 30 | .330 | 12.090 | 43.8 | 49.7 | 3.34 | 3.16 | 114.8 | 12.88 | 248.5 | 39.0 | 4.39 |
| | | API | .344 | 12.062 | 45.5 | 49.7 | 3.34 | 3.16 | 114.5 | 13.46 | 259.0 | 40.7 | 4.38 |
| | | Std. | .375 | 12.000 | 49.6 | 48.9 | 3.34 | 3.14 | 113.1 | 14.58 | 279.3 | 43.8 | 4.37 |
| | | 40 | .406 | 11.938 | 53.6 | 48.5 | 3.34 | 3.13 | 111.9 | 15.74 | 300.3 | 47.1 | 4.37 |
| | | API | .438 | 11.874 | 57.5 | 48.2 | 3.34 | 3.11 | 111.0 | 16.95 | 321.0 | 50.4 | 4.35 |
| | | X-Stg. | .500 | 11.750 | 65.4 | 46.9 | 3.34 | 3.08 | 108.4 | 19.24 | 361.5 | 56.7 | 4.33 |
| | | 60 | .562 | 11.626 | 73.2 | 46.0 | 3.34 | 3.04 | 106.2 | 21.52 | 400.5 | 62.8 | 4.31 |
| | | - - | .625 | 11.500 | 80.9 | 44.9 | 3.34 | 3.01 | 103.8 | 23.81 | 438.7 | 68.8 | 4.29 |
| | | 80 | .688 | 11.374 | 88.6 | 44.0 | 3.34 | 2.98 | 101.6 | 26.03 | 475.2 | 74.6 | 4.27 |
| | | - - | .750 | 11.250 | 96.2 | 43.1 | 3.34 | 2.94 | 99.4 | 28.27 | 510.7 | 80.1 | 4.25 |
| | | 100 | .844 | 11.062 | 108.0 | 41.6 | 3.34 | 2.90 | 96.1 | 31.53 | 561.8 | 88.1 | 4.22 |
| | | - - | .875 | 11.000 | 110.9 | 41.1 | 3.34 | 2.88 | 95.0 | 32.64 | 578.5 | 90.7 | 4.21 |
| | | 120 | 1.000 | 10.750 | 125.5 | 39.3 | 3.34 | 2.81 | 90.8 | 36.91 | 641.7 | 100.7 | 4.17 |
| | | 140 | 1.125 | 10.500 | 140.0 | 37.5 | 3.34 | 2.75 | 86.6 | 41.08 | 700.7 | 109.9 | 4.13 |
| | | - - | 1.250 | 10.250 | 153.6 | 35.8 | 3.34 | 2.68 | 82.5 | 45.16 | 755.5 | 118.5 | 4.09 |
| | | 160 | 1.312 | 10.126 | 161.0 | 34.9 | 3.34 | 2.65 | 80.5 | 47.14 | 781.3 | 122.6 | 4.07 |
| | | - - | 1.375 | 10.000 | 167.2 | 34.0 | 3.34 | 2.62 | 78.5 | 49.14 | 807.2 | 126.6 | 4.05 |
| | | - - | 1.500 | 9.750 | 180.4 | 32.4 | 3.34 | 2.55 | 74.7 | 53.01 | 853.8 | 133.9 | 4.01 |

APPENDIX 1 *continued*

| Nominal Pipe Size | Nominal Outside Diam. D | Designation | Wall Thickness | Inside Diam. d | Weight per Foot | Wt. of Water per Ft. of Pipe | Sq. Ft. Outside Surface per Ft. | Sq. Ft. Inside Surface per Ft. | Transverse Area in.$^2$ A | Area of Metal in.$^2$ A | Moment of Inertia in.$^4$ I | Section Modulus in.$^3$ Z | Radius of Gyration in. R |
|---|---|---|---|---|---|---|---|---|---|---|---|---|---|
| 14 | 14.000 | 10 Ga. | .134 | 13.732 | 20 | 64.2 | 3.67 | 3.59 | 148.1 | 5.84 | 140.4 | 20.1 | 4.90 |
| | | 8 Ga. | .164 | 13.672 | 24 | 63.6 | 3.67 | 3.58 | 146.8 | 7.13 | 170.7 | 24.4 | 4.89 |
| | | 6 Ga. | .194 | 13.612 | 29 | 63.1 | 3.67 | 3.56 | 145.5 | 8.41 | 200.6 | 28.7 | 4.88 |
| | | API | .210 | 13.580 | 31 | 62.8 | 3.67 | 3.55 | 144.8 | 9.10 | 216.2 | 30.9 | 4.87 |
| | | API | .219 | 13.562 | 32 | 62.6 | 3.67 | 3.55 | 144.5 | 9.48 | 225.1 | 32.2 | 4.87 |
| | | 3 Ga. | .239 | 13.522 | 35 | 62.3 | 3.67 | 3.54 | 143.6 | 10.33 | 244.9 | 35.0 | 4.87 |
| | | 10 | .250 | 13.500 | 37 | 62.1 | 3.67 | 3.54 | 143.0 | 10.82 | 256.0 | 36.6 | 4.86 |
| | | API | .281 | 13.438 | 41 | 61.5 | 3.67 | 3.52 | 141.8 | 12.11 | 285.2 | 40.7 | 4.85 |
| | | 20 | .312 | 13.375 | 46 | 60.8 | 3.67 | 3.50 | 140.5 | 13.44 | 314.9 | 45.0 | 4.84 |
| | | API | .344 | 13.312 | 50 | 60.3 | 3.67 | 3.48 | 139.2 | 14.76 | 344.3 | 49.2 | 4.83 |
| | | Std. | .375 | 13.250 | 55 | 59.7 | 3.67 | 3.47 | 137.9 | 16.05 | 372.8 | 53.2 | 4.82 |
| | | 40 | .438 | 13.124 | 63 | 58.5 | 3.67 | 3.44 | 135.3 | 18.66 | 429.6 | 61.4 | 4.80 |
| | | X-Stg. | .500 | 13.000 | 72 | 57.4 | 3.67 | 3.40 | 132.7 | 21.21 | 483.8 | 69.1 | 4.78 |
| | | 60 | .594 | 12.812 | 85 | 55.9 | 3.67 | 3.35 | 129.0 | 24.98 | 562.4 | 80.3 | 4.74 |
| | | - - | .625 | 12.750 | 89 | 55.3 | 3.67 | 3.34 | 127.7 | 26.26 | 588.5 | 84.1 | 4.73 |
| | | 80 | .750 | 12.500 | 107 | 51.2 | 3.67 | 3.27 | 122.7 | 31.22 | 687.5 | 98.2 | 4.69 |
| | | - - | .875 | 12.250 | 123 | 51.1 | 3.67 | 3.21 | 117.9 | 36.08 | 780.1 | 111.4 | 4.65 |
| | | 100 | .938 | 12.124 | 131 | 50.0 | 3.67 | 3.17 | 115.5 | 38.47 | 820.5 | 117.2 | 4.63 |
| | | - - | 1.000 | 12.000 | 139 | 49.0 | 3.67 | 3.14 | 113.1 | 40.84 | 868.0 | 124.0 | 4.61 |
| | | 120 | 1.094 | 11.812 | 151 | 47.5 | 3.67 | 3.09 | 109.6 | 44.32 | 929.8 | 132.8 | 4.58 |
| | | - - | 1.125 | 11.750 | 155 | 47.0 | 3.67 | 3.08 | 108.4 | 45.50 | 950.3 | 135.8 | 4.57 |
| | | 140 | 1.250 | 11.500 | 171 | 45.0 | 3.67 | 3.01 | 103.9 | 50.07 | 1027.5 | 146.8 | 4.53 |
| | | - - | 1.375 | 11.250 | 186 | 43.1 | 3.67 | 2.94 | 99.4 | 54.54 | 1099.5 | 157.1 | 4.49 |
| | | 160 | 1.406 | 11.188 | 190 | 42.6 | 3.67 | 2.93 | 98.3 | 55.63 | 1116.9 | 159.6 | 4.48 |
| | | - - | 1.500 | 11.000 | 200 | 41.2 | 3.67 | 2.88 | 95.0 | 58.90 | 1166.5 | 166.6 | 4.45 |

| | | | | | | | | | | | | | |
|---|---|---|---|---|---|---|---|---|---|---|---|---|---|
| 16 | 16.000 | 10 Ga. | .134 | 15.732 | 23 | 84.3 | 4.19 | 4.12 | 194.4 | 6.68 | 210 | 26.3 | 5.61 |
| | | 8 Ga. | .164 | 15.672 | 28 | 83.6 | 4.19 | 4.10 | 192.9 | 8.16 | 256 | 32.0 | 5.60 |
| | | - - | .188 | 15.624 | 32 | 83.3 | 4.19 | 4.09 | 192.0 | 9.39 | 294 | 36.7 | 5.59 |
| | | 6 Ga. | .194 | 15.612 | 33 | 83.0 | 4.19 | 4.09 | 191.4 | 9.63 | 301 | 37.6 | 5.59 |
| | | API | .219 | 15.562 | 37 | 82.5 | 4.19 | 4.07 | 190.2 | 10.86 | 338 | 42.3 | 5.58 |
| | | 3 Ga. | .239 | 15.522 | 40 | 82.0 | 4.19 | 4.06 | 189.2 | 11.83 | 368 | 45.9 | 5.57 |
| | | 10 | .250 | 15.500 | 42 | 82.1 | 4.19 | 4.06 | 189.0 | 12.40 | 385 | 48.1 | 5.57 |
| | | API | .281 | 15.438 | 47 | 81.2 | 4.19 | 4.04 | 187.0 | 13.90 | 430 | 53.8 | 5.56 |
| | | 20 | .312 | 15.375 | 52 | 80.1 | 4.19 | 4.03 | 185.6 | 15.40 | 474 | 59.2 | 5.55 |
| | | API | .344 | 15.312 | 57 | 80.0 | 4.19 | 4.01 | 184.1 | 16.94 | 519 | 64.9 | 5.54 |
| | | Std. | .375 | 15.250 | 63 | 79.1 | 4.19 | 4.00 | 182.6 | 18.41 | 562 | 70.3 | 5.53 |
| | | API | .438 | 15.124 | 73 | 78.2 | 4.19 | 3.96 | 180.0 | 21.42 | 650 | 81.2 | 5.51 |
| | | X-Stg. | .500 | 15.000 | 83 | 76.5 | 4.19 | 3.93 | 176.7 | 24.35 | 732 | 91.5 | 5.48 |
| | | - - | .625 | 14.750 | 103 | 74.1 | 4.19 | 3.86 | 170.9 | 30.19 | 893 | 111.7 | 5.44 |
| | | 60 | .656 | 14.688 | 108 | 73.4 | 4.19 | 3.85 | 169.4 | 31.62 | 933 | 116.6 | 5.43 |
| | | - - | .750 | 14.500 | 122 | 71.5 | 4.19 | 3.80 | 165.1 | 35.93 | 1047 | 130.9 | 5.40 |
| | | 80 | .844 | 14.312 | 137 | 69.7 | 4.19 | 3.75 | 160.9 | 40.14 | 1157 | 144.6 | 5.37 |
| | | - - | .875 | 14.250 | 141 | 69.1 | 4.19 | 3.73 | 159.5 | 41.58 | 1192 | 149.0 | 5.35 |
| | | - - | 1.000 | 14.000 | 160 | 66.7 | 4.19 | 3.66 | 153.9 | 47.12 | 1331 | 166.4 | 5.31 |
| | | 100 | 1.031 | 13.938 | 165 | 66.0 | 4.19 | 3.65 | 152.6 | 48.49 | 1366 | 170.7 | 5.30 |
| | | - - | 1.125 | 13.750 | 179 | 64.4 | 4.19 | 3.60 | 148.5 | 52.57 | 1463 | 182.9 | 5.27 |
| | | 120 | 1.219 | 13.562 | 193 | 62.6 | 4.19 | 3.55 | 144.5 | 56.56 | 1556 | 194.5 | 5.24 |
| | | - - | 1.250 | 13.500 | 197 | 62.1 | 4.19 | 3.53 | 143.1 | 57.92 | 1586 | 198.3 | 5.23 |
| | | - - | 1.375 | 13.250 | 215 | 59.8 | 4.19 | 3.47 | 137.9 | 63.17 | 1704 | 213.0 | 5.19 |
| | | 140 | 1.438 | 13.124 | 224 | 58.6 | 4.19 | 3.44 | 135.3 | 65.79 | 1761 | 220.1 | 5.17 |
| | | - - | 1.500 | 13.000 | 232 | 57.4 | 4.19 | 3.40 | 132.7 | 68.33 | 1816 | 227.0 | 5.15 |
| | | 160 | 1.594 | 12.812 | 245 | 55.9 | 4.19 | 3.35 | 129.0 | 72.10 | 1893 | 236.6 | 5.12 |

APPENDIX 1 *continued*

| Nominal Pipe Size | Nominal Outside Diam. D | Designation | Wall Thickness | Inside Diam. d | Weight per Foot | Wt. of Water per Ft. of Pipe | Sq. Ft. Outside Surface per Ft. | Sq. Ft. Inside Surface per Ft. | Transverse Area in.² A | Area of Metal in.² A | Moment of Inertia in.⁴ I | Section Modulus in.³ Z | Radius of Gyration in. R |
|---|---|---|---|---|---|---|---|---|---|---|---|---|---|
| 18 | 18.000 | 10 Ga | .134 | 17.732 | 26 | 107.1 | 4.71 | 4.64 | 246.9 | 7.52 | 300 | 33.4 | 6.32 |
| | | 8 Ga. | .164 | 17.672 | 31 | 106.3 | 4.71 | 4.63 | 245.3 | 9.19 | 366 | 40.6 | 6.31 |
| | | 6 Ga. | .194 | 17.612 | 37 | 105.6 | 4.71 | 4.61 | 243.6 | 10.85 | 430 | 47.8 | 6.29 |
| | | 3 Ga. | .239 | 17.522 | 45 | 104.5 | 4.71 | 4.59 | 241.1 | 13.34 | 526 | 58.4 | 6.28 |
| | | 10 | .250 | 17.500 | 47 | 104.6 | 4.71 | 4.58 | 241.0 | 13.96 | 550 | 61.1 | 6.28 |
| | | API | .281 | 17.438 | 49 | 104.0 | 4.71 | 4.56 | 240.0 | 14.49 | 570 | 63.4 | 6.27 |
| | | 20 | .312 | 17.375 | 59 | 102.5 | 4.71 | 4.55 | 237.1 | 17.36 | 679 | 75.5 | 6.25 |
| | | API | .344 | 17.312 | 65 | 102.0 | 4.71 | 4.53 | 235.4 | 19.08 | 744 | 82.6 | 6.24 |
| | | Std. | .375 | 17.250 | 71 | 101.2 | 4.71 | 4.51 | 233.7 | 20.76 | 807 | 89.6 | 6.23 |
| | | API | .406 | 17.188 | 76 | 100.6 | 4.71 | 4.50 | 232.0 | 22.44 | 869 | 96.6 | 6.22 |
| | | 30 | .438 | 17.124 | 82 | 99.5 | 4.71 | 4.48 | 229.5 | 24.95 | 963 | 107.0 | 6.21 |
| | | X-Stg. | .500 | 17.000 | 93 | 98.2 | 4.71 | 4.45 | 227.0 | 27.49 | 1053 | 117.0 | 6.19 |
| | | 40 | .562 | 16.876 | 105 | 97.2 | 4.71 | 4.42 | 224.0 | 30.85 | 1177 | 130.9 | 6.17 |
| | | - - | .625 | 16.750 | 116 | 95.8 | 4.71 | 4.39 | 220.5 | 34.15 | 1290 | 143.2 | 6.14 |
| | | 60 | .750 | 16.500 | 138 | 92.5 | 4.71 | 4.32 | 213.8 | 40.64 | 1515 | 168.3 | 6.10 |
| | | - - | .875 | 16.250 | 160 | 89.9 | 4.71 | 4.25 | 207.4 | 47.07 | 1730 | 192.3 | 6.06 |
| | | 80 | .938 | 16.124 | 171 | 88.5 | 4.71 | 4.22 | 204.2 | 50.23 | 1834 | 203.8 | 6.04 |
| | | - - | 1.000 | 16.000 | 182 | 87.2 | 4.71 | 4.19 | 201.1 | 53.41 | 1935 | 215.0 | 6.02 |
| | | - - | 1.125 | 15.750 | 203 | 84.5 | 4.71 | 4.12 | 194.8 | 59.64 | 2133 | 237.0 | 5.98 |
| | | 100 | 1.156 | 15.688 | 208 | 83.7 | 4.71 | 4.11 | 193.3 | 61.18 | 2182 | 242.3 | 5.97 |
| | | - - | 1.250 | 15.500 | 224 | 81.8 | 4.71 | 4.06 | 188.7 | 65.78 | 2319 | 257.7 | 5.94 |

| | | | | | | | | | | | | | |
|---|---|---|---|---|---|---|---|---|---|---|---|---|---|
| | | 120 | 1.375 | 15.250 | 244 | 79.2 | 4.71 | 3.99 | 182.7 | 71.82 | 2498 | 277.5 | 5.90 |
| | | - - | 1.500 | 15.000 | 265 | 76.6 | 4.71 | 3.93 | 176.7 | 77.75 | 2668 | 296.5 | 5.86 |
| | | 140 | 1.562 | 14.876 | 275 | 75.3 | 4.71 | 3.89 | 173.8 | 80.66 | 2750 | 305.5 | 5.84 |
| | | 160 | 1.781 | 14.438 | 309 | 71.0 | 4.71 | 3.78 | 163.7 | 90.75 | 3020 | 335.5 | 5.77 |
| **20** | **20.000** | 10 Ga. | .134 | 19.732 | 28 | 132.6 | 5.24 | 5.17 | 305.8 | 8.36 | 413 | 41.3 | 7.02 |
| | | 8 Ga. | .164 | 19.672 | 35 | 131.8 | 5.24 | 5.15 | 303.9 | 10.22 | 503 | 50.3 | 7.01 |
| | | 6 Ga. | .194 | 19.612 | 41 | 131.0 | 5.24 | 5.13 | 302.1 | 12.07 | 592 | 59.2 | 7.00 |
| | | 3 Ga. | .239 | 19.522 | 50 | 129.8 | 5.24 | 5.11 | 299.3 | 14.84 | 725 | 72.5 | 6.99 |
| | | 10 | .250 | 19.500 | 53 | 130.0 | 5.24 | 5.11 | 299.0 | 15.52 | 759 | 75.9 | 6.98 |
| | | API | .281 | 19.438 | 59 | 128.6 | 5.24 | 5.09 | 296.8 | 17.41 | 846 | 84.6 | 6.97 |
| | | API | .312 | 19.374 | 66 | 128.1 | 5.24 | 5.08 | 295.0 | 19.36 | 937 | 93.7 | 6.95 |
| | | API | .344 | 19.312 | 72 | 127.0 | 5.24 | 5.06 | 292.9 | 21.24 | 1026 | 102.6 | 6.95 |
| | | Std. | .375 | 19.250 | 79 | 126.0 | 5.24 | 5.04 | 291.1 | 23.12 | 1113 | 111.3 | 6.94 |
| | | API | .406 | 19.188 | 85 | 125.4 | 5.24 | 5.02 | 289.2 | 24.99 | 1200 | 120.0 | 6.93 |
| | | API | .438 | 19.124 | 92 | 125.1 | 5.24 | 5.01 | 288.0 | 26.95 | 1290 | 129.0 | 6.92 |
| | | X-Stg. | .500 | 19.000 | 105 | 122.8 | 5.24 | 4.97 | 283.5 | 30.63 | 1457 | 145.7 | 6.90 |
| | | 40 | .594 | 18.812 | 123 | 120.4 | 5.24 | 4.93 | 278.0 | 36.15 | 1704 | 170.4 | 6.86 |
| | | - - | .625 | 18.750 | 129 | 119.5 | 5.24 | 4.91 | 276.1 | 38.04 | 1787 | 178.7 | 6.85 |
| | | 60 | .812 | 18.376 | 167 | 114.9 | 5.24 | 4.81 | 265.2 | 48.95 | 2257 | 225.7 | 6.79 |
| | | - - | .875 | 18.250 | 179 | 113.2 | 5.24 | 4.78 | 261.6 | 52.57 | 2409 | 240.9 | 6.77 |
| | | - - | 1.000 | 18.000 | 203 | 110.3 | 5.24 | 4.71 | 254.5 | 59.69 | 2702 | 270.2 | 6.73 |
| | | 80 | 1.031 | 17.938 | 209 | 109.4 | 5.24 | 4.80 | 252.7 | 61.44 | 2771 | 277.1 | 6.72 |
| | | - - | 1.125 | 17.750 | 227 | 107.3 | 5.24 | 4.65 | 247.4 | 66.71 | 2981 | 298.1 | 6.68 |
| | | - - | 1.250 | 17.500 | 250 | 104.3 | 5.24 | 4.58 | 240.5 | 73.63 | 3249 | 324.9 | 6.64 |
| | | 100 | 1.281 | 17.438 | 256 | 103.4 | 5.24 | 4.56 | 238.8 | 75.34 | 3317 | 331.7 | 6.63 |
| | | - - | 1.375 | 17.250 | 274 | 101.3 | 5.24 | 4.52 | 233.7 | 80.45 | 3508 | 350.8 | 6.60 |
| | | 120 | 1.500 | 17.000 | 297 | 98.3 | 5.24 | 4.45 | 227.0 | 87.18 | 3755 | 375.5 | 6.56 |
| | | 140 | 1.750 | 16.500 | 342 | 92.6 | 5.24 | 4.32 | 213.8 | 100.33 | 4217 | 421.7 | 6.48 |
| | | 160 | 1.969 | 16.062 | 379 | 87.9 | 5.24 | 4.21 | 202.7 | 111.49 | 4586 | 458.6 | 6.41 |

APPENDIX 1 *continued*

| Nominal Pipe Size | Nominal Outside Diam. D | Designation | Wall Thickness | Inside Diam. d | Weight per Foot | Wt. of Water per Ft. of Pipe | Sq. Ft. Outside Surface per Ft. | Sq. Ft. Inside Surface per Ft. | Transverse Area in.$^2$ A | Area of Metal in.$^2$ A | Moment of Inertia in.$^4$ I | Section Modulus in.$^3$ Z | Radius of Gyration in. R |
|---|---|---|---|---|---|---|---|---|---|---|---|---|---|
| 22 | 22.000 | 8 Ga. | .164 | 21.672 | 38 | 159.9 | 5.76 | 5.67 | 368.9 | 11.25 | 671 | 61.0 | 7.72 |
| | | 6 Ga. | .194 | 21.612 | 45 | 159.0 | 5.76 | 5.66 | 366.8 | 13.29 | 790 | 71.8 | 7.71 |
| | | 3 Ga. | .239 | 21.522 | 56 | 157.7 | 5.76 | 5.63 | 363.8 | 16.34 | 967 | 87.9 | 7.69 |
| | | API | .250 | 21.500 | 58 | 157.4 | 5.76 | 5.63 | 363.1 | 17.18 | 1010 | 91.8 | 7.69 |
| | | API | .281 | 21.438 | 65 | 156.5 | 5.76 | 5.61 | 361.0 | 19.17 | 1131 | 102.8 | 7.68 |
| | | API | .312 | 21.376 | 72 | 155.6 | 5.76 | 5.60 | 358.9 | 21.26 | 1250 | 113.6 | 7.67 |
| | | API | .344 | 21.312 | 80 | 154.7 | 5.76 | 5.58 | 356.7 | 23.40 | 1373 | 124.8 | 7.66 |
| | | API | .375 | 21.250 | 87 | 153.7 | 5.76 | 5.56 | 354.7 | 25.48 | 1490 | 135.4 | 7.65 |
| | | API | .406 | 21.188 | 94 | 152.9 | 5.76 | 5.55 | 352.6 | 27.54 | 1607 | 146.1 | 7.64 |
| | | API | .438 | 21.124 | 101 | 151.9 | 5.76 | 5.53 | 350.5 | 29.67 | 1725 | 156.8 | 7.62 |
| | | API | .500 | 21.000 | 115 | 150.2 | 5.76 | 5.50 | 346.4 | 33.77 | 1953 | 177.5 | 7.61 |
| | | - - | .625 | 20.750 | 143 | 146.6 | 5.76 | 5.43 | 338.2 | 41.97 | 2400 | 218.2 | 7.56 |
| | | - - | .750 | 20.500 | 170 | 143.1 | 5.76 | 5.37 | 330.1 | 50.07 | 2829 | 257.2 | 7.52 |
| | | - - | .875 | 20.250 | 198 | 139.6 | 5.76 | 5.30 | 322.1 | 58.07 | 3245 | 295.0 | 7.47 |
| | | - - | 1.000 | 20.000 | 224 | 136.2 | 5.76 | 5.24 | 314.2 | 65.97 | 3645 | 331.4 | 7.43 |
| | | - - | 1.125 | 19.750 | 251 | 132.8 | 5.76 | 5.17 | 306.4 | 73.78 | 4029 | 366.3 | 7.39 |
| | | - - | 1.250 | 19.500 | 277 | 129.5 | 5.76 | 5.10 | 298.6 | 81.48 | 4400 | 400.0 | 7.35 |
| | | - - | 1.375 | 19.250 | 303 | 126.2 | 5.76 | 5.04 | 291.0 | 89.09 | 4758 | 432.6 | 7.31 |
| | | - - | 1.500 | 19.000 | 329 | 122.9 | 5.76 | 4.97 | 283.5 | 96.60 | 5103 | 463.9 | 7.27 |
| | | 8 Ga. | .164 | 23.672 | 42 | 190.8 | 6.28 | 6.20 | 440.1 | 12.28 | 872 | 72.7 | 8.43 |
| | | 6 Ga. | .194 | 23.612 | 49 | 189.8 | 6.28 | 6.18 | 437.9 | 14.51 | 1028 | 85.7 | 8.42 |
| | | 3 Ga. | .239 | 23.522 | 61 | 188.4 | 6.28 | 6.16 | 434.5 | 17.84 | 1260 | 105.0 | 8.40 |
| | | 10 | .250 | 23.500 | 63 | 189.0 | 6.28 | 6.15 | 435.0 | 18.67 | 1320 | 110.0 | 8.40 |
| | | API | .281 | 23.438 | 71 | 187.0 | 6.28 | 6.14 | 431.5 | 20.94 | 1472 | 122.7 | 8.38 |

| | | | | | | | | | | | | |
|---|---|---|---|---|---|---|---|---|---|---|---|---|
| 24 | 24.000 | API | .312 | 23.376 | 79 | 186.9 | 6.28 | 6.12 | 430.0 | 23.20 | 1630 | 136.0 | 8.38 |
| | | API | .344 | 23.312 | 87 | 185.0 | 6.28 | 6.10 | 426.8 | 25.57 | 1789 | 149.1 | 8.36 |
| | | Std. | .375 | 23.250 | 95 | 183.8 | 6.28 | 6.09 | 424.6 | 27.83 | 1942 | 161.9 | 8.35 |
| | | API | .406 | 23.188 | 102 | 183.1 | 6.28 | 6.07 | 422.3 | 30.09 | 2095 | 174.6 | 8.34 |
| | | API | .438 | 23.124 | 110 | 182.1 | 6.28 | 6.05 | 420.0 | 32.42 | 2252 | 187.7 | 8.33 |
| | | X-Stg. | .500 | 23.000 | 125 | 181.0 | 6.28 | 6.02 | 416.0 | 36.90 | 2550 | 213.0 | 8.31 |
| | | 30 | .562 | 22.876 | 141 | 178.5 | 6.28 | 5.99 | 411.0 | 41.40 | 2840 | 237.0 | 8.28 |
| | | - - | .625 | 22.750 | 156 | 175.9 | 6.28 | 5.96 | 406.5 | 45.90 | 3137 | 261.4 | 8.27 |
| | | 40 | .688 | 22.624 | 171 | 174.2 | 6.28 | 5.92 | 402.1 | 50.30 | 3422 | 285.2 | 8.25 |
| | | - - | .750 | 22.500 | 186 | 172.1 | 6.28 | 5.89 | 397.6 | 54.78 | 3705 | 308.8 | 8.22 |
| | | - - | .875 | 22.250 | 216 | 168.6 | 6.28 | 5.82 | 388.8 | 63.57 | 4257 | 354.7 | 8.18 |
| | | 60 | .969 | 22.062 | 238 | 165.8 | 6.28 | 5.78 | 382.3 | 70.04 | 4652 | 387.7 | 8.15 |
| | | - - | 1.000 | 22.000 | 246 | 164.8 | 6.28 | 5.76 | 380.1 | 72.26 | 4788 | 399.0 | 8.14 |
| | | - - | 1.125 | 21.750 | 275 | 161.1 | 6.28 | 5.69 | 371.5 | 80.85 | 5302 | 441.8 | 8.10 |
| | | 80 | 1.219 | 21.562 | 297 | 158.2 | 6.28 | 5.65 | 365.2 | 87.17 | 5673 | 472.8 | 8.07 |
| | | - - | 1.250 | 21.500 | 304 | 157.4 | 6.28 | 5.63 | 363.1 | 89.34 | 5797 | 483.0 | 8.05 |
| | | - - | 1.375 | 21.250 | 332 | 153.8 | 6.28 | 5.56 | 354.7 | 97.73 | 6275 | 522.9 | 8.01 |
| | | - - | 1.500 | 21.000 | 361 | 150.2 | 6.28 | 5.50 | 346.4 | 106.03 | 6740 | 561.7 | 7.97 |
| | | 100 | 1.531 | 20.938 | 367 | 149.3 | 6.28 | 5.48 | 344.3 | 108.07 | 6847 | 570.6 | 7.96 |
| | | 120 | 1.812 | 20.376 | 429 | 141.4 | 6.28 | 5.33 | 326.1 | 126.30 | 7823 | 651.9 | 7.87 |
| | | 140 | 2.062 | 19.876 | 484 | 134.4 | 6.28 | 5.20 | 310.3 | 142.10 | 8627 | 718.9 | 7.79 |
| | | 160 | 2.344 | 19.312 | 542 | 127.0 | 6.28 | 5.06 | 293.1 | 159.40 | 9457 | 788.1 | 7.70 |
| 26 | 26.000 | 8 Ga. | .164 | 25.672 | 45 | 224.4 | 6.81 | 6.72 | 517.6 | 13.31 | 1111 | 85.4 | 9.13 |
| | | 6 Ga. | .194 | 25.612 | 54 | 223.4 | 6.81 | 6.70 | 515.2 | 15.73 | 1310 | 100.7 | 9.12 |
| | | 3 Ga. | .239 | 25.522 | 66 | 221.8 | 6.81 | 6.68 | 511.6 | 19.34 | 1605 | 123.4 | 9.11 |
| | | API | .250 | 25.500 | 67 | 221.4 | 6.81 | 6.68 | 510.7 | 19.85 | 1646 | 126.6 | 9.10 |
| | | API | .281 | 25.438 | 77 | 220.3 | 6.81 | 6.66 | 508.2 | 22.70 | 1877 | 144.4 | 9.09 |
| | | API | .312 | 25.376 | 84 | 219.2 | 6.81 | 6.64 | 505.8 | 25.18 | 2076 | 159.7 | 9.08 |

APPENDIX 1 *continued*

| Nominal Pipe Size | Nominal Outside Diam. D | Designation | Wall Thickness | Inside Diam. d | Weight per Foot | Wt. of Water per Ft. of Pipe | Sq. Ft. Outside Surface per Ft. | Sq. Ft. Inside Surface per Ft. | Transverse Area in.² A | Area of Metal in.² A | Moment of Inertia in.⁴ I | Section Modulus in.³ Z | Radius of Gyration in. R |
|---|---|---|---|---|---|---|---|---|---|---|---|---|---|
| 26 cont. | 26.000 | API | .344 | 25.312 | 94 | 218.2 | 6.81 | 6.63 | 503.2 | 27.73 | 2280 | 175.4 | 9.07 |
| | | API | .375 | 25.250 | 103 | 217.1 | 6.81 | 6.61 | 500.7 | 30.19 | 2478 | 190.6 | 9.06 |
| | | API | .406 | 25.188 | 111 | 216.0 | 6.81 | 6.59 | 498.3 | 32.64 | 2673 | 205.6 | 9.05 |
| | | API | .438 | 25.124 | 120 | 214.9 | 6.81 | 6.58 | 495.8 | 35.17 | 2874 | 221.1 | 9.04 |
| | | API | .500 | 25.000 | 136 | 212.8 | 6.81 | 6.54 | 490.9 | 40.06 | 3259 | 250.7 | 9.02 |
| | | - - | .625 | 24.750 | 169 | 208.6 | 6.81 | 6.48 | 481.1 | 49.82 | 4013 | 308.7 | 8.98 |
| | | - - | .750 | 24.500 | 202 | 204.4 | 6.81 | 6.41 | 471.4 | 59.49 | 4744 | 364.9 | 8.93 |
| | | - - | .875 | 24.250 | 235 | 200.2 | 6.81 | 6.35 | 461.9 | 69.07 | 5458 | 419.9 | 8.89 |
| | | - - | 1.000 | 24.000 | 267 | 196.1 | 6.81 | 6.28 | 452.4 | 78.54 | 6149 | 473.0 | 8.85 |
| | | - - | 1.125 | 23.750 | 299 | 192.1 | 6.81 | 6.22 | 443.0 | 87.91 | 6813 | 524.1 | 8.80 |
| | | - - | 1.375 | 23.250 | 362 | 184.1 | 6.81 | 6.09 | 424.6 | 106.37 | 8088 | 622.2 | 8.72 |
| | | - - | 1.500 | 23.000 | 393 | 180.1 | 6.81 | 6.02 | 415.5 | 115.45 | 8695 | 668.8 | 8.68 |
| | | 8 Ga. | .164 | 29.672 | 52 | 299.9 | 7.85 | 7.77 | 691.4 | 15.37 | 1711 | 114.0 | 10.55 |
| | | 6 Ga. | .194 | 29.612 | 62 | 298.6 | 7.85 | 7.75 | 688.6 | 18.17 | 2017 | 134.4 | 10.53 |
| | | 3 Ga. | .239 | 29.522 | 76 | 296.7 | 7.85 | 7.73 | 684.4 | 22.35 | 2474 | 165.0 | 10.52 |
| | | API | .250 | 29.500 | 79 | 296.3 | 7.85 | 7.72 | 683.4 | 23.37 | 2585 | 172.3 | 10.52 |
| | | API | .281 | 29.438 | 89 | 295.1 | 7.85 | 7.70 | 680.5 | 26.24 | 2897 | 193.1 | 10.51 |
| | | 10 | .312 | 29.376 | 99 | 293.7 | 7.85 | 7.69 | 677.8 | 29.19 | 3201 | 213.4 | 10.50 |
| | | API | .344 | 29.312 | 109 | 292.6 | 7.85 | 7.67 | 674.8 | 32.04 | 3524 | 235.0 | 10.49 |
| | | API | .375 | 29.250 | 119 | 291.2 | 7.85 | 7.66 | 672.0 | 34.90 | 3823 | 254.8 | 10.48 |
| | | API | .406 | 29.188 | 130 | 290.7 | 7.85 | 7.64 | 669.0 | 37.75 | 4132 | 275.5 | 10.46 |

| | | | | | | | | | | | | |
|---|---|---|---|---|---|---|---|---|---|---|---|---|
| 30 | 30.000 | API | .438 | 29.124 | 138 | 288.8 | 7.85 | 7.62 | 666.1 | 40.68 | 4442 | 296.2 | 10.45 |
| | | 20 | .500 | 29.000 | 158 | 286.2 | 7.85 | 7.59 | 660.5 | 46.34 | 5033 | 335.5 | 10.43 |
| | | 30 | .625 | 28.750 | 196 | 281.3 | 7.85 | 7.53 | 649.2 | 57.68 | 6213 | 414.2 | 10.39 |
| | | - - | .750 | 28.500 | 234 | 276.6 | 7.85 | 7.46 | 637.9 | 68.92 | 7371 | 491.4 | 10.34 |
| | | - - | .875 | 28.250 | 272 | 271.8 | 7.85 | 7.39 | 620.7 | 80.06 | 8494 | 566.2 | 10.30 |
| | | - - | 1.000 | 28.000 | 310 | 267.0 | 7.85 | 7.33 | 615.7 | 91.11 | 9591 | 639.4 | 10.26 |
| | | - - | 1.125 | 27.750 | 347 | 262.2 | 7.85 | 7.26 | 604.7 | 102.05 | 10653 | 710.2 | 10.22 |
| | | - - | 1.250 | 27.500 | 384 | 257.5 | 7.85 | 7.20 | 593.9 | 112.90 | 11682 | 778.8 | 10.17 |
| | | - - | 1.375 | 27.250 | 421 | 252.9 | 7.85 | 7.13 | 583.1 | 123.65 | 12694 | 846.2 | 10.13 |
| | | - - | 1.500 | 27.000 | 457 | 248.2 | 7.85 | 7.07 | 572.5 | 134.30 | 13673 | 911.5 | 10.09 |
| 32 | 32.000 | API | .250 | 31.500 | 85 | 337.8 | 8.38 | 8.25 | 779.2 | 24.93 | 3141 | 196.3 | 11.22 |
| | | API | .281 | 31.438 | 95 | 336.5 | 8.38 | 8.23 | 776.2 | 28.04 | 3525 | 220.3 | 11.21 |
| | | API | .312 | 31.376 | 106 | 335.2 | 8.38 | 8.21 | 773.2 | 31.02 | 3891 | 243.2 | 11.20 |
| | | API | .344 | 31.312 | 116 | 333.8 | 8.38 | 8.20 | 770.0 | 34.24 | 4287 | 268.0 | 11.19 |
| | | API | .375 | 31.250 | 127 | 332.5 | 8.38 | 8.18 | 766.9 | 37.25 | 4656 | 291.0 | 11.18 |
| | | API | .406 | 31.188 | 137 | 331.2 | 8.38 | 8.16 | 764.0 | 40.29 | 5025 | 314.1 | 11.17 |
| | | API | .438 | 31.124 | 148 | 329.8 | 8.38 | 8.15 | 760.8 | 43.43 | 5407 | 337.9 | 11.16 |
| | | API | .500 | 31.000 | 168 | 327.2 | 8.38 | 8.11 | 754.7 | 49.48 | 6140 | 383.8 | 11.14 |
| | | - - | .625 | 30.750 | 209 | 321.9 | 8.38 | 8.05 | 742.5 | 61.59 | 7578 | 473.6 | 11.09 |
| | | - - | .750 | 30.500 | 250 | 316.7 | 8.38 | 7.98 | 730.5 | 73.63 | 8990 | 561.9 | 11.05 |
| | | - - | .875 | 30.250 | 291 | 311.5 | 8.38 | 7.92 | 718.6 | 85.53 | 10368 | 648.0 | 11.01 |
| | | - - | 1.000 | 30.000 | 331 | 306.4 | 8.38 | 7.85 | 706.8 | 97.38 | 11680 | 730.0 | 10.95 |
| | | - - | 1.125 | 29.750 | 371 | 301.3 | 8.38 | 7.79 | 695.0 | 109.0 | 13003 | 812.7 | 10.92 |
| | | - - | 1.250 | 29.500 | 410 | 296.3 | 8.38 | 7.72 | 680.5 | 120.7 | 14398 | 899.9 | 10.88 |
| | | - - | 1.375 | 29.250 | 450 | 291.2 | 8.38 | 7.66 | 671.9 | 132.2 | 15526 | 970.4 | 10.84 |
| | | - - | 1.500 | 29.000 | 489 | 286.3 | 8.38 | 7.59 | 660.5 | 143.7 | 16752 | 1047.0 | 10.80 |

APPENDIX 1 *continued*

| Nominal Pipe Size | Nominal Outside Diam. D | Designation | Wall Thickness | Inside Diam. d | Weight per Foot | Wt. of Water per Ft. of Pipe | Sq. Ft. Outside Surface per Ft. | Sq. Ft. Inside Surface per Ft. | Transverse Area in.² A | Area of Metal in.² A | Moment of Inertia in.⁴ I | Section Modulus in.³ Z | Radius of Gyration in. R |
|---|---|---|---|---|---|---|---|---|---|---|---|---|---|
| 34 | 34.000 | API | .250 | 33.500 | 90 | 382.0 | 8.90 | 8.77 | 881.2 | 26.50 | 3773 | 221.9 | 11.93 |
| | | API | .281 | 33.438 | 101 | 380.7 | 8.90 | 8.75 | 878.2 | 29.77 | 4230 | 248.8 | 11.92 |
| | | API | .312 | 33.376 | 112 | 379.3 | 8.90 | 8.74 | 874.9 | 32.99 | 4680 | 275.3 | 11.91 |
| | | API | .344 | 33.312 | 124 | 377.8 | 8.90 | 8.72 | 871.6 | 36.36 | 5147 | 302.8 | 11.90 |
| | | API | .375 | 33.250 | 135 | 376.2 | 8.90 | 8.70 | 867.8 | 39.61 | 5597 | 329.2 | 11.89 |
| | | API | .406 | 33.188 | 146 | 375.0 | 8.90 | 8.69 | 865.0 | 42.88 | 6047 | 355.7 | 11.87 |
| | | API | .438 | 33.124 | 157 | 373.6 | 8.90 | 8.67 | 861.7 | 46.18 | 6501 | 382.4 | 11.86 |
| | | API | .500 | 33.000 | 179 | 370.8 | 8.90 | 8.64 | 855.3 | 52.62 | 7385 | 434.4 | 11.85 |
| | | - - | .625 | 32.750 | 223 | 365.0 | 8.90 | 8.57 | 841.9 | 65.53 | 9124 | 536.7 | 11.80 |
| | | - - | .750 | 32.500 | 266 | 359.5 | 8.90 | 8.51 | 829.3 | 78.34 | 10829 | 637.0 | 11.76 |
| | | - - | .875 | 32.250 | 308 | 354.1 | 8.90 | 8.44 | 816.8 | 90.66 | 12442 | 731.9 | 11.71 |
| | | - - | 1.000 | 32.000 | 353 | 348.6 | 8.90 | 8.38 | 804.2 | 103.6 | 14114 | 830.2 | 11.67 |
| | | - - | 1.125 | 31.750 | 395 | 343.2 | 8.90 | 8.31 | 791.6 | 116.1 | 15703 | 923.7 | 11.63 |
| | | - - | 1.250 | 31.500 | 437 | 337.8 | 8.90 | 8.25 | 779.2 | 128.5 | 17246 | 1014.5 | 11.58 |
| | | - - | 1.375 | 31.250 | 479 | 332.4 | 8.90 | 8.18 | 766.9 | 140.9 | 18770 | 1104.1 | 11.54 |
| | | - - | 1.500 | 31.000 | 521 | 327.2 | 8.90 | 8.11 | 754.7 | 153.1 | 20247 | 1191.0 | 11.50 |
| | | - - | .164 | 35.672 | 63 | 433.2 | 9.42 | 9.34 | 999.3 | 18.53 | 2975 | 165.3 | 12.67 |
| | | - - | .194 | 35.612 | 74 | 431.8 | 9.42 | 9.32 | 996.0 | 21.83 | 3499 | 194.4 | 12.66 |
| | | - - | .239 | 35.522 | 91 | 429.6 | 9.42 | 9.30 | 991.0 | 26.86 | 4293 | 238.5 | 12.64 |
| | | API | .250 | 35.500 | 96 | 429.1 | 9.42 | 9.29 | 989.7 | 28.11 | 4491 | 249.5 | 12.64 |
| | | API | .281 | 35.438 | 107 | 427.6 | 9.42 | 9.28 | 986.4 | 31.49 | 5023 | 279.1 | 12.63 |
| | | API | .312 | 35.376 | 119 | 426.1 | 9.42 | 9.26 | 982.9 | 34.95 | 5565 | 309.1 | 12.62 |

| | | | | | | | | | | | | | |
|---|---|---|---|---|---|---|---|---|---|---|---|---|---|
| 36 | 36.000 | API | .344 | 35.312 | 131 | 424.6 | 9.42 | 9.24 | 979.3 | 38.56 | 6127 | 340.4 | 12.60 |
| | | API | .375 | 35.250 | 143 | 423.1 | 9.42 | 9.23 | 975.8 | 42.01 | 6664 | 370.2 | 12.59 |
| | | API | .406 | 35.188 | 154 | 421.6 | 9.42 | 9.21 | 972.5 | 45.40 | 7191 | 399.5 | 12.58 |
| | | API | .438 | 35.124 | 166 | 420.1 | 9.42 | 9.19 | 968.9 | 48.93 | 7737 | 429.9 | 12.57 |
| | | API | .500 | 35.000 | 190 | 417.1 | 9.42 | 9.16 | 962.1 | 55.76 | 8785 | 488.1 | 12.55 |
| | | - - | .625 | 34.750 | 236 | 411.1 | 9.42 | 9.10 | 948.3 | 69.50 | 10872 | 604.0 | 12.51 |
| | | - - | .750 | 34.500 | 282 | 405.3 | 9.42 | 9.03 | 934.7 | 83.01 | 12898 | 716.5 | 12.46 |
| | | - - | .875 | 34.250 | 329 | 399.4 | 9.42 | 8.97 | 921.2 | 96.60 | 14906 | 828.1 | 12.42 |
| | | - - | 1.000 | 34.000 | 374 | 393.6 | 9.42 | 8.90 | 907.9 | 109.9 | 16851 | 936.2 | 12.38 |
| | | - - | 1.125 | 33.750 | 419 | 387.8 | 9.42 | 8.83 | 894.5 | 123.3 | 18766 | 1042.6 | 12.34 |
| | | - - | 1.250 | 33.500 | 464 | 382.1 | 9.42 | 8.77 | 881.3 | 136.5 | 20624 | 1145.8 | 12.29 |
| | | - - | 1.375 | 33.250 | 509 | 376.4 | 9.42 | 8.70 | 868.2 | 149.6 | 22451 | 1247.3 | 12.25 |
| | | - - | 1.500 | 33.000 | 553 | 370.8 | 9.42 | 8.64 | 855.3 | 162.6 | 24237 | 1346.5 | 12.21 |
| 42 | 42.000 | - - | .250 | 41.500 | 112 | 586.4 | 10.99 | 10.86 | 1352.6 | 32.82 | 7126 | 339.3 | 14.73 |
| | | - - | .375 | 41.250 | 167 | 579.3 | 10.99 | 10.80 | 1336.3 | 49.08 | 10627 | 506.1 | 14.71 |
| | | - - | .500 | 41.000 | 222 | 572.3 | 10.99 | 10.73 | 1320.2 | 65.18 | 14037 | 668.4 | 14.67 |
| | | - - | .625 | 40.750 | 276 | 565.4 | 10.99 | 10.67 | 1304.1 | 81.28 | 17373 | 827.3 | 14.62 |
| | | - - | .750 | 40.500 | 331 | 558.4 | 10.99 | 10.60 | 1288.2 | 97.23 | 20689 | 985.2 | 14.59 |
| | | - - | .875 | 40.250 | 385 | 551.6 | 10.99 | 10.54 | 1272.3 | 113.0 | 23896 | 1137.9 | 14.54 |
| | | - - | 1.000 | 40.000 | 438 | 544.8 | 10.99 | 10.47 | 1256.6 | 128.8 | 27080 | 1289.5 | 14.50 |
| | | - - | 1.125 | 39.750 | 492 | 537.9 | 10.99 | 10.41 | 1240.9 | 144.5 | 30193 | 1437.8 | 14.45 |
| | | - - | 1.250 | 39.500 | 544 | 531.2 | 10.99 | 10.34 | 1225.3 | 160.0 | 33233 | 1582.5 | 14.41 |
| | | - - | 1.375 | 39.250 | 597 | 524.4 | 10.99 | 10.27 | 1209.9 | 175.5 | 36240 | 1725.7 | 14.37 |
| | | - - | 1.500 | 39.000 | 649 | 517.9 | 10.99 | 10.21 | 1194.5 | 190.8 | 39181 | 1865.7 | 14.33 |

# APPENDIX 2

## Flow of water through schedule 40 steel pipe

*(Courtesy Crane Co.)*

| Discharge | | Pressure Drop per 100 feet and Velocity in Schedule 40 Pipe for Water at 60 F. | | | | | | | | | | | | | | | |
|---|---|---|---|---|---|---|---|---|---|---|---|---|---|---|---|---|---|
| | | Veloc-ity | Press. Drop | Veloc-ity | Press. Drop | Veloc-ity | Press. Drop | Veloc-ity | Press. Drop | Veloc-ity | Press. Drop | Veloc-ity | Press. Drop | Veloc-ity | Press. Drop | Veloc-ity | Press. Drop |
| Gallons per Minute | Cubic Ft. per Second | Feet per Second | Lbs. per Sq. In. | Feet per Second | Lbs. per Sq. In. | Feet per Second | Lbs. per Sq. In. | Feet per Second | Lbs. per Sq. In. | Feet per Second | Lbs. per Sq. In. | Feet per Second | Lbs. per Sq. In. | Feet per Second | Lbs. per Sq. In. | Feet per Second | Lbs. per Sq. In. |
| | | 1/8″ | | 1/4″ | | 3/8″ | | 1/2″ | | | | | | | | | |
| .2 | 0.000446 | 1.13 | 1.86 | 0.616 | 0.359 | | | | | | | | | | | | |
| .3 | 0.000668 | 1.69 | 4.22 | 0.924 | 0.903 | 0.504 | 0.159 | 0.317 | 0.061 | 3/4″ | | | | | | | |
| .4 | 0.000891 | 2.26 | 6.98 | 1.23 | 1.61 | 0.672 | 0.345 | 0.422 | 0.086 | | | | | | | | |
| .5 | 0.00111 | 2.82 | 10.5 | 1.54 | 2.39 | 0.840 | 0.539 | 0.528 | 0.167 | 0.301 | 0.033 | | | | | | |
| .6 | 0.00134 | 3.39 | 14.7 | 1.85 | 3.29 | 1.01 | 0.751 | 0.633 | 0.240 | 0.361 | 0.041 | | | | | | |
| .8 | 0.00178 | 4.52 | 25.0 | 2.46 | 5.44 | 1.34 | 1.25 | 0.844 | 0.408 | 0.481 | 0.102 | 1″ | | 1 1/4″ | | | |
| 1 | 0.00223 | 5.65 | 37.2 | 3.08 | 8.28 | 1.68 | 1.85 | 1.06 | 0.600 | 0.602 | 0.155 | 0.371 | 0.048 | | | 1 1/2″ | |
| 2 | 0.00446 | 11.29 | 134.4 | 6.16 | 30.1 | 3.36 | 6.58 | 2.11 | 2.10 | 1.20 | 0.526 | 0.743 | 0.164 | 0.429 | 0.044 | | |
| 3 | 0.00668 | | | 9.25 | 64.1 | 5.04 | 13.9 | 3.17 | 4.33 | 1.81 | 1.09 | 1.114 | 0.336 | 0.644 | 0.090 | 0.473 | 0.043 |
| 4 | 0.00891 | | | 12.33 | 111.2 | 6.72 | 23.9 | 4.22 | 7.42 | 2.41 | 1.83 | 1.49 | 0.565 | 0.858 | 0.150 | 0.630 | 0.071 |
| 5 | 0.01114 | 2″ | | | | 8.40 | 36.7 | 5.28 | 11.2 | 3.01 | 2.75 | 1.86 | 0.835 | 1.073 | 0.223 | 0.788 | 0.104 |
| 6 | 0.01337 | 0.574 | 0.044 | 2 1/2″ | | 10.08 | 51.9 | 6.33 | 15.8 | 3.61 | 3.84 | 2.23 | 1.17 | 1.29 | 0.309 | 0.946 | 0.145 |
| 8 | 0.01782 | 0.765 | 0.073 | | | 13.44 | 91.1 | 8.45 | 27.7 | 4.81 | 6.60 | 2.97 | 1.99 | 1.72 | 0.518 | 1.26 | 0.241 |
| 10 | 0.02228 | 0.956 | 0.108 | 0.670 | 0.046 | | | 10.56 | 42.4 | 6.02 | 9.99 | 3.71 | 2.99 | 2.15 | 0.774 | 1.58 | 0.361 |
| 15 | 0.03342 | 1.43 | 0.224 | 1.01 | 0.094 | 3″ | | | | 9.03 | 21.6 | 5.57 | 6.36 | 3.22 | 1.63 | 2.37 | 0.755 |
| 20 | 0.04456 | 1.91 | 0.375 | 1.34 | 0.158 | 0.868 | 0.056 | 3 1/2″ | | 12.03 | 37.8 | 7.43 | 10.9 | 4.29 | 2.78 | 3.16 | 1.28 |

| | | | | | | | | | | | | | | | | | |
|---|---|---|---|---|---|---|---|---|---|---|---|---|---|---|---|---|---|
| 25 | 0.05570 | 2.39 | 0.561 | 1.68 | 0.234 | 1.09 | 0.083 | 0.812 | 0.041 | 4″ | | 9.28 | 16.7 | 5.37 | 4.22 | 3.94 | 1.93 |
| 30 | 0.06684 | 2.87 | 0.786 | 2.01 | 0.327 | 1.30 | 0.114 | 0.974 | 0.056 | | | 11.14 | 23.8 | 6.44 | 5.92 | 4.73 | 2.72 |
| 35 | 0.07798 | 3.35 | 1.05 | 2.35 | 0.436 | 1.52 | 0.151 | 1.14 | 0.071 | 0.882 | 0.041 | 12.99 | 32.2 | 7.51 | 7.90 | 5.52 | 3.64 |
| 40 | 0.08912 | 3.83 | 1.35 | 2.68 | 0.556 | 1.74 | 0.192 | 1.30 | 0.095 | 1.01 | 0.052 | 14.85 | 41.5 | 8.59 | 10.24 | 6.30 | 4.65 |
| 45 | 0.1003 | 4.30 | 1.67 | 3.02 | 0.668 | 1.95 | 0.239 | 1.46 | 0.117 | 1.13 | 0.064 | | | 9.67 | 12.80 | 7.09 | 5.85 |
| 50 | 0.1114 | 4.78 | 2.03 | 3.35 | 0.839 | 2.17 | 0.288 | 1.62 | 0.142 | 1.26 | 0.076 | 5″ | | 10.74 | 15.66 | 7.88 | 7.15 |
| 60 | 0.1337 | 5.74 | 2.87 | 4.02 | 1.18 | 2.60 | 0.406 | 1.95 | 0.204 | 1.51 | 0.107 | | | 12.89 | 22.2 | 9.47 | 10.21 |
| 70 | 0.1560 | 6.70 | 3.84 | 4.69 | 1.59 | 3.04 | 0.540 | 2.27 | 0.261 | 1.76 | 0.143 | 1.12 | 0.047 | | | 11.05 | 13.71 |
| 80 | 0.1782 | 7.65 | 4.97 | 5.36 | 2.03 | 3.47 | 0.687 | 2.60 | 0.334 | 2.02 | 0.180 | 1.28 | 0.060 | | | 12.62 | 17.59 |
| 90 | 0.2005 | 8.60 | 6.20 | 6.03 | 2.53 | 3.91 | 0.861 | 2.92 | 0.416 | 2.27 | 0.224 | 1.44 | 0.074 | 6″ | | 14.20 | 22.0 |
| 100 | 0.2228 | 9.56 | 7.59 | 6.70 | 3.09 | 4.34 | 1.05 | 3.25 | 0.509 | 2.52 | 0.272 | 1.60 | 0.090 | 1.11 | 0.036 | 15.78 | 26.9 |
| 125 | 0.2785 | 11.97 | 11.76 | 8.38 | 4.71 | 5.43 | 1.61 | 4.06 | 0.769 | 3.15 | 0.415 | 2.01 | 0.135 | 1.39 | 0.055 | 19.72 | 41.4 |
| 150 | 0.3342 | 14.36 | 16.70 | 10.05 | 6.69 | 6.51 | 2.24 | 4.87 | 1.08 | 3.78 | 0.580 | 2.41 | 0.190 | 1.67 | 0.077 | | |
| 175 | 0.3899 | 16.75 | 22.3 | 11.73 | 8.97 | 7.60 | 3.00 | 5.68 | 1.44 | 4.41 | 0.744 | 2.81 | 0.253 | 1.94 | 0.102 | | |
| 200 | 0.4456 | 19.14 | 28.8 | 13.42 | 11.68 | 8.68 | 3.87 | 6.49 | 1.85 | 5.04 | 0.985 | 3.21 | 0.323 | 2.22 | 0.130 | 8″ | |
| 225 | 0.5013 | ... | ... | 15.09 | 14.63 | 9.77 | 4.83 | 7.30 | 2.32 | 5.67 | 1.23 | 3.61 | 0.401 | 2.50 | 0.162 | 1.44 | 0.043 |
| 250 | 0.557 | ... | ... | ... | ... | 10.85 | 5.93 | 8.12 | 2.84 | 6.30 | 1.46 | 4.01 | 0.495 | 2.78 | 0.195 | 1.60 | 0.051 |
| 275 | 0.6127 | ... | ... | ... | ... | 11.94 | 7.14 | 8.93 | 3.40 | 6.93 | 1.79 | 4.41 | 0.583 | 3.05 | 0.234 | 1.76 | 0.061 |
| 300 | 0.6684 | ... | ... | ... | ... | 13.00 | 8.36 | 9.74 | 4.02 | 7.56 | 2.11 | 4.81 | 0.683 | 3.33 | 0.275 | 1.92 | 0.072 |
| 325 | 0.7241 | ... | ... | ... | ... | 14.12 | 9.89 | 10.53 | 4.09 | 8.19 | 2.47 | 5.21 | 0.797 | 3.61 | 0.320 | 2.08 | 0.083 |
| 350 | 0.7798 | | | ... | ... | ... | ... | 11.36 | 5.41 | 8.82 | 2.84 | 5.62 | 0.919 | 3.89 | 0.367 | 2.24 | 0.095 |
| 375 | 0.8355 | | | ... | ... | ... | ... | 12.17 | 6.18 | 9.45 | 3.25 | 6.02 | 1.05 | 4.16 | 0.416 | 2.40 | 0.108 |
| 400 | 0.8912 | | | ... | ... | ... | ... | 12.98 | 7.03 | 10.08 | 3.68 | 6.42 | 1.19 | 4.44 | 0.471 | 2.56 | 0.121 |
| 425 | 0.9469 | | | ... | ... | ... | ... | 13.80 | 7.89 | 10.71 | 4.12 | 6.82 | 1.33 | 4.72 | 0.529 | 2.73 | 0.136 |
| 450 | 1.003 | 10″ | | ... | ... | ... | ... | 14.61 | 8.80 | 11.34 | 4.60 | 7.22 | 1.48 | 5.00 | 0.590 | 2.89 | 0.151 |
| 475 | 1.059 | 1.93 | 0.054 | | | ... | ... | ... | ... | 11.97 | 5.12 | 7.62 | 1.64 | 5.27 | 0.653 | 3.04 | 0.166 |
| 500 | 1.114 | 2.03 | 0.059 | | | ... | ... | ... | ... | 12.60 | 5.65 | 8.02 | 1.81 | 5.55 | 0.720 | 3.21 | 0.182 |
| 550 | 1.225 | 2.24 | 0.071 | | | ... | ... | ... | ... | 13.85 | 6.79 | 8.82 | 2.17 | 6.11 | 0.861 | 3.53 | 0.219 |
| 600 | 1.337 | 2.44 | 0.083 | | | ... | ... | ... | ... | 15.12 | 8.04 | 9.63 | 2.55 | 6.66 | 1.02 | 3.85 | 0.258 |
| 650 | 1.448 | 2.64 | 0.097 | 12″ | | ... | ... | ... | ... | ... | ... | 10.43 | 2.98 | 7.22 | 1.18 | 4.17 | 0.301 |
| 700 | 1.560 | 2.85 | 0.112 | 2.01 | 0.047 | | | ... | ... | ... | ... | 11.23 | 3.43 | 7.78 | 1.35 | 4.49 | 0.343 |
| 750 | 1.671 | 3.05 | 0.127 | 2.15 | 0.054 | 14″ | | ... | ... | ... | ... | 12.03 | 3.92 | 8.33 | 1.55 | 4.81 | 0.392 |
| 800 | 1.782 | 3.25 | 0.143 | 2.29 | 0.061 | | | ... | ... | ... | ... | 12.83 | 4.43 | 8.88 | 1.75 | 5.13 | 0.443 |
| 850 | 1.894 | 3.46 | 0.160 | 2.44 | 0.068 | 2.02 | 0.042 | ... | ... | ... | ... | 13.64 | 5.00 | 9.44 | 1.96 | 5.45 | 0.497 |
| 900 | 2.005 | 3.66 | 0.179 | 2.58 | 0.075 | 2.13 | 0.047 | ... | ... | ... | ... | 14.44 | 5.58 | 9.99 | 2.18 | 5.77 | 0.554 |

## APPENDIX 2 *continued*

| Discharge | | Pressure Drop per 100 feet and Velocity in Schedule 40 Pipe for Water at 60 F. | | | | | | | | | | | | | | | |
|---|---|---|---|---|---|---|---|---|---|---|---|---|---|---|---|---|---|
| | | Veloc-ity | Press. Drop | Veloc-ity | Press. Drop | Veloc-ity | Press. Drop | Veloc-ity | Press. Drop | Veloc-ity | Press. Drop | Veloc-ity | Press. Drop | Veloc-ity | Press. Drop | Veloc-ity | Press. Drop |
| Gallons per Minute | Cubic Ft. per Second | Feet per Second | Lbs. per Sq. In. | Feet per Second | Lbs. per Sq. In. | Feet per Second | Lbs. per Sq. In. | Feet per Second | Lbs. per Sq. In. | Feet per Second | Lbs. per Sq. In. | Feet per Second | Lbs. per Sq. In. | Feet per Second | Lbs. per Sq. In. | Feet per Second | Lbs. per Sq. In. |
| 950 | 2.117 | 3.86 | 0.198 | 2.72 | 0.083 | 2.25 | 0.052 | | | ... | ... | 15.24 | 6.21 | 10.55 | 2.42 | 6.09 | 0.613 |
| 1 000 | 2.228 | 4.07 | 0.218 | 2.87 | 0.091 | 2.37 | 0.057 | 16″ | | ... | ... | 16.04 | 6.84 | 11.10 | 2.68 | 6.41 | 0.675 |
| 1 100 | 2.451 | 4.48 | 0.260 | 3.15 | 0.110 | 2.61 | 0.068 | | | ... | ... | 17.65 | 8.23 | 12.22 | 3.22 | 7.05 | 0.807 |
| 1 200 | 2.674 | 4.88 | 0.306 | 3.44 | 0.128 | 2.85 | 0.080 | 2.18 | 0.042 | ... | ... | ... | ... | 13.33 | 3.81 | 7.70 | 0.948 |
| 1 300 | 2.896 | 5.29 | 0.355 | 3.73 | 0.150 | 3.08 | 0.093 | 2.36 | 0.048 | ... | ... | ... | ... | 14.43 | 4.45 | 8.33 | 1.11 |
| 1 400 | 3.119 | 5.70 | 0.409 | 4.01 | 0.171 | 3.32 | 0.107 | 2.54 | 0.055 | | | | | 15.55 | 5.13 | 8.98 | 1.28 |
| 1 500 | 3.342 | 6.10 | 0.466 | 4.30 | 0.195 | 3.56 | 0.122 | 2.72 | 0.063 | 18″ | | | | 16.66 | 5.85 | 9.62 | 1.46 |
| 1 600 | 3.565 | 6.51 | 0.527 | 4.59 | 0.219 | 3.79 | 0.138 | 2.90 | 0.071 | | | | | 17.77 | 6.61 | 10.26 | 1.65 |
| 1 800 | 4.010 | 7.32 | 0.663 | 5.16 | 0.276 | 4.27 | 0.172 | 3.27 | 0.088 | 2.58 | 0.050 | | | 19.99 | 8.37 | 11.54 | 2.08 |
| 2 000 | 4.456 | 8.14 | 0.808 | 5.73 | 0.339 | 4.74 | 0.209 | 3.63 | 0.107 | 2.87 | 0.060 | 20″ | | 22.21 | 10.3 | 12.82 | 2.55 |
| 2 500 | 5.570 | 10.17 | 1.24 | 7.17 | 0.515 | 5.93 | 0.321 | 4.54 | 0.163 | 3.59 | 0.091 | | | | | 16.03 | 3.94 |
| 3 000 | 6.684 | 12.20 | 1.76 | 8.60 | 0.731 | 7.11 | 0.451 | 5.45 | 0.232 | 4.30 | 0.129 | 3.46 | 0.075 | 24″ | | 19.24 | 5.59 |
| 3 500 | 7.798 | 14.24 | 2.38 | 10.03 | 0.982 | 8.30 | 0.607 | 6.35 | 0.312 | 5.02 | 0.173 | 4.04 | 0.101 | | | 22.44 | 7.56 |
| 4 000 | 8.912 | 16.27 | 3.08 | 11.47 | 1.27 | 9.48 | 0.787 | 7.26 | 0.401 | 5.74 | 0.222 | 4.62 | 0.129 | 3.19 | 0.052 | 22.65 | 9.80 |
| 4 500 | 10.03 | 18.31 | 3.87 | 12.90 | 1.60 | 10.67 | 0.990 | 8.17 | 0.503 | 6.46 | 0.280 | 5.20 | 0.162 | 3.59 | 0.065 | 28.87 | 12.2 |
| 5 000 | 11.14 | 20.35 | 4.71 | 14.33 | 1.95 | 11.85 | 1.21 | 9.08 | 0.617 | 7.17 | 0.340 | 5.77 | 0.199 | 3.99 | 0.079 | ... | ... |
| 6 000 | 13.37 | 24.41 | 6.74 | 17.20 | 2.77 | 14.23 | 1.71 | 10.89 | 0.877 | 8.61 | 0.483 | 6.93 | 0.280 | 4.79 | 0.111 | ... | ... |
| 7 000 | 15.60 | 28.49 | 9.11 | 20.07 | 3.74 | 16.60 | 2.31 | 12.71 | 1.18 | 10.04 | 0.652 | 8.08 | 0.376 | 5.59 | 0.150 | ... | ... |
| 8 000 | 17.82 | ... | ... | 22.93 | 4.84 | 18.96 | 2.99 | 14.52 | 1.51 | 11.47 | 0.839 | 9.23 | 0.488 | 6.38 | 0.192 | ... | ... |
| 9 000 | 20.05 | ... | ... | 25.79 | 6.09 | 21.34 | 3.76 | 16.34 | 1.90 | 12.91 | 1.05 | 10.39 | 0.608 | 7.18 | 0.242 | ... | |
| 10 000 | 22.28 | ... | ... | 28.66 | 7.46 | 23.71 | 4.61 | 18.15 | 2.34 | 14.34 | 1.28 | 11.54 | 0.739 | 7.98 | 0.294 | ... | ... |
| 12 000 | 26.74 | ... | ... | 34.40 | 10.7 | 28.45 | 6.59 | 21.79 | 3.33 | 17.21 | 1.83 | 13.85 | 1.06 | 9.58 | 0.416 | ... | ... |
| 14 000 | 31.19 | ... | ... | ... | ... | 33.19 | 8.89 | 25.42 | 4.49 | 20.08 | 2.45 | 16.16 | 1.43 | 11.17 | 0.562 | ... | ... |
| 16 000 | 35.65 | ... | ... | ... | ... | ... | ... | 29.05 | 5.83 | 22.95 | 3.18 | 18.47 | 1.85 | 12.77 | 0.723 | ... | ... |
| 18 000 | 40.10 | ... | ... | ... | ... | ... | ... | 32.68 | 7.31 | 25.82 | 4.03 | 20.77 | 2.32 | 14.36 | 0.907 | ... | |
| 20 000 | 44.56 | ... | ... | ... | ... | ... | ... | 36.31 | 9.03 | 28.69 | 4.93 | 23.08 | 2.86 | 15.96 | 1.12 | ... | |

For pipe lengths other than 100 feet, the pressure drop is proportional to the length. Thus, for 50 feet of pipe, the pressure drop is approximately one-half the value given in the table for 300 feet, three times the given value etc. Velocity is a function of the cross sectional flow area; thus it is constant for a given flow rate and is independent of pipe length.

# APPENDIX 3

# Size, capacity and weight of A.P.I storage tanks

*(Courtesy General American Transportation Corp.)*

| Size | 42 Gallon Barrels | U.S. Gallons | Tank Weight | Size | 42 Gallon Barrels | U.S. Gallons | Tank Weight |
|---|---|---|---|---|---|---|---|
| 10x12 | 168 | 7056 | 4920 | 35x12 | 2052 | 86184 | 32485 |
| 10x18 | 252 | 10584 | 6390 | 35x18 | 3078 | 129276 | 38715 |
| 10x24 | 336 | 14112 | 7870 | 35x24 | 4104 | 172368 | 43985 |
| | | | | 35x30 | 5130 | 215460 | 49340 |
| 15x12 | 372 | 15624 | 9470 | 35x36 | 6158 | 258636 | 56495 |
| 15x18 | 558 | 23436 | 11770 | 35x42 | 7182 | 301644 | 63645 |
| 15x24 | 774 | 32508 | 14060 | 35x48 | 8208 | 344736 | 71015 |
| | | | | | | | |
| 18x12 | 540 | 22680 | 12205 | 36x12 | 2172 | 91224 | 34635 |
| 18x18 | 810 | 34020 | 14950 | 36x18 | 3258 | 136836 | 40020 |
| 18x24 | 1080 | 45360 | 18590 | 36x24 | 4344 | 182448 | 45430 |
| 18x30 | 1350 | 56700 | 20500 | 36x30 | 5430 | 228060 | 50935 |
| 18x36 | 1620 | 68040 | 23510 | 36x36 | 6516 | 273672 | 58185 |
| | | | | 36x42 | 7602 | 319284 | 65630 |
| 20x12 | 672 | 28224 | 12805 | 36x48 | 8688 | 364896 | 73405 |
| 20x18 | 1008 | 42336 | 16845 | | | | |
| 20x24 | 1344 | 56448 | 19875 | 40x12 | 2688 | 112896 | 41010 |
| 20x30 | 1680 | 70560 | 22980 | 40x18 | 4032 | 169344 | 46970 |
| 20x36 | 2016 | 84672 | 26285 | 40x24 | 5376 | 225792 | 52955 |
| | | | | 40x30 | 6720 | 282240 | 60960 |
| 24x12 | 972 | 40824 | 18080 | 40x36 | 8064 | 338688 | 69070 |
| 24x18 | 1458 | 61236 | 21715 | 40x42 | 9408 | 395136 | 77180 |
| 24x24 | 1944 | 81648 | 25335 | 40x48 | 10752 | 451584 | 86765 |
| 24x30 | 2430 | 102060 | 29030 | | | | |
| 24x36 | 2916 | 122472 | 32925 | 42′6″x12 | 3036 | 127512 | 45665 |
| | | | | 42′6″x18 | 4554 | 191268 | 52000 |
| 25x12 | 1044 | 43848 | 19815 | 42′6″x24 | 6072 | 255024 | 58350 |
| 25x18 | 1566 | 65772 | 23585 | 42′6″x32 | 8096 | 340032 | 69715 |
| 25x24 | 2088 | 87696 | 27350 | 42′6″x40 | 10120 | 425040 | 81650 |
| 25x30 | 2610 | 109620 | 31195 | | | | |
| 25x36 | 3132 | 131544 | 35240 | 42x12 | 3386 | 142632 | 50685 |
| 25x42 | 3654 | 153468 | 39225 | 45x18 | 5094 | 213948 | 57435 |
| | | | | 45x24 | 6792 | 285264 | 64180 |
| 30x12 | 1512 | 63504 | 25820 | 45x30 | 8490 | 356580 | 73890 |
| 30x18 | 2268 | 95256 | 30325 | 45x36 | 10188 | 427896 | 82290 |
| 30x24 | 3024 | 127008 | 34830 | 45x42 | 11886 | 499212 | 92875 |
| 30x30 | 3780 | 158760 | 39450 | 45x48 | 13584 | 570528 | 104585 |
| 30x36 | 4536 | 190512 | 44190 | | | | |
| 30x42 | 5292 | 222264 | 50370 | 48x12 | 3864 | 162288 | 56300 |
| 30x48 | 6048 | 254016 | 56765 | 48x18 | 5796 | 243432 | 63470 |

## APPENDIX 3 *continued*

| Size | 42 Gallon Barrels | U.S. Gallons | Tank Weight |
|---|---|---|---|
| 48x24 | 7728 | 324576 | 70665 |
| 48x30 | 9660 | 405720 | 80265 |
| 48x36 | 11592 | 486864 | 90230 |
| 48x42 | 13524 | 568008 | 101995 |
| 48x48 | 15456 | 649152 | 115760 |
| 50x16 | 5600 | 235200 | 73155 |
| 50x24 | 8400 | 352800 | 86280 |
| 50x32 | 11200 | 470400 | 100075 |
| 50x40 | 14000 | 588000 | 116635 |
| 50x48 | 16800 | 705600 | 136455 |
| 50x56 | 19600 | 823200 | 159630 |
| 60x16 | 8048 | 338016 | 98970 |
| 60x24 | 12072 | 507024 | 115245 |
| 60x32 | 16096 | 676032 | 135095 |
| 60x40 | 20120 | 845040 | 159185 |
| 60x48 | 24144 | 1014048 | 190390 |
| 60x56 | 28168 | 1183056 | 229710 |
| 67x16 | 10048 | 422016 | 126410 |
| 67x24 | 15072 | 633024 | 144550 |
| 67x32 | 20096 | 844032 | 168435 |
| 67x40 | 25120 | 1055040 | 198600 |
| 67x48 | 30144 | 1266048 | 236145 |
| 67x56 | 35618 | 1495956 | 283520 |
| 70x16 | 10960 | 460320 | 133515 |
| 70x24 | 16440 | 690480 | 153075 |
| 70x32 | 21920 | 920640 | 179125 |
| 70x40 | 27400 | 1150800 | 211645 |
| 70x48 | 32880 | 1380960 | 251780 |
| 70x56 | 38360 | 1611120 | 301985 |
| 78x16 | 13616 | 571872 | 159120 |
| 78x24 | 20424 | 857808 | 182220 |
| 78x32 | 27232 | 1143744 | 215100 |
| 78x40 | 34040 | 1429680 | 254705 |
| 78x48 | 40848 | 1715616 | 304070 |
| 78x56 | 47656 | 2001552 | 365465 |
| 80x16 | 14320 | 601440 | 164275 |
| 80x24 | 21480 | 902160 | 189290 |
| 80x32 | 28640 | 1202880 | 223045 |
| 80x40 | 35800 | 1503600 | 264950 |
| 80x48 | 42960 | 1804320 | 316850 |
| 80x56 | 50120 | 2105040 | 379625 |
| 90x16 | 18128 | 761376 | 199840 |
| 90x24 | 27192 | 1142064 | 230940 |
| 90x32 | 36256 | 1522752 | 273345 |
| 90x40 | 45320 | 1903440 | 326365 |
| 90x48 | 54384 | 2284128 | 390735 |
| 90x56 | 63448 | 2664816 | 470650 |
| 100x16 | 22384 | 940128 | 245930 |
| 100x24 | 33576 | 1410192 | 284420 |
| 100x32 | 44768 | 1880256 | 337875 |
| 100x40 | 55960 | 2350320 | 403305 |
| 100x48 | 67152 | 2820384 | 486795 |
| 100x56 | 78344 | 3290448 | 591005 |
| 102x16 | 23280 | 977760 | 255450 |
| 102x24 | 34920 | 1466640 | 296420 |
| 102x32 | 46560 | 1955520 | 350820 |
| 102x40 | 58200 | 2444400 | 419220 |
| 102x48 | 69840 | 2933280 | 504130 |
| 102x56 | 81480 | 3422160 | 611875 |
| 110x16 | 27072 | 1137024 | 293890 |
| 110x24 | 40608 | 1705536 | 340020 |
| 110x32 | 54144 | 2274048 | 402775 |
| 110x40 | 67680 | 2842650 | 482290 |
| 110x48 | 81216 | 3411072 | 582325 |
| 110x56 | 94752 | 3979584 | 708390 |
| 120x16 | 32224 | 1353408 | 339160 |
| 120x24 | 48336 | 2030112 | 395145 |
| 120x32 | 64448 | 2706816 | 471155 |
| 120x40 | 80560 | 3383520 | 565410 |
| 120x48 | 96672 | 4060224 | 682880 |
| 120x56 | 112784 | 4736928 | 831065 |
| 130x16 | 37824 | 1588608 | 416840 |
| 130x24 | 56736 | 2382912 | 481860 |
| 130x32 | 75648 | 3177216 | 568945 |
| 130x40 | 94560 | 3971520 | 678475 |
| 130x48 | 113472 | 4765824 | 815315 |
| 130x56 | 132384 | 5560128 | 985855 |
| 134x16 | 40176 | 1687392 | 426600 |
| 134x24 | 60264 | 2531088 | 496425 |
| 134x32 | 80352 | 3374784 | 589505 |
| 134x40 | 100440 | 4218480 | 705745 |
| 134x48 | 120528 | 5062176 | 851770 |
| 134x56 | 140616 | 5905872 | 1031380 |
| 140x16 | 43856 | 1841952 | 465750 |
| 140x24 | 65784 | 2762928 | 540195 |
| 140x32 | 87712 | 3683904 | 640275 |
| 140x40 | 109640 | 4604880 | 766625 |
| 140x48 | 131568 | 5525856 | 922845 |
| 140x56 | 153496 | 6446832 | 1118365 |
| 144x16 | 46416 | 1949472 | 487385 |

# APPENDIX 4

## Capacities of cylinders and spheres

*(Courtesy Buffalo Tank Corp.)*

| Diam. in. Feet | Cu. Ft. per Foot of Cylinder | Gallons per Foot of Cylinder | 42 Gallon Barrels per Foot of Cylinder | Sphere Surface in Sq. Ft. | Sphere Volume in Cu. Ft | Diam. in. Feet | Cu. Ft. per Foot of Cylinder | Gallons per Foot of Cylinder | 42 Gallon Barrels per Foot of Cylinder | Sphere Surface in Sq. Ft. | Sphere Volume in Cu. Ft |
|---|---|---|---|---|---|---|---|---|---|---|---|
| 1/64 | .0002 | .00143 | .000034 | .00077 | .000002 | 23/32 | .4057 | 3.0351 | .07227 | 1.6230 | .19442 |
| 1/32 | .0008 | .00574 | .000137 | .00307 | .000016 | 3/4 | .4418 | 3.3048 | .07869 | 1.7671 | .22089 |
| 1/16 | .0031 | .02295 | .000546 | .01227 | .000128 | 25/32 | .4794 | 3.5859 | .08538 | 1.9175 | .24967 |
| 3/32 | .0069 | .05164 | .00123 | .02761 | .000431 | 13/16 | .5185 | 3.8785 | .09235 | 2.0739 | .28085 |
| 1/8 | .0123 | .09180 | .00219 | .04909 | .00102 | 27/32 | .5591 | 4.1826 | .09959 | 2.2365 | .31451 |
| 5/32 | .0192 | .14344 | .00342 | .07670 | .00200 | 7/8 | .6013 | 4.4982 | .10710 | 2.4053 | .35077 |
| 3/16 | .0276 | .20655 | .00492 | .11045 | .00345 | 29/32 | .6450 | 4.8250 | .11489 | 2.5802 | .38971 |
| 7/32 | .0376 | .28114 | .00669 | .15033 | .00548 | 15/16 | .6903 | 5.1637 | .12295 | 2.7612 | .43143 |
| 1/4 | .0491 | .36720 | .00874 | .19635 | .00818 | 31/32 | .7371 | 5.5137 | .13128 | 2.9483 | .47603 |
| 9/32 | .0621 | .46474 | .01107 | .24850 | .01165 | | | | | | |
| 5/16 | .0767 | .57375 | .01366 | .30680 | .01598 | 1 | .7854 | 5.8752 | .13989 | 3.1416 | .52360 |
| 11/32 | .0928 | .69424 | .01653 | .37122 | .02127 | 1 1/16 | .8866 | 6.6325 | .15792 | 3.5466 | .62804 |
| 3/8 | .1104 | .82620 | .01967 | .44179 | .02761 | 1 1/8 | .9940 | 7.4358 | .17704 | 3.9761 | .74551 |
| 13/32 | .1296 | .96964 | .02309 | .51849 | .03511 | 1 3/16 | 1.1075 | 8.2849 | .19726 | 4.4301 | .87680 |
| 7/16 | .1503 | 1.1245 | .02677 | .60132 | .04385 | 1 1/4 | 1.2272 | 9.1800 | .21857 | 4.9087 | 1.0227 |
| 15/32 | .1726 | 1.2909 | .03074 | .69029 | .05393 | 1 5/16 | 1.3530 | 10.121 | .24097 | 5.4119 | 1.1838 |
| 1/2 | .1963 | 1.4688 | .03497 | .78540 | .06545 | 1 3/8 | 1.4849 | 11.108 | .26447 | 5.9396 | 1.3612 |
| 17/32 | .2217 | 1.6581 | .03948 | .88664 | .07850 | 1 7/16 | 1.6230 | 12.141 | .28906 | 6.4918 | 1.5553 |
| 9/16 | .2485 | 1.8589 | .04426 | .99402 | .09319 | 1 1/2 | 1.7671 | 13.219 | .31474 | 7.0686 | 1.7671 |
| 19/32 | .2769 | 2.0712 | .04932 | 1.1075 | .10960 | 1 9/16 | 1.9175 | 14.344 | .34152 | 7.6699 | 1.9974 |
| 5/8 | .3068 | 2.2950 | .05464 | 1.2272 | .12783 | 1 5/8 | 2.0739 | 15.514 | .36938 | 8.2958 | 2.2468 |
| 21/32 | .3382 | 2.5302 | .06024 | 1.3530 | .14798 | 1 11/16 | 2.2365 | 16.731 | .39835 | 8.9462 | 2.5161 |
| 11/16 | .3712 | 2.7769 | .06612 | 1.4849 | .17014 | 1 3/4 | 2.4053 | 17.993 | .42840 | 9.6211 | 2.8062 |

## APPENDIX 4 *continued*

| Diam. in. Feet | Cu. Ft. per Foot of Cylinder | Gallons per Foot of Cylinder | 42 Gallon Barrels per Foot of Cylinder | Sphere Surface in Sq. Ft. | Sphere Volume in Cu. Ft |
|---|---|---|---|---|---|
| $1\frac{13}{16}$ | 2.5802 | 19.301 | .45955 | 10.321 | 3.1177 |
| $1\frac{7}{8}$ | 2.7612 | 20.655 | .49178 | 11.045 | .34515 |
| $1\frac{15}{16}$ | 2.9483 | 22.055 | .52512 | 11.793 | 3.8082 |
| 2 | 3.1416 | 23.501 | .55954 | 12.566 | 4.1888 |
| $2\frac{1}{16}$ | 3.3410 | 24.992 | .59505 | 13.364 | 4.5939 |
| $2\frac{1}{8}$ | 3.5466 | 26.530 | .63167 | 14.186 | 5.0243 |
| $2\frac{3}{16}$ | 3.7583 | 28.114 | .66937 | 15.033 | 5.4808 |
| $2\frac{1}{4}$ | 3.9761 | 29.743 | .70817 | 15.904 | 5.9641 |
| $2\frac{5}{16}$ | 4.2000 | 31.418 | .74806 | 16.800 | 6.4751 |
| $2\frac{3}{8}$ | 4.4301 | 33.140 | .78904 | 17.721 | 7.0144 |
| $2\frac{7}{16}$ | 4.6664 | 34.907 | .83112 | 18.665 | 7.5829 |
| $2\frac{1}{2}$ | 4.9087 | 36.720 | .87428 | 19.635 | 8.1812 |
| $2\frac{9}{16}$ | 5.1572 | 38.579 | .91854 | 20.629 | 8.8103 |
| $2\frac{5}{8}$ | 5.4119 | 40.484 | .96390 | 21.648 | 9.4708 |
| $2\frac{11}{16}$ | 5.6727 | 42.434 | 1.0103 | 22.691 | 10.164 |
| $2\frac{3}{4}$ | 5.9396 | 44.431 | 1.0578 | 23.758 | 10.889 |
| $2\frac{13}{16}$ | 6.2126 | 46.474 | 1.1065 | 24.850 | 11.649 |
| $2\frac{7}{8}$ | 6.4918 | 48.562 | 1.1562 | 25.967 | 12.443 |
| $2\frac{15}{16}$ | 6.7771 | 50.696 | 1.2071 | 27.109 | 13.272 |
| 3 | 7.0686 | 52.877 | 1.2590 | 28.272 | 14.137 |
| $3\frac{1}{16}$ | 7.3662 | 55.103 | 1.3120 | 29.465 | 15.039 |
| $3\frac{1}{8}$ | 7.6699 | 57.375 | 1.3661 | 30.680 | 15.979 |
| $3\frac{3}{16}$ | 7.9798 | 59.693 | 1.4213 | 31.919 | 16.957 |
| $3\frac{1}{4}$ | 8.2958 | 62.057 | 1.4775 | 33.183 | 17.974 |
| $3\frac{5}{16}$ | 8.6179 | 64.466 | 1.5349 | 34.472 | 19.031 |
| $3\frac{3}{8}$ | 8.9462 | 66.922 | 1.5934 | 35.785 | 20.129 |
| $3\frac{7}{16}$ | 9.2806 | 69.424 | 1.6529 | 37.122 | 21.268 |

| Diam. in. Feet | Cu. Ft. per Foot of Cylinder | Gallons per Foot of Cylinder | 42 Gallon Barrels per Foot of Cylinder | Sphere Surface in Sq. Ft. | Sphere Volume in Cu. Ft |
|---|---|---|---|---|---|
| $3\frac{1}{2}$ | 9.6211 | 71.971 | 1.7136 | 38.485 | 22.449 |
| $3\frac{5}{8}$ | 10.321 | 77.204 | 1.8382 | 41.282 | 24.942 |
| $3\frac{3}{4}$ | 11.045 | 82.620 | 1.9671 | 44.179 | 27.612 |
| $3\frac{7}{8}$ | 11.793 | 88.220 | 2.1005 | 47.173 | 30.466 |
| 4 | 12.566 | 94.003 | 2.2382 | 50.265 | 33.510 |
| $4\frac{1}{8}$ | 13.364 | 99.970 | 2.3802 | 53.456 | 36.751 |
| $4\frac{1}{4}$ | 14.186 | 106.12 | 2.5267 | 56.745 | 40.194 |
| $4\frac{3}{8}$ | 15.033 | 112.45 | 2.6775 | 60.132 | 43.846 |
| $4\frac{1}{2}$ | 15.904 | 118.97 | 2.8327 | 63.617 | 47.713 |
| $4\frac{5}{8}$ | 16.800 | 125.67 | 2.9922 | 67.201 | 51.800 |
| $4\frac{3}{4}$ | 17.721 | 132.56 | 3.1562 | 70.882 | 56.115 |
| $4\frac{7}{8}$ | 18.665 | 139.63 | 3.3245 | 74.662 | 60.663 |
| 5 | 19.635 | 146.88 | 3.4971 | 78.540 | 65.450 |
| $5\frac{1}{8}$ | 20.629 | 154.32 | 3.6742 | 82.516 | 70.482 |
| $5\frac{1}{4}$ | 21.648 | 161.93 | 3.8556 | 86.590 | 75.766 |
| $5\frac{3}{8}$ | 22.691 | 169.74 | 4.0414 | 90.763 | 81.308 |
| $5\frac{1}{2}$ | 23.758 | 177.72 | 4.2315 | 95.033 | 87.114 |
| $5\frac{5}{8}$ | 24.850 | 185.89 | 4.4261 | 99.402 | 93.189 |
| $5\frac{3}{4}$ | 25.967 | 194.25 | 4.6250 | 103.87 | 99.541 |
| $5\frac{7}{8}$ | 27.109 | 202.79 | 4.8282 | 108.43 | 106.17 |
| 6 | 28.274 | 211.51 | 5.0359 | 113.10 | 113.10 |
| $6\frac{1}{8}$ | 29.465 | 220.41 | 5.2479 | 117.86 | 120.31 |
| $6\frac{1}{4}$ | 30.680 | 229.50 | 5.4643 | 122.72 | 127.83 |
| $6\frac{3}{8}$ | 31.919 | 238.77 | 5.6850 | 127.68 | 135.66 |
| $6\frac{1}{2}$ | 33.183 | 248.23 | 5.9102 | 132.73 | 143.79 |
| $6\frac{5}{8}$ | 34.472 | 257.87 | 6.1397 | 137.89 | 152.25 |
| $6\frac{3}{4}$ | 35.785 | 267.69 | 6.3735 | 143.14 | 161.03 |

| | | | | | |
|---|---|---|---|---|---|
| 6 7/8 | 37.122 | 277.69 | 6.6118 | 148.49 | 170.14 |
| 7 | 38.485 | 287.88 | 6.8544 | 153.94 | 179.59 |
| 7 1/8 | 39.871 | 298.26 | 7.1014 | 159.48 | 189.39 |
| 7 1/4 | 41.282 | 308.81 | 7.3527 | 165.13 | 199.53 |
| 7 3/8 | 42.781 | 319.56 | 7.6085 | 170.87 | 210.03 |
| 7 1/2 | 44.179 | 330.48 | 7.8686 | 176.71 | 220.89 |
| 7 5/8 | 45.664 | 341.59 | 8.1330 | 182.65 | 232.12 |
| 7 3/4 | 47.173 | 352.88 | 8.4019 | 188.69 | 243.73 |
| 7 7/8 | 48.707 | 364.35 | 8.6751 | 194.83 | 255.71 |
| 8 | 50.265 | 376.01 | 8.9527 | 201.06 | 268.08 |
| 8 1/8 | 51.849 | 387.85 | 9,2346 | 207.39 | 280.85 |
| 8 1/4 | 53.456 | 399.88 | 9.5209 | 213.82 | 294.01 |
| 8 3/8 | 55.088 | 412.09 | 9.8116 | 220.35 | 307.58 |
| 8 1/2 | 56.745 | 424.48 | 10.107 | 226.98 | 321.56 |
| 8 5/8 | 58.426 | 437.06 | 10.406 | 233.71 | 335.95 |
| 8 3/4 | 60.132 | 449.82 | 10.710 | 240.53 | 350.77 |
| 8 7/8 | 61.862 | 462.76 | 11.018 | 247.45 | 366.02 |
| 9 | 63.617 | 475.89 | 11.331 | 254.47 | 381.70 |
| 9 1/8 | 65.397 | 489.20 | 11.648 | 261.59 | 397.83 |
| 9 1/4 | 67.201 | 502.70 | 11.969 | 268.80 | 414.40 |
| 9 3/8 | 69.029 | 516.37 | 12.295 | 276.12 | 431.43 |
| 9 1/2 | 70.882 | 530.24 | 12.625 | 283.53 | 448.92 |
| 9 5/8 | 72.760 | 544.28 | 12.959 | 291.04 | 466.88 |
| 9 3/4 | 74.662 | 558.51 | 13.298 | 298.65 | 485.30 |
| 9 7/8 | 76.598 | 572.92 | 13.641 | 306.35 | 504.21 |
| 10 | 78.540 | 587.52 | 13.989 | 314.16 | 523.60 |
| 10 1/4 | 82.516 | 617.26 | 14.697 | 330.06 | 563.86 |
| 10 1/2 | 86.590 | 647.74 | 15.422 | 346.36 | 606.13 |
| 10 3/4 | 90.763 | 678.95 | 16.166 | 363.05 | 650.47 |
| 11 | 95.033 | 710.90 | 16.926 | 380.13 | 696.91 |
| 11 1/4 | 99.402 | 743.58 | 17.704 | 397.61 | 745.51 |
| 11 1/2 | 103.87 | 776.99 | 18.500 | 415.48 | 796.33 |
| 11 3/4 | 108.43 | 811.14 | 19.313 | 433.74 | 849.40 |
| 12 | 113.10 | 846.03 | 20.143 | 452.39 | 904.78 |
| 12 1/4 | 117.86 | 881.65 | 20.992 | 471.44 | 962.51 |

| | | | | | |
|---|---|---|---|---|---|
| 12 1/2 | 122.72 | 918.00 | 21.857 | 490.87 | 1022.7 |
| 12 3/4 | 127.68 | 955.08 | 22.740 | 510.71 | 1085.2 |
| 13 | 132.73 | 992.91 | 23.641 | 530.93 | 1150.3 |
| 13 1/4 | 137.98 | 1031.5 | 24.559 | 551.55 | 1218.0 |
| 13 1/2 | 143.14 | 1070.8 | 25.494 | 572.56 | 1288.2 |
| 13 3/4 | 148.49 | 1110.8 | 26.447 | 593.96 | 1461.2 |
| 14 | 153.94 | 1151.5 | 27.418 | 615.75 | 1436.8 |
| 14 1/4 | 159.48 | 1193.0 | 28.405 | 637.94 | 1515.1 |
| 14 1/2 | 165.13 | 1235.3 | 29.411 | 660.52 | 1596.3 |
| 14 3/4 | 170.87 | 1278.2 | 30.434 | 683.49 | 1680.3 |
| 15 | 176.71 | 1321.9 | 31.474 | 706.86 | 1757.1 |
| 15 1/4 | 182.65 | 1366.3 | 32.532 | 730.62 | 1857.0 |
| 15 1/2 | 188.69 | 1411.5 | 33.607 | 754.77 | 1949.8 |
| 15 3/4 | 194.83 | 1457.4 | 34.700 | 779.31 | 2045.7 |
| 16 | 201.06 | 1504.0 | 35.811 | 804.25 | 2144.7 |
| 16 1/4 | 207.39 | 1551.4 | 36.938 | 829.58 | 2246.8 |
| 16 1/2 | 213.82 | 1599.5 | 38.084 | 855.30 | 2353.1 |
| 16 3/4 | 220.35 | 1648.4 | 39.247 | 881.41 | 2460.6 |
| 17 | 226.98 | 1697.9 | 40.427 | 907.92 | 2572.4 |
| 17 1/4 | 233.71 | 1748.2 | 41.625 | 934.82 | 2687.6 |
| 17 1/2 | 240.53 | 1799.3 | 42.840 | 962.11 | 2806.2 |
| 17 3/4 | 247.45 | 1851.1 | 44.073 | 989.80 | 2928.2 |
| 18 | 254.47 | 1903.6 | 45.323 | 1017.9 | 3053.6 |
| 18 1/4 | 261.59 | 1956.8 | 46.591 | 1046.3 | 3182.6 |
| 18 1/2 | 268.80 | 2010.8 | 47.876 | 1075.2 | 3315.2 |
| 18 3/4 | 276.12 | 2065.5 | 49.178 | 1104.5 | 3451.5 |
| 19 | 282.53 | 2120.9 | 50.499 | 1134.1 | 3591.4 |
| 19 1/4 | 291.04 | 2177.1 | 51.836 | 1164.2 | 3735.0 |
| 19 1/2 | 298.65 | 2234.0 | 53.191 | 1194.6 | 3882.4 |
| 19 3/4 | 306.35 | 2291.7 | 54.564 | 1225.4 | 4033.7 |
| 20 | 314.16 | 2350.1 | 55.954 | 1256.6 | 4188.8 |
| 20 1/4 | 322.06 | 2409.2 | 57.362 | 1288.2 | 4347.8 |
| 20 1/2 | 330.06 | 2469.0 | 58.787 | 1320.3 | 4510.9 |

## APPENDIX 4 *continued*

| Diam. in. Feet | Cu. Ft. per Foot of Cylinder | Gallons per Foot of Cylinder | 42 Gallon Barrels per Foot of Cylinder | Sphere Surface in Sq. Ft. | Sphere Volume in Cu. Ft |
|---|---|---|---|---|---|
| 20 3/4 | 338.16 | 2529.6 | 60.229 | 1352.7 | 4677.9 |
| 21 | 346.36 | 2591.0 | 61.689 | 1385.4 | 4849.0 |
| 21 1/4 | 354.66 | 2653.0 | 63.167 | 1418.6 | 5024.3 |
| 21 1/2 | 363.05 | 2715.8 | 64.662 | 1452.2 | 5203.7 |
| 21 3/4 | 371.54 | 2779.3 | 66.175 | 1486.2 | 5387.4 |
| 22 | 380.13 | 2843.6 | 67.705 | 1520.5 | 5575.3 |
| 22 1/4 | 388.82 | 2908.6 | 69.252 | 1555.3 | 5767.5 |
| 22 1/2 | 397.61 | 2974.3 | 70.817 | 1590.4 | 5964.1 |
| 22 3/4 | 406.49 | 3040.8 | 72.399 | 1626.0 | 6165.1 |
| 23 | 415.48 | 3108.0 | 73.999 | 1661.9 | 6370.6 |
| 23 1/4 | 424.56 | 3175.9 | 75.617 | 1698.2 | 6580.6 |
| 23 1/2 | 433.74 | 3244.6 | 77.252 | 1734.9 | 6795.2 |
| 23 3/4 | 443.01 | 3314.0 | 78.904 | 1772.1 | 7014.4 |
| 24 | 452.39 | 3384.1 | 80.574 | 1809.6 | 7238.2 |
| 24 1/4 | 461.86 | 3455.0 | 82.261 | 1847.5 | 7466.8 |
| 24 1/2 | 471.44 | 3526.6 | 83.966 | 1885.7 | 7700.1 |
| 24 3/4 | 481.11 | 3598.9 | 85.689 | 1924.4 | 7938.2 |
| 25 | 490.87 | 3672.0 | 87.428 | 1963.5 | 8181.2 |
| 25 1/4 | 500.74 | 3745.8 | 89.186 | 2003.0 | 8429.1 |
| 25 1/2 | 510.71 | 3820.3 | 90.960 | 2042.8 | 8682.0 |
| 25 3/4 | 520.77 | 3895.6 | 92.753 | 2083.1 | 8939.9 |
| 26 | 530.93 | 3971.6 | 94.563 | 2123.7 | 9202.8 |
| 26 1/4 | 541.19 | 4048.4 | 96.390 | 2164.8 | 9470.8 |
| 26 1/2 | 551.55 | 4125.8 | 98.235 | 2206.2 | 9744.0 |

| Diam. in. Feet | Cu. Ft. per Foot of Cylinder | Gallons per Foot of Cylinder | 42 Callon Barrels per Foot of Cylinder | Sphere Surface in Sq. Ft. | Sphere Volume in Cu. Ft |
|---|---|---|---|---|---|
| 26 3/4 | 562.00 | 4204.1 | 100.10 | 2248.0 | 10022 |
| 27 | 572.56 | 4283.0 | 101.98 | 2290.2 | 10306 |
| 27 1/4 | 583.21 | 4362.7 | 103.87 | 2332.8 | 10595 |
| 27 1/2 | 593.96 | 4443.1 | 105.79 | 2375.8 | 10889 |
| 27 3/4 | 604.81 | 4524.3 | 107.72 | 2419.2 | 11189 |
| 28 | 615.75 | 4606.1 | 109.67 | 2463.0 | 11494 |
| 28 1/4 | 626.80 | 4688.8 | 111.64 | 2507.2 | 11805 |
| 28 1/2 | 637.94 | 4772.1 | 113.62 | 2551.8 | 12121 |
| 28 3/4 | 649.18 | 4856.2 | 115.62 | 2596.7 | 12443 |
| 29 | 660.52 | 4941.0 | 117.64 | 2642.1 | 12770 |
| 29 1/4 | 671.96 | 5026.6 | 119.68 | 2687.8 | 13103 |
| 29 1/2 | 683.49 | 5112.9 | 121.74 | 2734.0 | 13442 |
| 29 3/4 | 695.13 | 5199.9 | 123.81 | 2780.5 | 13787 |
| 30 | 706.86 | 5287.7 | 125.90 | 2827.4 | 14137 |
| 30 1/4 | 718.69 | 5376.2 | 128.00 | 2874.8 | 14494 |
| 30 1/2 | 730.62 | 5465.4 | 130.13 | 2922.5 | 14856 |
| 30 3/4 | 742.64 | 5555.4 | 132.27 | 2970.6 | 15224 |
| 31 | 754.77 | 5646.1 | 134.43 | 3019.1 | 15599 |
| 31 1/3 | 766.99 | 5737.5 | 136.61 | 3068.0 | 15979 |
| 31 1/2 | 779.31 | 5829.7 | 138.80 | 3117.2 | 16366 |
| 31 3/4 | 791.73 | 5922.6 | 141.01 | 3166.9 | 16758 |
| 32 | 804.25 | 6016.2 | 143.24 | 3217.0 | 17157 |
| 32 1/4 | 816.86 | 6110.6 | 145.49 | 3267.5 | 17563 |
| 32 1/2 | 829.58 | 6205.7 | 147.75 | 3318.3 | 17974 |
| 32 3/4 | 842.39 | 6301.5 | 150.04 | 3369.6 | 18392 |

| | | | | | |
|---|---|---|---|---|---|
| 33 | 855.30 | 6398.1 | 152.34 | 3421.2 | 18817 |
| 33 1/4 | 868.31 | 6495.4 | 154.65 | 3473.2 | 19247 |
| 33 1/2 | 881.41 | 6593.4 | 156.99 | 3525.7 | 19685 |
| 33 3/4 | 894.62 | 6692.2 | 159.34 | 3578.5 | 20129 |
| 34 | 907.92 | 6791.7 | 161.71 | 3631.7 | 20580 |
| 34 1/4 | 921.32 | 6892.0 | 164.09 | 3685.3 | 21037 |
| 34 1/2 | 934.82 | 6992.9 | 166.50 | 3739.3 | 21501 |
| 34 3/4 | 948.42 | 7094.7 | 168.92 | 3793.7 | 21972 |
| 35 | 962.11 | 7197.1 | 171.36 | 3848.5 | 22449 |
| 35 1/4 | 975.91 | 7300.3 | 173.82 | 3903.6 | 22934 |
| 35 1/2 | 989.80 | 7404.2 | 176.29 | 3959.2 | 23425 |
| 35 3/4 | 1003.8 | 7508.9 | 178.78 | 4015.2 | 23924 |
| 36 | 1017.9 | 7614.2 | 181.29 | 4071.5 | 24429 |
| 36 1/4 | 1032.1 | 7720.4 | 183.82 | 4128.2 | 24942 |
| 36 1/2 | 1046.3 | 7827.2 | 186.36 | 4185.4 | 25461 |
| 36 3/4 | 1060.7 | 7934.8 | 188.92 | 4242.9 | 25988 |
| 37 | 1075.2 | 8043.1 | 191.50 | 4300.8 | 26522 |
| 37 1/4 | 1089.8 | 8152.2 | 194.10 | 4359.2 | 27063 |
| 37 1/2 | 1104.5 | 8262.0 | 196.71 | 4417.9 | 27612 |
| 37 3/4 | 1119.2 | 8372.5 | 199.35 | 4477.0 | 28168 |
| 38 | 1134.1 | 8483.8 | 201.99 | 4536.5 | 28731 |
| 38 1/4 | 1149.1 | 8585.8 | 204.65 | 4596.3 | 29302 |
| 38 1/2 | 1164.2 | 8708.5 | 207.35 | 4656.6 | 29880 |
| 38 3/4 | 1179.3 | 8822.0 | 210.05 | 4717.3 | 30466 |
| 39 | 1194.6 | 8936.2 | 212.77 | 4778.4 | 31059 |
| 39 1/4 | 1210.0 | 9051.1 | 215.50 | 4839.8 | 31660 |
| 39 1/2 | 1225.4 | 9166.8 | 218.26 | 4901.7 | 32269 |
| 39 3/4 | 1241.0 | 9283.2 | 221.03 | 4963.9 | 32886 |
| 40 | 1256.6 | 9400.3 | 223.82 | 5026.5 | 33510 |
| 40 1/4 | 1272.4 | 9518.2 | 226.62 | 5089.6 | 34143 |
| 40 1/2 | 1288.2 | 9636.8 | 229.45 | 5153.0 | 34783 |
| 40 3/4 | 1304.2 | 9756.1 | 232.29 | 5216.8 | 35431 |
| 41 | 1320.3 | 9876.2 | 235.15 | 5281.0 | 36087 |
| 41 1/4 | 1336.4 | 9997.0 | 238.02 | 5345.6 | 36751 |
| 41 1/2 | 1352.7 | 10119. | 240.92 | 5410.6 | 37423 |
| 41 3/4 | 1369.0 | 10241. | 243.83 | 5476.0 | 38104 |
| 42 | 1385.4 | 10364 | 246.76 | 5541.8 | 38792 |
| 42 1/4 | 1402.0 | 10488 | 249.70 | 5607.9 | 39489 |
| 42 1/2 | 1418.6 | 10612 | 252.67 | 5674.5 | 40194 |
| 42 3/4 | 1435.4 | 10737 | 255.65 | 5714.5 | 40908 |
| 43 | 1452.2 | 10863 | 258.65 | 5808.8 | 41630 |
| 43 1/4 | 1469.1 | 10990 | 261.66 | 5876.5 | 42360 |
| 43 1/2 | 1486.2 | 11117 | 264.70 | 5944.7 | 43099 |
| 43 3/4 | 1503.3 | 11245 | 267.75 | 6013.2 | 43846 |
| 44 | 1520.5 | 11374 | 270.82 | 6082.1 | 44602 |
| 44 1/4 | 1537.9 | 11504 | 273.90 | 6151.4 | 45367 |
| 44 1/2 | 1555.3 | 11634 | 277.01 | 6221.1 | 46140 |
| 44 3/4 | 1572.8 | 11765 | 280.13 | 6291.2 | 46922 |
| 45 | 1590.4 | 11897 | 283.27 | 6361.7 | 47713 |
| 45 1/4 | 1608.2 | 12030 | 286.42 | 6432.6 | 48513 |
| 45 1/2 | 1626.0 | 12163 | 289.60 | 6503.9 | 49321 |
| 45 3/4 | 1643.9 | 12297 | 292.79 | 6575.5 | 50139 |
| 46 | 1661.9 | 12432 | 296.00 | 6647.6 | 50965 |
| 46 1/4 | 1680.0 | 12567 | 299.22 | 6720.1 | 51800 |
| 46 1/2 | 1698.2 | 12704 | 302.47 | 6792.9 | 52645 |
| 46 3/4 | 1716.5 | 12841 | 305.73 | 6866.1 | 53499 |
| 47 | 1734.9 | 12978 | 309.01 | 6939.8 | 54362 |
| 47 1/4 | 1753.5 | 13117 | 312.30 | 7013.8 | 55234 |
| 47 1/2 | 1772.1 | 13256 | 315.62 | 7088.2 | 56115 |
| 47 3/4 | 1790.8 | 13396 | 318.95 | 7163.0 | 57006 |
| 48 | 1809.6 | 13536 | 322.30 | 7238.2 | 57906 |
| 48 1/4 | 1828.5 | 13678 | 325.66 | 7313.8 | 58815 |
| 48 1/2 | 1847.5 | 13820 | 329.05 | 7389.8 | 59734 |
| 48 3/4 | 1866.5 | 13963 | 332.45 | 7466.2 | 60663 |
| 49 | 1885.7 | 14106 | 335.86 | 7543.0 | 61601 |
| 49 1/4 | 1905.0 | 14251 | 339.30 | 7620.1 | 62549 |

## APPENDIX 4 *continued*

| Diam. in. Feet | Cu. Ft. per Foot of Cylinder | Gallons per Foot of Cylinder | 42 Gallon Barrels per Foot of Cylinder | Sphere Surface in Sq. Ft. | Sphere Volume in Cu. Ft |
|---|---|---|---|---|---|
| 49 1/2 | 1924.4 | 14396 | 342.75 | 7697.7 | 63506 |
| 49 3/4 | 1943.9 | 14541 | 346.23 | 7775.6 | 64473 |
| 50 | 1963.5 | 14688 | 349.71 | 7854.0 | 65450 |
| 50 1/4 | 1983.2 | 14835 | 353.22 | 7932.7 | 66437 |
| 50 1/2 | 2003.0 | 14983 | 356.74 | 8011.8 | 67433 |
| 50 3/4 | 2022.8 | 15132 | 360.28 | 8091.4 | 68439 |
| 51 | 2042.8 | 15281 | 363.84 | 8171.3 | 69456 |
| 51 1/4 | 2062.9 | 15432 | 367.42 | 8251.6 | 70482 |
| 51 1/2 | 2083.1 | 15582 | 371.01 | 8332.3 | 71519 |
| 51 3/4 | 2103.3 | 15734 | 374.62 | 8413.4 | 72565 |
| 52 | 2123.7 | 15887 | 378.25 | 8494.9 | 73622 |
| 52 1/4 | 2144.2 | 16040 | 381.90 | 8576.7 | 74689 |
| 52 1/2 | 2164.8 | 16193 | 385.56 | 8659.0 | 75766 |
| 52 3/4 | 2185.4 | 16348 | 389.24 | 8741.7 | 76854 |
| 53 | 2206.2 | 16503 | 392.94 | 8824.7 | 77952 |
| 53 1/4 | 2227.0 | 16659 | 396.65 | 8908.2 | 79060 |
| 53 1/2 | 2248.0 | 16816 | 400.39 | 8992.0 | 80179 |
| 53 3/4 | 2269.1 | 16974 | 404.14 | 9076.3 | 81308 |
| 54 | 2290.2 | 17132 | 407.91 | 9160.9 | 82448 |
| 54 1/4 | 2311.5 | 17291 | 411.59 | 9245.9 | 83598 |
| 54 1/2 | 2332.8 | 17451 | 415.49 | 9331.3 | 84759 |
| 54 3/4 | 2354.3 | 17611 | 419.32 | 9417.1 | 85931 |
| 55 | 2375.8 | 17772 | 423.15 | 9503.3 | 87114 |
| 55 1/4 | 2397.5 | 17934 | 427.01 | 9589.9 | 88307 |
| 55 1/2 | 2419.2 | 18097 | 430.88 | 9676.9 | 89511 |
| 55 3/4 | 2441.1 | 18260 | 434.77 | 9764.3 | 90726 |
| 56 | 2463.0 | 18245 | 438.68 | 9852.0 | 91952 |
| 56 1/4 | 2485.0 | 18589 | 442.61 | 9940.2 | 93189 |
| 56 1/2 | 2507.2 | 18755 | 446.55 | 10029 | 94437 |
| 56 3/4 | 2529.4 | 18921 | 450.51 | 10118 | 95697 |
| 57 | 2551.8 | 19088 | 454.49 | 10207 | 96967 |
| 57 1/4 | 2574.2 | 19256 | 458.48 | 10297 | 98248 |
| 57 1/2 | 2596.7 | 19425 | 462.50 | 10387 | 99541 |
| 57 3/4 | 2619.4 | 19594 | 466.53 | 10477 | 100845 |
| 58 | 2642.1 | 19764 | 470.57 | 10568 | 102160 |
| 58 1/4 | 2664.9 | 19935 | 474.64 | 10660 | 103487 |
| 58 1/2 | 2687.8 | 20106 | 478.72 | 10751 | 104825 |
| 58 3/4 | 2710.9 | 20279 | 482.82 | 10843 | 106175 |
| 59 | 2734.0 | 20452 | 486.94 | 10936 | 107536 |
| 59 1/4 | 2757.2 | 20625 | 491.08 | 11029 | 108909 |
| 59 1/2 | 2780.5 | 20800 | 495.23 | 11122 | 110293 |
| 59 3/4 | 2803.9 | 20975 | 499.40 | 11216 | 111690 |
| 60 | 2827.4 | 21151 | 503.59 | 11310 | 113097 |
| 60 1/4 | 2851.0 | 21327 | 507.79 | 11404 | 114517 |
| 60 1/2 | 2874.8 | 21505 | 512.02 | 11499 | 115948 |
| 60 3/4 | 2898.6 | 21683 | 516.26 | 11594 | 117392 |
| 61 | 2922.5 | 21862 | 520.51 | 11690 | 118847 |
| 61 1/4 | 2946.5 | 22041 | 524.79 | 11786 | 120314 |
| 61 1/2 | 2970.6 | 22221 | 529.08 | 11882 | 121793 |
| 61 3/4 | 2994.8 | 22402 | 533.39 | 11979 | 123285 |

| | | | | | |
|---|---|---|---|---|---|
| 62 | 3019.1 | 22584 | 537.72 | 12076 | 124788 |
| 62¼ | 3043.4 | 22767 | 542.06 | 12174 | 126304 |
| 62½ | 3068.0 | 22950 | 546.43 | 12272 | 127832 |
| 62¾ | 3092.6 | 23134 | 550.81 | 12370 | 129372 |
| 63 | 3117.2 | 23319 | 555.21 | 12469 | 130924 |
| 63¼ | 3142.0 | 23504 | 559.62 | 12568 | 132489 |
| 63½ | 3166.9 | 23690 | 564.05 | 12668 | 134066 |
| 63¾ | 3191.9 | 23877 | 568.50 | 12768 | 135656 |
| 64 | 3217.0 | 24065 | 572.97 | 12868 | 137258 |
| 64¼ | 3242.2 | 24253 | 577.46 | 12969 | 138873 |
| 64½ | 3267.5 | 24442 | 581.96 | 13070 | 140500 |
| 64¾ | 3292.8 | 24632 | 586.48 | 13171 | 142141 |
| 65 | 3318.3 | 24823 | 591.02 | 13273 | 143793 |
| 65¼ | 3343.9 | 25014 | 595.57 | 13376 | 145459 |
| 65½ | 3369.6 | 25206 | 600.14 | 13478 | 147137 |
| 65¾ | 3395.3 | 25399 | 604.73 | 13581 | 148828 |
| 66 | 3421.2 | 25592 | 609.34 | 13685 | 150533 |
| 66¼ | 3447.2 | 25787 | 613.97 | 13789 | 152250 |
| 66½ | 3473.2 | 25982 | 618.61 | 13893 | 153980 |
| 66¾ | 3499.4 | 26177 | 623.27 | 13998 | 155723 |
| 67 | 3525.7 | 26374 | 627.95 | 14103 | 157479 |
| 67¼ | 3552.0 | 26571 | 632.64 | 14208 | 159249 |
| 67¼ | 3578.5 | 26769 | 637.35 | 14314 | 161031 |
| 67¾ | 3605.0 | 26967 | 642.08 | 14420 | 162827 |
| 68 | 3631.7 | 27167 | 646.83 | 14527 | 164636 |
| 68¼ | 3658.4 | 27367 | 651.59 | 14634 | 166459 |
| 68½ | 3685.3 | 27568 | 656.38 | 14741 | 168295 |
| 68¾ | 3712.2 | 27769 | 661.18 | 14849 | 170144 |
| 69 | 3739.3 | 27972 | 665.99 | 14957 | 172007 |
| 69¼ | 3766.4 | 28175 | 670.83 | 15066 | 173883 |
| 69½ | 3793.7 | 28379 | 675.68 | 15175 | 175773 |
| 69¾ | 3821.0 | 28583 | 680.55 | 15284 | 177677 |
| 70 | 3848.5 | 28788 | 685.44 | 15394 | 179594 |
| 70¼ | 3876.0 | 28994 | 690.34 | 15504 | 181525 |
| 70½ | 3903.6 | 29201 | 695.27 | 15615 | 183470 |
| 70¾ | 3931.4 | 29409 | 700.21 | 15725 | 185429 |
| 71 | 3959.2 | 29617 | 705.16 | 15837 | 187402 |
| 71¼ | 3987.1 | 29826 | 710.14 | 15948 | 189388 |
| 71½ | 4015.2 | 30035 | 715.13 | 16061 | 191389 |
| 71¾ | 4043.3 | 30246 | 720.14 | 16173 | 193404 |
| 72 | 4071.5 | 30457 | 725.17 | 16286 | 195432 |
| 72¼ | 4099.8 | 30669 | 730.21 | 16399 | 197475 |
| 72½ | 4128.2 | 30881 | 735.27 | 16513 | 199532 |
| 72¾ | 4156.8 | 31095 | 740.35 | 16627 | 201603 |
| 73 | 4185.4 | 31309 | 745.45 | 16742 | 203689 |
| 73¼ | 4214.1 | 31524 | 750.56 | 16856 | 205789 |
| 73½ | 4242.9 | 31739 | 755.70 | 16972 | 207903 |
| 73¾ | 4271.8 | 31956 | 760.85 | 17087 | 210032 |
| 74 | 4300.8 | 32173 | 766.01 | 17203 | 212175 |
| 74¼ | 4329.9 | 32390 | 771.20 | 17320 | 214332 |
| 74½ | 4359.2 | 32609 | 776.40 | 17437 | 216505 |
| 74¾ | 4388.5 | 32828 | 781.62 | 17554 | 218692 |
| 75 | 4417.9 | 33048 | 786.86 | 17671 | 220893 |
| 75½ | 4477.0 | 33490 | 797.38 | 17908 | 225341 |
| 75¾ | 4506.7 | 33712 | 802.67 | 18027 | 227587 |
| 76 | 4536.5 | 33935 | 807.98 | 18146 | 229847 |
| 76¼ | 4566.4 | 34159 | 813.30 | 18265 | 232123 |
| 76½ | 4596.3 | 34383 | 818.64 | 18385 | 234414 |
| 76¾ | 4626.4 | 34608 | 824.00 | 18506 | 236719 |
| 77 | 4656.6 | 34834 | 829.38 | 18627 | 239040 |
| 77¼ | 4686.9 | 35061 | 834.77 | 18748 | 241376 |
| 77½ | 4717.3 | 35288 | 840.19 | 18869 | 243727 |
| 77¾ | 4747.8 | 35516 | 845.62 | 18991 | 246093 |
| 78 | 4778.4 | 35745 | 851.06 | 19113 | 248475 |
| 78¼ | 4809.0 | 35974 | 856.53 | 19236 | 250872 |
| 78½ | 4839.8 | 36204 | 862.01 | 19359 | 253284 |

## APPENDIX 4 *continued*

| Diam. in. Feet | Cu. Ft. per Foot of Cylinder | Gallons per Foot of Cylinder | 42 Gallon Barrels per Foot of Cylinder | Sphere Surface in Sq. Ft. | Sphere Volume in Cu. Ft |
|---|---|---|---|---|---|
| $78^3/_4$ | 4870.7 | 36435 | 867.51 | 19483 | 255712 |
| 79 | 4901.7 | 36667 | 873.02 | 19607 | 258155 |
| $79^1/_4$ | 4932.7 | 36899 | 878.56 | 19731 | 260613 |
| $79^1/_2$ | 4963.9 | 37133 | 884.11 | 19856 | 263087 |
| $79^3/_4$ | 4995.2 | 37367 | 889.68 | 19981 | 265577 |
| 80 | 5026.5 | 37601 | 895.27 | 20106 | 268083 |
| $80^1/_4$ | 5058.0 | 37837 | 900.87 | 200232 | 270604 |
| $80^1/_2$ | 5089.6 | 38073 | 906.49 | 20358 | 273141 |
| $80^3/_4$ | 5121.2 | 38310 | 912.13 | 20485 | 275693 |
| 81 | 5153.0 | 38547 | 917.79 | 20612 | 278262 |
| $81^1/_4$ | 5184.9 | 38785 | 923.46 | 20739 | 280846 |
| $81^1/_2$ | 5216.8 | 39024 | 929.15 | 20867 | 283447 |
| $81^3/_4$ | 528.9 | 39264 | 934.86 | 20995 | 286063 |
| 82 | 5281.0 | 39505 | 940.59 | 21124 | 288696 |
| $82^1/_4$ | 5313.3 | 39746 | 946.33 | 21253 | 291344 |
| $82^1/_2$ | 5345.6 | 39988 | 952.09 | 21382 | 294009 |
| $82^3/_4$ | 5378.1 | 40231 | 957.87 | 21512 | 296690 |
| 83 | 5410.6 | 40474 | 963.67 | 21642 | 299387 |
| $83^1/_4$ | 5443.3 | 40718 | 969.48 | 21773 | 302100 |
| $83^1/_2$ | 5476.0 | 40963 | 975.32 | 21904 | 304830 |
| $83^3/_4$ | 5508.8 | 41209 | 981.16 | 22035 | 307576 |
| 84 | 5541.8 | 41455 | 987.03 | 22167 | 310339 |
| $84^1/_4$ | 5574.8 | 41702 | 992.92 | 22299 | 313118 |
| $84^1/_2$ | 5607.9 | 41950 | 998.82 | 22432 | 315914 |
| $84^3/_4$ | 5641.2 | 42199 | 1004.7 | 22565 | 318726 |

| Diam. in. Feet | Cu. Ft. per Foot of Cylinder | Gallons per Foot of Cylinder | 42 Gallon Barrels per Foot of Cylinder | Sphere Surface in Sq. Ft. | Sphere Volume in Cu. Ft |
|---|---|---|---|---|---|
| 85 | 5674.5 | 42448 | 1010.7 | 22698 | 321555 |
| $85^1/_4$ | 5707.9 | 42698 | 1016.6 | 22832 | 324401 |
| $85^1/_2$ | 5741.5 | 42949 | 1022.6 | 22966 | 327263 |
| $85^3/_4$ | 5775.1 | 43201 | 1028.6 | 23100 | 330142 |
| 86 | 5808.8 | 43453 | 1034.6 | 23235 | 333038 |
| $86^1/_4$ | 5842.6 | 43706 | 1040.6 | 23371 | 335951 |
| $86^1/_2$ | 5876.5 | 43960 | 1046.7 | 23506 | 338881 |
| $86^3/_4$ | 5910.6 | 44214 | 1052.7 | 23642 | 341828 |
| 87 | 5944.7 | 44469 | 1058.8 | 23779 | 344791 |
| $87^1/_4$ | 5978.9 | 44725 | 1064.9 | 23916 | 347772 |
| $87^1/_2$ | 6013.2 | 44982 | 1071.0 | 24053 | 350770 |
| $87^3/_4$ | 6047.6 | 45239 | 1077.1 | 24190 | 353785 |
| 88 | 6082.1 | 45497 | 1083.3 | 24328 | 356818 |
| $88^1/_4$ | 6116.7 | 45756 | 1089.4 | 24467 | 359868 |
| $88^1/_2$ | 6151.4 | 46016 | 1095.6 | 24606 | 362935 |
| $88^3/_4$ | 6186.2 | 46276 | 1101.8 | 24745 | 366019 |
| 89 | 6221.1 | 46537 | 1108.0 | 24885 | 369121 |
| $89^1/_4$ | 6256.1 | 46799 | 1114.3 | 25025 | 372240 |
| $89^1/_2$ | 6291.2 | 47062 | 1120.5 | 25165 | 375377 |
| $89^3/_4$ | 6326.4 | 47325 | 1126.8 | 25306 | 378531 |
| 90 | 6361.7 | 47589 | 1133.1 | 25447 | 381704 |
| $90^1/_4$ | 6397.1 | 47854 | 1139.4 | 25588 | 384893 |
| $90^1/_2$ | 6432.6 | 48119 | 1145.7 | 25730 | 388101 |
| $90^3/_4$ | 6468.2 | 48385 | 1152.0 | 25873 | 391326 |
| 91 | 6503.9 | 48652 | 1158.4 | 26016 | 394569 |

| | | | | | |
|---|---|---|---|---|---|
| 91¼ | 6539.7 | 48920 | 1164.8 | 26159 | 397830 |
| 91½ | 6575.5 | 49189 | 1171.2 | 26302 | 401109 |
| 91¾ | 6611.5 | 49458 | 1177.6 | 26446 | 404405 |
| 92 | 6647.6 | 49728 | 1184.0 | 26590 | 407720 |
| 92¼ | 6683.8 | 49998 | 1190.4 | 26735 | 411053 |
| 92½ | 6720.1 | 50270 | 1196.9 | 26880 | 414404 |
| 92¾ | 6756.4 | 50542 | 1203.4 | 27026 | 417773 |
| 93 | 6792.9 | 50814 | 1209.9 | 27172 | 421160 |
| 93¼ | 6829.5 | 51088 | 1216.4 | 27318 | 424566 |
| 93½ | 6866.1 | 51362 | 1222.9 | 27465 | 427990 |
| 93¾ | 6902.9 | 51637 | 1229.5 | 27612 | 431432 |
| 94 | 6939.8 | 51913 | 1236.0 | 27759 | 434893 |
| 94¼ | 6976.7 | 52190 | 1242.6 | 27907 | 438372 |
| 94½ | 7013.8 | 52467 | 1249.2 | 28055 | 441870 |
| 94¾ | 7051.0 | 52745 | 1255.8 | 28204 | 445386 |
| 95 | 7088.2 | 53024 | 1262.5 | 28353 | 448920 |
| 95¼ | 7125.6 | 53303 | 1269.1 | 28502 | 452474 |
| 95½ | 7163.0 | 53583 | 1275.8 | 28652 | 456046 |
| 95¾ | 7200.6 | 53864 | 1282.5 | 28802 | 459637 |
| 96 | 7238.2 | 54146 | 1289.2 | 28953 | 463247 |
| 96¼ | 7276.0 | 54428 | 1295.9 | 29104 | 466875 |
| 96½ | 7313.8 | 54711 | 1302.6 | 29255 | 470523 |
| 96¾ | 7351.8 | 54995 | 1309.4 | 29407 | 474189 |
| 97 | 7389.8 | 55280 | 1316.2 | 29559 | 477874 |
| 97¼ | 7428.0 | 55565 | 1323.0 | 29712 | 481579 |
| 97½ | 7466.2 | 55851 | 1329.8 | 29865 | 485302 |
| 97¾ | 7504.5 | 56138 | 1336.6 | 30018 | 489045 |
| 98 | 7543.0 | 56425 | 1343.5 | 30172 | 492807 |
| 98¼ | 7581.5 | 56714 | 1350.3 | 30326 | 496588 |
| 98½ | 7620.1 | 57003 | 1357.2 | 30481 | 500388 |
| 98¾ | 7658.9 | 57292 | 1364.1 | 30635 | 504208 |
| 99 | 7697.7 | 57583 | 1371.0 | 30791 | 508047 |
| 99¼ | 7736.6 | 57874 | 1377.9 | 30946 | 511906 |
| 99½ | 7775.6 | 58166 | 1384.9 | 31103 | 515784 |
| 99¾ | 7814.8 | 58458 | 1391.9 | 31259 | 519682 |
| 100 | 7854.0 | 58752 | 1398.9 | 31416 | 523599 |
| 100¼ | 7893.3 | 59046 | 1405.9 | 31573 | 527536 |
| 100½ | 7932.7 | 59341 | 1412.9 | 31731 | 531492 |
| 100¾ | 7972.2 | 59636 | 1419.9 | 31889 | 535468 |
| 101 | 8011.8 | 59933 | 1427.0 | 32047 | 539464 |
| 101¼ | 8051.6 | 60230 | 1434.0 | 32206 | 543480 |
| 101½ | 8091.4 | 60528 | 1441.1 | 32365 | 547516 |
| 101¾ | 8131.3 | 60826 | 1448.2 | 32525 | 551572 |
| 102 | 8171.3 | 61125 | 1455.4 | 32685 | 555647 |
| 102¼ | 8211.4 | 61425 | 1462.5 | 32846 | 559743 |
| 102½ | 8251.6 | 61726 | 1469.7 | 33006 | 563859 |
| 102¾ | 8291.9 | 62028 | 1476.8 | 33168 | 567994 |
| 103 | 8332.3 | 62330 | 1484.0 | 33329 | 572151 |
| 103¼ | 8372.8 | 62633 | 1491.3 | 33491 | 576327 |
| 103½ | 8413.4 | 62936 | 1498.5 | 33654 | 580523 |
| 103¾ | 8454.1 | 63241 | 1505.7 | 33816 | 584740 |
| 104 | 8494.9 | 63546 | 1513.0 | 33979 | 588977 |
| 104¼ | 8535.8 | 63852 | 1520.3 | 34143 | 593235 |
| 104½ | 8576.7 | 64159 | 1527.6 | 34307 | 597513 |
| 104¾ | 8617.8 | 64466 | 1534.9 | 34471 | 601812 |
| 105 | 8659.0 | 64774 | 1542.2 | 34536 | 606131 |
| 105¼ | 8700.3 | 65083 | 1549.6 | 34801 | 610471 |
| 105½ | 8741.7 | 65392 | 1557.0 | 34967 | 614831 |
| 105¾ | 8783.2 | 65703 | 1564.3 | 35133 | 619213 |
| 106 | 8824.7 | 66014 | 1571.8 | 35299 | 623615 |
| 106¼ | 8866.4 | 66325 | 1579.2 | 35466 | 628037 |
| 106½ | 8908.2 | 66638 | 1586.6 | 35633 | 6322481 |
| 106¾ | 8950.1 | 66951 | 1594.1 | 35800 | 636945 |
| 107 | 8992.0 | 67265 | 1601.5 | 35968 | 641431 |
| 107¼ | 9034.1 | 67580 | 1609.0 | 36136 | 645938 |
| 107½ | 9076.3 | 67895 | 1616.6 | 36305 | 650465 |

## APPENDIX 4 *continued*

| Diam. in. Feet | Cu. Ft. per Foot of Cylinder | Gallons per Foot of Cylinder | 42 Gallon Barrels per Foot of Cylinder | Sphere Surface in Sq. Ft. | Sphere Volume in Cu. Ft |
|---|---|---|---|---|---|
| $107\frac{3}{4}$ | 9118.5 | 68211 | 1624.1 | 36474 | 655014 |
| 108 | 9160.9 | 68528 | 1631.6 | 36644 | 659584 |
| $108\frac{1}{4}$ | 9203.3 | 68846 | 1639.2 | 36813 | 664175 |
| $108\frac{1}{2}$ | 9245.9 | 69164 | 1646.8 | 36984 | 668787 |
| $108\frac{3}{4}$ | 9288.6 | 69483 | 1654.5 | 37154 | 673421 |
| 109 | 9331.3 | 69803 | 1662.0 | 37325 | 678076 |
| $109\frac{1}{4}$ | 9374.2 | 70124 | 1669.6 | 37497 | 682752 |
| $109\frac{1}{2}$ | 9417.1 | 70445 | 1677.3 | 37668 | 687450 |
| $109\frac{3}{4}$ | 9460.2 | 70767 | 1684.9 | 37841 | 692169 |
| 110 | 9503.3 | 71090 | 1692.6 | 38013 | 696910 |
| $110\frac{1}{4}$ | 9546.6 | 71413 | 1700.3 | 38186 | 701672 |
| $110\frac{1}{2}$ | 9589.9 | 71737 | 1708.0 | 38360 | 706457 |
| $110\frac{3}{4}$ | 9633.4 | 72062 | 1715.8 | 38533 | 711262 |
| 111 | 9676.9 | 72388 | 1723.5 | 38708 | 716090 |
| $111\frac{1}{4}$ | 9720.5 | 72715 | 1731.3 | 38882 | 720939 |
| $111\frac{1}{2}$ | 9764.3 | 73042 | 1739.1 | 39057 | 725810 |
| $111\frac{3}{4}$ | 9808.1 | 73370 | 1746.9 | 39232 | 730704 |
| 112 | 9852.0 | 73698 | 1754.7 | 39408 | 735619 |
| $112\frac{1}{4}$ | 9896.1 | 74028 | 1762.6 | 39584 | 740556 |
| $112\frac{1}{2}$ | 9940.2 | 74358 | 1770.4 | 39761 | 745515 |
| $112\frac{3}{4}$ | 9984.4 | 74689 | 1778.3 | 39938 | 750496 |
| 113 | 10029 | 75020 | 1786.2 | 40115 | 755499 |
| $113\frac{1}{4}$ | 10073 | 75353 | 1794.1 | 40293 | 760525 |
| $113\frac{1}{2}$ | 10118 | 75686 | 1802.0 | 40471 | 765572 |
| $113\frac{3}{4}$ | 10162 | 76019 | 1810.0 | 40649 | 770642 |

| Diam. in. Feet | Cu. Ft. per Foot of Cylinder | Gallons per Foot of Cylinder | 42 Gallon Barrels per Foot of Cylinder | Sphere Surface in Sq. Ft. | Sphere Volume in Cu. Ft |
|---|---|---|---|---|---|
| 114 | 10207 | 76354 | 1818.0 | 40828 | 775735 |
| $114\frac{1}{4}$ | 10252 | 76689 | 1825.9 | 41007 | 780849 |
| $114\frac{1}{2}$ | 10297 | 77025 | 1833.9 | 41187 | 785986 |
| $114\frac{3}{4}$ | 10342 | 77362 | 1841.9 | 41367 | 791146 |
| 115 | 10387 | 77699 | 1850.0 | 41548 | 796328 |
| $115\frac{1}{4}$ | 10432 | 78038 | 1858.0 | 41728 | 801533 |
| $115\frac{1}{2}$ | 10477 | 78376 | 1866.1 | 41910 | 806760 |
| $115\frac{3}{4}$ | 10523 | 78716 | 1874.2 | 42091 | 812010 |
| 116 | 10568 | 79057 | 1882.3 | 42273 | 817283 |
| $116\frac{1}{4}$ | 10614 | 79398 | 1890.4 | 42456 | 822579 |
| $116\frac{1}{2}$ | 10660 | 79739 | 1898.6 | 42638 | 827897 |
| $116\frac{3}{4}$ | 10705 | 80082 | 1906.7 | 42822 | 833238 |
| 117 | 10751 | 80425 | 1914.9 | 43005 | 838603 |
| $117\frac{1}{4}$ | 10797 | 80769 | 1923.1 | 43189 | 843990 |
| $117\frac{1}{2}$ | 10843 | 81114 | 1931.3 | 43374 | 849400 |
| $117\frac{3}{4}$ | 10890 | 81460 | 1939.5 | 43558 | 854833 |
| 118 | 10936 | 81806 | 1947.8 | 43744 | 860290 |
| $118\frac{1}{4}$ | 10982 | 82153 | 1956.0 | 43929 | 865769 |
| $118\frac{1}{2}$ | 11029 | 82501 | 1964.3 | 44115 | 871272 |
| $118\frac{3}{4}$ | 11075 | 82849 | 1972.6 | 44301 | 876798 |
| 119 | 11122 | 83199 | 1980.9 | 44488 | 882347 |
| $119\frac{1}{4}$ | 11169 | 83548 | 1989.2 | 44675 | 887920 |
| $119\frac{1}{2}$ | 11216 | 83899 | 1997.6 | 44863 | 893516 |
| $119\frac{3}{4}$ | 11263 | 84251 | 2006.0 | 45051 | 899136 |
| 120 | 11310 | 84603 | 2014.3 | 45239 | 904779 |

| | | | | | |
|---|---|---|---|---|---|
| 120¼ | 11357 | 84956 | 2022.8 | 45428 | 910445 |
| 120½ | 11404 | 85309 | 2031.2 | 45617 | 916136 |
| 120¾ | 11452 | 85664 | 2039.6 | 45806 | 921850 |
| 121 | 11499 | 86019 | 2048.1 | 45996 | 927587 |
| 121¼ | 11547 | 86374 | 2056.5 | 46186 | 933349 |
| 121½ | 11594 | 86731 | 2065.0 | 46377 | 939134 |
| 121¾ | 11642 | 87088 | 2073.5 | 46568 | 944943 |
| 122 | 11690 | 87446 | 2082.1 | 46759 | 950776 |
| 122¼ | 11738 | 87805 | 2090.6 | 46951 | 966633 |
| 122½ | 11786 | 88165 | 2099.2 | 47144 | 962514 |
| 122¾ | 11834 | 88525 | 2107.7 | 47336 | 968419 |
| 123 | 11882 | 88886 | 2116.3 | 47529 | 974348 |
| 123¼ | 11931 | 89247 | 2124.9 | 47723 | 980301 |
| 123½ | 11979 | 89610 | 2133.6 | 47916 | 986278 |
| 123¾ | 12028 | 89973 | 2142.2 | 48111 | 992280 |
| 124 | 12076 | 90337 | 2150.9 | 48305 | 998306 |
| 124¼ | 12125 | 90701 | 2159.6 | 48500 | 1004356 |
| 124½ | 12174 | 91067 | 2168.3 | 48695 | 1010431 |
| 124¾ | 12223 | 91433 | 2177.0 | 48891 | 1016530 |
| 125 | 12272 | 91800 | 2185.7 | 49087 | 1022654 |
| 125¼ | 12321 | 92167 | 2194.5 | 49284 | 1028802 |
| 125½ | 12370 | 92536 | 2203.2 | 49481 | 1034975 |
| 125¾ | 12420 | 92905 | 2212.0 | 49678 | 1041172 |
| 126 | 12469 | 93274 | 2220.8 | 49876 | 1047394 |
| 126¼ | 12519 | 93645 | 2229.6 | 50074 | 1053641 |
| 126½ | 12568 | 94016 | 2238.5 | 50273 | 1059913 |
| 126¾ | 12618 | 94388 | 2247.3 | 50471 | 1066209 |
| 127 | 12668 | 94761 | 2256.2 | 50671 | 1072531 |
| 127¼ | 12718 | 95134 | 2265.1 | 50870 | 1078877 |
| 127½ | 12768 | 95508 | 2274.0 | 51071 | 1085248 |
| 127¾ | 12818 | 95883 | 2282.9 | 51271 | 1091654 |
| 128 | 12868 | 96259 | 2291.9 | 51472 | 1098066 |
| 128¼ | 12918 | 96635 | 2300.8 | 51673 | 1104513 |
| 128½ | 12969 | 97013 | 2309.8 | 51875 | 1110985 |
| 128¾ | 13019 | 97390 | 2318.8 | 52077 | 1117481 |
| 129 | 13070 | 97769 | 2327.8 | 52279 | 1124004 |
| 129¼ | 13121 | 98148 | 2336.9 | 52482 | 1130551 |
| 129½ | 13171 | 98528 | 2345.9 | 52685 | 1137124 |
| 129¾ | 13222 | 98909 | 2355.0 | 52889 | 1143723 |
| 130 | 13273 | 99291 | 2364.1 | 53093 | 1150347 |
| 130¼ | 13324 | 99673 | 2373.2 | 53297 | 1156996 |
| 130½ | 13376 | 100056 | 2382.3 | 53502 | 1163671 |
| 130¾ | 13427 | 100440 | 2391.4 | 53707 | 1170371 |
| 131 | 13478 | 100824 | 2400.6 | 53913 | 1177098 |
| 131¼ | 13530 | 101209 | 2409.7 | 54119 | 1183850 |
| $131^{2}/_{2}$ | 13581 | 101595 | 2418.9 | 54325 | 1190627 |
| 131¾ | 13633 | 101982 | 2428.1 | 54532 | 1197431 |
| 132 | 13685 | 102369 | 2437.4 | 54739 | 1204260 |
| 132¼ | 13737 | 102757 | 2446.6 | 54947 | 1211116 |
| 132½ | 13789 | 103146 | 2455.9 | 55155 | 1217997 |
| 132¾ | 13841 | 103536 | 2465.1 | 55363 | 1224904 |
| 133 | 13983 | 103926 | 2474.4 | 55572 | 1231838 |
| 133¼ | 13945 | 104317 | 2483.7 | 55781 | 1238797 |
| 133½ | 13998 | 104709 | 2493.1 | 55990 | 1245783 |
| 133¾ | 14050 | 105102 | 2502.4 | 56200 | 1252795 |
| 134 | 14103 | 105495 | 2511.8 | 56410 | 1259833 |
| 134¼ | 14155 | 105889 | 2521.2 | 56621 | 1266898 |
| 134½ | 14208 | 106284 | 2530.6 | 56832 | 1273988 |
| 134¾ | 14261 | 106679 | 2540.0 | 57044 | 1281106 |
| 135 | 14314 | 107075 | 2549.4 | 57256 | 1288249 |
| 135¼ | 14367 | 107472 | 2558.9 | 57468 | 1295420 |
| 135½ | 14420 | 107870 | 2568.3 | 57680 | 1302616 |
| 135¾ | 14473 | 108268 | 2577.8 | 57893 | 1309840 |
| 136 | 14527 | 108667 | 2587.3 | 58107 | 1317090 |
| 136¼ | 14580 | 109067 | 2596.8 | 59321 | 1324366 |
| 136½ | 14634 | 109468 | 2606.4 | 58535 | 1331670 |

## APPENDIX 4 *continued*

| Diam. in. Feet | Cu. Ft. per Foot of Cylinder | Gallons per Foot of Cylinder | 42 Gallon Barrels per Foot of Cylinder | Sphere Surface in Sq. Ft. | Sphere Volume in Cu. Ft |
|---|---|---|---|---|---|
| 136¾ | 14687 | 109869 | 2615.9 | 58750 | 1339000 |
| 137 | 14741 | 110271 | 2625.5 | 58965 | 1346357 |
| 137¼ | 14795 | 110674 | 2635.1 | 59180 | 1353741 |
| 137½ | 14849 | 111078 | 2644.7 | 59396 | 1361152 |
| 137¾ | 14903 | 111482 | 2654.3 | 59612 | 1368590 |
| 138 | 14957 | 111887 | 2664.0 | 59828 | 1376055 |
| 138¼ | 15011 | 112293 | 2673.6 | 60045 | 1383547 |
| 138½ | 15066 | 112699 | 2683.3 | 60263 | 1391067 |
| 138¾ | 15120 | 113107 | 2693.0 | 60481 | 1398613 |
| 139 | 15175 | 113514 | 2702.7 | 60699 | 1406187 |
| 139¼ | 15229 | 113923 | 2712.5 | 60917 | 1413788 |
| 139½ | 15284 | 114333 | 2722.2 | 61136 | 1421416 |
| 139¾ | 15339 | 114743 | 2732.0 | 61356 | 1429072 |
| 140 | 15394 | 115154 | 2741.8 | 61575 | 1436755 |
| 140¼ | 15449 | 115565 | 2751.6 | 61795 | 1444466 |
| 140½ | 15504 | 115978 | 2761.4 | 62016 | 1452204 |
| 140¾ | 15559 | 116391 | 2771.2 | 62237 | 1459970 |
| 141 | 15615 | 116805 | 2781 1 | 62458 | 1467763 |
| 141¼ | 15670 | 117219 | 2790.9 | 62680 | 1475584 |
| 141½ | 15725 | 117634 | 2800.8 | 62902 | 1483433 |
| 141¾ | 15781 | 118050 | 2810.7 | 63124 | 1491310 |
| 142 | 15837 | 118467 | 2820.6 | 63347 | 1499214 |
| 142¼ | 15893 | 118885 | 2830.6 | 63570 | 1507146 |
| 142½ | 15948 | 119303 | 2840.5 | 63794 | 1515107 |
| 142¾ | 16005 | 119722 | 2850.5 | 64018 | 1523095 |
| 143 | 16061 | 120142 | 2860.5 | 64242 | 1531111 |

| Diam. in. Feet | Cu. Ft. per Foot of Cylinder | Gallons per Foot of Cylinder | 42 Gallon Barrels per Foot of Cylinder | Sphere Surface in Sq. Ft. | Sphere Volume in Cu. Ft |
|---|---|---|---|---|---|
| 143½ | 16173 | 120983 | 2880.6 | 64692 | 1547228 |
| 143¾ | 16230 | 121405 | 2890.6 | 64918 | 1555329 |
| 144 | 16286 | 121828 | 2900.7 | 65144 | 1563458 |
| 144¼ | 16343 | 122251 | 2910.7 | 65370 | 1571615 |
| 144½ | 16399 | 122675 | 2920.8 | 65597 | 1579800 |
| 144¾ | 16456 | 123100 | 2931.0 | 65824 | 1588014 |
| 145 | 16513 | 123526 | 2941.1 | 66052 | 1596256 |
| 145¼ | 16570 | 123952 | 2951.2 | 66280 | 1604527 |
| 145½ | 16627 | 124379 | 2961.4 | 66508 | 1612826 |
| 145¾ | 16684 | 124807 | 2971.6 | 66737 | 1621154 |
| 146 | 16742 | 125235 | 2981.8 | 66966 | 1629511 |
| 146½ | 16799 | 125665 | 2992.0 | 67196 | 1637896 |
| 146½ | 16856 | 126095 | 3002.3 | 67426 | 1646310 |
| 146¾ | 16914 | 126525 | 3012.5 | 67656 | 1654752 |
| 147 | 16972 | 126957 | 3022.8 | 67887 | 1663224 |
| 147¼ | 17029 | 127389 | 3033.1 | 68118 | 1671724 |
| 147½ | 17087 | 127822 | 3043.4 | 68349 | 1680253 |
| 147¾ | 17145 | 128256 | 3053.7 | 68581 | 1688811 |
| 148 | 17203 | 128690 | 3064.0 | 68813 | 1697398 |
| 148¼ | 17262 | 129125 | 3074.4 | 69046 | 1706015 |
| 148½ | 17320 | 129561 | 3084.8 | 69279 | 1714660 |
| 148¾ | 17378 | 129998 | 3095.2 | 69513 | 1723334 |
| 149 | 17437 | 130435 | 3105.6 | 69746 | 1732038 |
| 149¼ | 17495 | 130873 | 3116.0 | 69981 | 1740771 |
| 149½ | 17554 | 131312 | 3126.5 | 70215 | 1749533 |
| 149¾ | 17613 | 131571 | 3136.9 | 70450 | 1758325 |

# APPENDIX 5

## Weights of rings and discs

*(Courtesy Taylor Forge Group, Gulf & Western Mfg. Co.)*

The table below provides a convenient means for determining the weights of rectangular section rings and discs in diameters from 1/2″ up to and including 150″ by size increments of 1/8″.

The column at the left gives the "even-inch" diameters and columns across to the right give weights of that diameter and of added fractional diameters.

Computed weights are based on steel weighing .283 lbs per cu. in. Thus, for other materials, weights may be readily calculated merely by application of the proper factor.

For determining rectangular ring weights, proceed as in the following example:

Ring size 96″ O.D. x 80″ I.D. x 6″ thickness.

| | |
|---|---|
| Weight of 96″ disc | 2051 lbs. |
| Weight of 80″ disc | 1424 lbs. |
| Ring weight per 1″ thickness | 627 lbs. |
| Times 6″ thickness | 6 |
| Weight of ring | 3762 lbs. |

| Diameter Inches | WEIGHTS OF STEEL DISCS PER INCH OF THICKNESS | | | | | | | |
|---|---|---|---|---|---|---|---|---|
| | Added Fractional Increments 0″ | 1/8″ | 1/4″ | 3/8″ | 1/2″ | 5/8″ | 3/4″ | 7/8″ |
| 0 | - - | - - | - - | - - | .05 | .09 | .13 | .17 |
| 1 | .22 | .28 | .35 | .42 | .50 | .59 | .68 | .78 |
| 2 | .89 | 1.01 | 1.13 | 1.26 | 1.39 | 1.54 | 1.69 | 1.84 |

| Diameter Inches | WEIGHTS OF STEEL DISCS PER INCH OF THICKNESS | | | | | | | |
|---|---|---|---|---|---|---|---|---|
| | Added Fractional Increments | | | | | | | |
| | 0″ | 1/8″ | 1/4″ | 3/8″ | 1/2″ | 5/8″ | 3/4″ | 7/8″ |
| 3 | 2.01 | 2.17 | 2.35 | 2.53 | 2.72 | 2.92 | 3.13 | 3.34 |
| 4 | 3.56 | 3.78 | 4.01 | 4.25 | 4.50 | 4.75 | 5.02 | 5.28 |
| 5 | 5.56 | 5.84 | 6.13 | 6.42 | 6.72 | 7.03 | 7.35 | 7.67 |
| 6 | 8.00 | 8.35 | 8.69 | 9.04 | 9.40 | 9.77 | 10.14 | 10.52 |
| 7 | 10.90 | 11.30 | 11.70 | 12.10 | 12.52 | 12.94 | 13.36 | 13.80 |
| 8 | 14.24 | 14.69 | 15.14 | 15.61 | 16.08 | 16.55 | 17.04 | 17.53 |
| 9 | 18.02 | 18.53 | 19.04 | 19.56 | 20.08 | 20.61 | 21.15 | 21.70 |
| 10 | 22.25 | 22.81 | 23.38 | 23.95 | 24.53 | 25.12 | 25.71 | 26.32 |
| 11 | 26.92 | 27.54 | 28.16 | 28.79 | 29.43 | 30.07 | 30.72 | 31.38 |
| 12 | 32.04 | 32.71 | 33.39 | 34.08 | 34.77 | 35.47 | 36.17 | 36.88 |
| 13 | 37.60 | 38.33 | 39.06 | 39.80 | 40.55 | 41.31 | 42.07 | 42.84 |
| 14 | 43.62 | 44.39 | 45.18 | 45.98 | 46.78 | 47.59 | 48.41 | 49.23 |
| 15 | 50.06 | 50.90 | 51.75 | 52.60 | 53.46 | 54.32 | 55.20 | 56.08 |
| 16 | 56.96 | 57.86 | 58.76 | 59.66 | 60.58 | 61.50 | 62.43 | 63.36 |
| 17 | 64.30 | 65.25 | 66.21 | 67.17 | 68.14 | 69.12 | 70.10 | 71.09 |
| 18 | 72.09 | 73.10 | 74.11 | 75.13 | 76.15 | 77.19 | 78.22 | 79.27 |
| 19 | 80.32 | 81.39 | 82.45 | 83.53 | 84.61 | 85.70 | 86.79 | 87.89 |
| 20 | 89.00 | 90.12 | 91.24 | 92.37 | 93.51 | 94.65 | 95.80 | 96.96 |
| 21 | 98.13 | 99.30 | 100.48 | 101.66 | 102.85 | 104.05 | 105.26 | 106.47 |
| 22 | 107.69 | 108.92 | 110.15 | 111.40 | 112.64 | 113.90 | 115.16 | 116.43 |
| 23 | 117.71 | 118.99 | 120.28 | 121.58 | 122.88 | 124.19 | 125.51 | 126.83 |
| 24 | 128.16 | 129.50 | 130.85 | 132.20 | 133.57 | 134.93 | 136.30 | 137.68 |
| 25 | 139.07 | 140.46 | 141.86 | 143.27 | 144.68 | 146.11 | 147.54 | 148.97 |
| 26 | 150.41 | 151.86 | 153.32 | 154.78 | 156.25 | 157.73 | 159.22 | 160.71 |

| | | | | | | | | |
|---|---|---|---|---|---|---|---|---|
| **27** | 162.21 | 163.71 | 165.22 | 166.74 | 168.27 | 169.80 | 171.34 | 172.89 |
| **28** | 174.44 | 176.01 | 177.57 | 179.15 | 180.73 | 182.32 | 183.91 | 185.52 |
| **29** | 187.13 | 188.74 | 190.37 | 192.00 | 193.64 | 195.28 | 196.93 | 198.59 |
| **30** | 200.25 | 201.93 | 203.61 | 205.29 | 206.99 | 208.69 | 210.39 | 212.11 |
| **31** | 213.83 | 215.56 | 217.29 | 219.03 | 220.78 | 222.54 | 224.30 | 226.07 |
| **32** | 227.85 | 229.63 | 231.42 | 233.22 | 235.02 | 236.83 | 238.65 | 240.48 |
| **33** | 242.31 | 244.15 | 245.99 | 247.85 | 249.71 | 251.57 | 253.45 | 255.33 |
| **34** | 257.22 | 259.11 | 261.01 | 262.92 | 264.84 | 266.76 | 268.69 | 270.63 |
| **35** | 272.57 | 274.52 | 276.48 | 278.44 | 280.41 | 282.39 | 284.38 | 286.37 |
| **36** | 288.37 | 290.38 | 292.39 | 294.41 | 296.42 | 298.46 | 300.50 | 302.55 |
| **37** | 304.60 | 306.67 | 308.74 | 310.81 | 312.90 | 314.97 | 317.07 | 319.19 |
| **38** | 321.29 | 323.42 | 325.54 | 327.66 | 329.82 | 331.94 | 334.10 | 336.25 |
| **39** | 338.4 | 340.6 | 342.8 | 345.0 | 347.2 | 349.4 | 351.6 | 353.8 |
| **40** | 356.0 | 358.2 | 360.5 | 362.7 | 365.0 | 367.2 | 369.5 | 371.7 |
| **41** | 374.0 | 376.3 | 378.6 | 380.9 | 383.2 | 385.5 | 387.8 | 390.2 |
| **42** | 392.5 | 394.8 | 397.2 | 399.5 | 401.9 | 404.3 | 406.6 | 409.0 |
| **43** | 411.4 | 413.8 | 416.2 | 418.6 | 421.0 | 423.2 | 425.9 | 428.3 |
| **44** | 430.8 | 433.2 | 435.7 | 438.1 | 440.6 | 443.1 | 445.6 | 448.1 |
| **45** | 450.6 | 453.1 | 455.6 | 458.1 | 460.6 | 463.2 | 465.7 | 468.3 |
| **46** | 470.8 | 473.4 | 475.9 | 478.5 | 481.1 | 483.7 | 486.3 | 488.9 |
| **47** | 491.5 | 494.1 | 496.8 | 499.4 | 502.0 | 504.7 | 507.3 | 510.0 |
| **48** | 512.7 | 515.3 | 518.0 | 520.7 | 523.4 | 526.1 | 528.8 | 531.5 |
| **49** | 534.2 | 537.0 | 539.7 | 542.4 | 545.2 | 548.0 | 550.7 | 553.5 |
| **50** | 556.3 | 559.0 | 561.8 | 564.7 | 567.4 | 570.2 | 573.1 | 575.9 |
| **51** | 578.7 | 581.6 | 584.4 | 587.3 | 590.1 | 593.0 | 595.9 | 598.7 |
| **52** | 601.6 | 604.5 | 607.4 | 610.4 | 613.3 | 616.2 | 619.1 | 622.1 |
| **53** | 625.0 | 628.0 | 630.9 | 633.9 | 636.9 | 639.8 | 642.8 | 645.8 |

| Diameter Inches | WEIGHTS OF STEEL DISCS PER INCH OF THICKNESS | | | | | | | |
|---|---|---|---|---|---|---|---|---|
| | Added Fractional Increments 0" | 1/8" | 1/4" | 3/8" | 1/2" | 5/8" | 3/4" | 7/8" |
| 54 | 648.8 | 651.8 | 654.8 | 657.8 | 660.9 | 663.9 | 667.0 | 670.0 |
| 55 | 673.1 | 676.1 | 679.2 | 682.3 | 685.4 | 688.4 | 691.6 | 694.6 |
| 56 | 697.8 | 700.9 | 704.0 | 707.1 | 710.3 | 713.4 | 716.6 | 719.7 |
| 57 | 722.9 | 726.1 | 729.3 | 732.4 | 735.6 | 738.8 | 742.1 | 745.3 |
| 58 | 748.5 | 751.7 | 755.0 | 758.2 | 761.4 | 764.7 | 768.0 | 771.3 |
| 59 | 774.5 | 777.8 | 781.1 | 784.4 | 787.7 | 791.0 | 794.3 | 797.7 |
| 60 | 801.0 | 804.3 | 807.7 | 811.1 | 814.4 | 817.8 | 821.2 | 824.5 |
| 61 | 827.9 | 831.3 | 834.7 | 838.1 | 841.6 | 845.0 | 848.4 | 851.8 |
| 62 | 855.3 | 858.8 | 862.2 | 865.7 | 869.2 | 872.6 | 876.1 | 879.6 |
| 63 | 883.1 | 886.6 | 890.1 | 893.7 | 897.2 | 900.7 | 904.3 | 907.8 |
| 64 | 911.4 | 914.9 | 918.5 | 922.1 | 925.7 | 929.2 | 932.8 | 936.5 |
| 65 | 940.1 | 943.7 | 947.3 | 950.9 | 954.6 | 958.2 | 961.9 | 965.5 |
| 66 | 969.2 | 972.9 | 976.6 | 980.3 | 984.0 | 987.7 | 991.4 | 995.1 |
| 67 | 998.8 | 1002.5 | 1006.3 | 1010.0 | 1013.8 | 1017.5 | 1021.3 | 1025.1 |
| 68 | 1028.9 | 1032.6 | 1036.4 | 1040.2 | 1044.0 | 1047.8 | 1051.7 | 1055.5 |
| 69 | 1059.3 | 1063.2 | 1067.0 | 1070.9 | 1074.7 | 1078.6 | 1082.5 | 1086.4 |
| 70 | 1090.3 | 1094.2 | 1098.1 | 1102.0 | 1105.9 | 1109.8 | 1113.8 | 1117.7 |
| 71 | 1121.6 | 1125.6 | 1129.5 | 1133.5 | 1137.5 | 1141.5 | 1145.5 | 1149.5 |
| 72 | 1153.5 | 1157.5 | 1161.5 | 1165.5 | 1169.5 | 1173.6 | 1177.6 | 1181.7 |
| 73 | 1185.7 | 1189.8 | 1193.8 | 1197.9 | 1202.0 | 1206.1 | 1210.2 | 1214.3 |
| 74 | 1218.4 | 1222.5 | 1226.7 | 1230.8 | 1235.0 | 1239.1 | 1243.3 | 1247.4 |
| 75 | 1251.6 | 1255.8 | 1259.9 | 1264.1 | 1268.3 | 1272.5 | 1276.7 | 1280.9 |
| 76 | 1285.2 | 1289.4 | 1293.7 | 1297.9 | 1302.1 | 1306.4 | 1310.7 | 1314.9 |
| 77 | 1319.2 | 1323.5 | 1327.8 | 1332.1 | 1336.4 | 1340.7 | 1345.0 | 1349.4 |

| | | | | | | | | |
|---|---|---|---|---|---|---|---|---|
| **78** | 1353.7 | 1358.1 | 1362.4 | 1366.7 | 1371.1 | 1375.5 | 1379.9 | 1384.3 |
| **79** | 1388.6 | 1393.0 | 1397.4 | 1401.8 | 1406.3 | 1410.7 | 1415.1 | 1419.6 |
| **80** | 1424.0 | 1428.5 | 1432.9 | 1437.4 | 1441.9 | 1446.4 | 1450.8 | 1455.3 |
| **81** | 1459.8 | 1464.3 | 1468.9 | 1473.4 | 1477.9 | 1482.4 | 1487.0 | 1491.5 |
| **82** | 1496.1 | 1500.7 | 1505.3 | 1509.8 | 1514.4 | 1519.0 | 1523.6 | 1528.2 |
| **83** | 1532.8 | 1537.4 | 1542.1 | 1546.7 | 1551.3 | 1556.0 | 1560.6 | 1565.3 |
| **84** | 1570.0 | 1574.7 | 1579.3 | 1584.0 | 1588.7 | 1593.4 | 1598.1 | 1602.8 |
| **85** | 1607.6 | 1612.3 | 1617.0 | 1621.8 | 1626.6 | 1631.3 | 1636.1 | 1640.8 |
| **86** | 1645.6 | 1650.4 | 1655.2 | 1660.0 | 1664.8 | 1669.6 | 1674.5 | 1679.3 |
| **87** | 1684.1 | 1689.0 | 1693.8 | 1698.7 | 1703.5 | 1708.4 | 1713.3 | 1718.2 |
| **88** | 1723.1 | 1728.0 | 1732.9 | 1737.8 | 1742.7 | 1747.6 | 1752.5 | 1757.5 |
| **89** | 1762.4 | 1767.4 | 1772.3 | 1777.3 | 1782.3 | 1787.3 | 1792.3 | 1797.3 |
| **90** | 1802.3 | 1807.3 | 1812.3 | 1817.3 | 1822.4 | 1827.4 | 1832.4 | 1837.5 |
| **91** | 1842.5 | 1847.6 | 1852.7 | 1857.8 | 1862.8 | 1867.9 | 1873.0 | 1878.2 |
| **92** | 1883.3 | 1888.4 | 1893.5 | 1898.6 | 1903.8 | 1908.9 | 1914.1 | 1919.3 |
| **93** | 1924 | 1930 | 1935 | 1940 | 1945 | 1950 | 1956 | 1961 |
| **94** | 1966 | 1971 | 1976 | 1982 | 1987 | 1992 | 1998 | 2003 |
| **95** | 2008 | 2013 | 2019 | 2024 | 2029 | 2035 | 2040 | 2045 |
| **96** | 2051 | 2056 | 2061 | 2067 | 2072 | 2077 | 2083 | 2088 |
| **97** | 2094 | 2099 | 2104 | 2110 | 2115 | 2121 | 2126 | 2131 |
| **98** | 2137 | 2142 | 2148 | 2153 | 2159 | 2164 | 2170 | 2175 |
| **99** | 2181 | 2186 | 2192 | 2197 | 2203 | 2208 | 2214 | 2219 |
| **100** | 2225 | 2231 | 2236 | 2242 | 2248 | 2253 | 2259 | 2264 |
| **101** | 2270 | 2276 | 2281 | 2287 | 2293 | 2298 | 2304 | 2309 |
| **102** | 2315 | 2321 | 2327 | 2332 | 2338 | 2344 | 2349 | 2355 |
| **103** | 2361 | 2367 | 2372 | 2378 | 2384 | 2390 | 2395 | 2401 |
| **104** | 2407 | 2413 | 2418 | 2424 | 2430 | 2436 | 2442 | 2448 |
| **105** | 2453 | 2459 | 2465 | 2471 | 2477 | 2483 | 2489 | 2494 |
| **106** | 2500 | 2506 | 2512 | 2518 | 2524 | 2530 | 2536 | 2542 |
| **107** | 2548 | 2554 | 2560 | 2566 | 2572 | 2578 | 2584 | 2590 |

## WEIGHTS OF STEEL DISCS PER INCH OF THICKNESS

| Diameter Inches | Added Fractional Increments 0" | 1/8" | 1/4" | 3/8" | 1/2" | 5/8" | 3/4" | 7/8" |
|---|---|---|---|---|---|---|---|---|
| 108 | 2596 | 2602 | 2608 | 2614 | 2620 | 2626 | 2632 | 2638 |
| 109 | 2644 | 2650 | 2656 | 2662 | 2668 | 2674 | 2680 | 2686 |
| 110 | 2692 | 2698 | 2704 | 2711 | 2717 | 2723 | 2729 | 2735 |
| 111 | 2741 | 2748 | 2757 | 2760 | 2766 | 2772 | 2779 | 2785 |
| 112 | 2791 | 2797 | 2804 | 2810 | 2816 | 2822 | 2829 | 2835 |
| 113 | 2841 | 2847 | 2854 | 2860 | 2866 | 2873 | 2879 | 2885 |
| 114 | 2892 | 2898 | 2904 | 2911 | 2917 | 2923 | 2930 | 2936 |
| 115 | 2943 | 2949 | 2955 | 2962 | 2968 | 2975 | 2981 | 2988 |
| 116 | 2994 | 3000 | 3007 | 3013 | 3020 | 3026 | 3033 | 3039 |
| 117 | 3046 | 3052 | 3059 | 3065 | 3072 | 3078 | 3085 | 3092 |
| 118 | 3098 | 3105 | 3111 | 3118 | 3124 | 3131 | 3138 | 3144 |
| 119 | 3151 | 3158 | 3164 | 3171 | 3177 | 3184 | 3191 | 3197 |
| 120 | 3204 | 3211 | 3217 | 3226 | 3231 | 3238 | 3244 | 3251 |
| 121 | 3258 | 3264 | 3271 | 3278 | 3285 | 3291 | 3298 | 3305 |
| 122 | 3312 | 3319 | 3325 | 3332 | 3339 | 3346 | 3353 | 3359 |
| 123 | 3366 | 3373 | 3380 | 3387 | 3394 | 3401 | 3407 | 3414 |
| 124 | 3421 | 3428 | 3435 | 3442 | 3449 | 3456 | 3463 | 3470 |
| 125 | 3477 | 3484 | 3491 | 3498 | 3505 | 3511 | 3518 | 3525 |
| 126 | 3532 | 3539 | 3546 | 3554 | 3561 | 3568 | 3575 | 3582 |
| 127 | 3589 | 3596 | 3603 | 3610 | 3617 | 3624 | 3631 | 3638 |
| 128 | 3646 | 3653 | 3660 | 3667 | 3674 | 3681 | 3688 | 3696 |
| 129 | 3703 | 3710 | 3717 | 3724 | 3731 | 3739 | 3746 | 3753 |
| 130 | 3760 | 3768 | 3775 | 3782 | 3789 | 3797 | 3804 | 3811 |
| 131 | 3818 | 3826 | 3833 | 3840 | 3848 | 3855 | 3862 | 3870 |

| | | | | | | | | |
|---|---|---|---|---|---|---|---|---|
| 132 | 3877 | 3884 | 3892 | 3899 | 3906 | 3914 | 3921 | 3928 |
| 133 | 3936 | 3943 | 3951 | 3958 | 3966 | 3973 | 3980 | 3988 |
| 134 | 3995 | 4003 | 4010 | 4018 | 4025 | 4033 | 4040 | 4048 |
| 135 | 4055 | 4063 | 4070 | 4078 | 4085 | 4093 | 4100 | 4108 |
| 136 | 4115 | 4123 | 4131 | 4138 | 4146 | 4153 | 4161 | 4169 |
| 137 | 4176 | 4184 | 4191 | 4199 | 4207 | 4214 | 4222 | 4230 |
| 138 | 4237 | 4245 | 4253 | 4260 | 4268 | 4276 | 4283 | 4291 |
| 139 | 4299 | 4307 | 4314 | 4322 | 4330 | 4338 | 4345 | 4353 |
| 140 | 4361 | 4369 | 4377 | 4384 | 4392 | 4400 | 4408 | 4416 |
| 141 | 4424 | 4431 | 4439 | 4447 | 4455 | 4463 | 4471 | 4479 |
| 142 | 4487 | 4494 | 4502 | 4510 | 4518 | 4526 | 4534 | 4542 |
| 143 | 4550 | 4558 | 4566 | 4574 | 4582 | 4590 | 4598 | 4606 |
| 144 | 4614 | 4622 | 4630 | 4638 | 4646 | 4654 | 4662 | 4670 |
| 145 | 4678 | 4686 | 4694 | 4702 | 4710 | 4719 | 4727 | 4735 |
| 146 | 4743 | 4751 | 4759 | 4767 | 4775 | 4784 | 4792 | 4800 |
| 147 | 4808 | 4816 | 4824 | 4833 | 4841 | 4849 | 4857 | 4865 |
| 148 | 4874 | 4882 | 4890 | 4898 | 4907 | 4915 | 4923 | 4932 |
| 149 | 4940 | 4948 | 4956 | 4965 | 4973 | 4981 | 4990 | 4998 |
| 150 | 5006 | 5015 | 5023 | 5031 | 5040 | 5048 | 5057 | 5065 |

# APPENDIX 6

# Specifications and standards for dimensional guidance

*(Reproduced by permission of American Society of Mechanical Engineers)*

NOTES (referring to ANSI B31.1):
(1) ASME SA, SB, and SFA Specifications shall be used for boiler external piping as defined in Para. 100.1.2(A).
(2) For all other piping, materials conforming to an ASME SA or SB Specification may be used interchangeably with material specified to an ASTM A or B Specification of the same number listed in Table 126.1.
(3) The approved year of issue of the Specifications and Standards is not given in this Table. This information is given in Appendix F of this Code.

ASTM Ferrous Material Specifications

Pipe—Seamless for High Temp. Service

| | |
|---|---|
| A 106 | Carbon Steel |
| A 335 | Ferritic Alloy Steel |
| A 376 | Austenitic Central Station Service |
| A 405 | Ferritic Alloy Special Heat Treated |

Pipe—Seamless & Welded

| | |
|---|---|
| A 53 | Carbon Steel |
| A 120 | Black & Zinc Coated (Ordinary Use) |
| A 312 | Austenitic Stainless Steel |
| A 333 | Carbon Steel (Low Temp. Service) |
| A 530 | General Requirements for Specialized Carbon & Alloy Steel Pipe |

Pipe—Welded

| | |
|---|---|
| A 134 | Arc-Welded Steel Plate 16 in. & over |
| A 135 | Electric-Resistance Welded Steel |
| A 139 | Arc-Welded Steel 4 in. & over |
| A 155 | Arc-Welded Steel for H.T. Service |
| A 211 | Spiral Welded Steel or Iron |
| A 358 | Arc-Welded Cr-Ni Alloy H.T. Service |

Pipe—Forged & Bored H.T. Service

| | |
|---|---|
| A 369 | Ferritic Alloy Steel |
| A 430 | Austenitic Stainless Steel |

Pipe—Centrifugally Cast H.T. Service

A 426 Ferritic Alloys
A 451 Austenitic Alloys
A 452 Austenitic Cold-Wrought Alloys

Tube—Seamless

A 179 Low Carbon Steel for Heat Exchanger & Condensers
A 192 Carbon Steel for Boilers at H.T. Service
A 199 Cold Drawn Intermediate Alloy-Steel
A 210 Medium-Carbon Steel for Boilers
A 213 Ferritic & Austenitic Alloys for Boilers

Tube—Seamless & Welded

A 268 Ferritic Stainless Steel for General Use
A 450 General Requirements for Carbon, Ferritic, & Austenitic Alloy

Tube—Welded

A 178 Arc-welded Carbon Steel Boiler Tubes
A 214 Electric Resistance Welded Carbon Steel
A 226 Carbon Steel High Pressure Service
A 249 Austenitic Stainless Steel
A 254 Copper Brazed Steel

Plates for Pressure Vessels

A 240 Stainless, Chromium and Chrome-Nickel
A 285 Low & Intermediates T.S. Carbon Steel

Plates for Pressure Vessels (Cont'd)

A 299 Carbon-Manganese-Silicon Steel
A 387 Chromium-Molybdenum Alloy Steel
A 515 Carbon Steel for Intermediate & H.T. Service
A 516 Carbon Steel for Moderate & L.T. Service

Welding Fittings—Factory Made

A 234 Wrought Carbon & Ferritic Alloy Steel
A 403 Austenitic Steel
A 420 Piping Fittings of Wrought Carbon Steel and Alloy Steel for L.T. Service

Forgings

A 105 Carbon Steel for H.T. Service
A 181 Carbon Steel for General Service
A 182 Alloy Steel for H.T. Service
A 350 Forgings, Carbon and Low Alloy Steel Requiring Toughness Testing for Piping Components

Bolts, Nuts, & Studs

A 193 Alloy & Stainless Steel Bolts for H.T. Service
A 194 Carbon & Alloy Steel Nuts for H.T. Service
A 307 Low Carbon Steel Threaded Fasteners
A 320 Alloy Steel Bolting Materials for L.T. Service
A 354 Quenched & Tempered Alloy Steel Bolts & Studs
A 437 Alloy Steel Turbine-Type Material for H.T. Service
A 453 Material with Exp. Coeff. Comparable to Austenitic Steel

**APPENDIX 6** *continued*

Bolts, Nuts, & Studs (Cont'd)

| | |
|---|---|
| A 564 | Hot Rolled & Cold Finished Stainless & Heat Resisting |

Castings

| | |
|---|---|
| A 47 | Malleable Iron |
| A 48 | Gray Iron |
| A 126 | Gray Iron for Valves, Flanges & Pipe Fittings |
| A 197 | Cupola Malleable Iron |
| A 216 | Carbon Steel for Welding & H.T. Service |
| A 217 | Alloy Steel for Pressure & H.T. Parts |
| A 278 | Gray Iron for Pressure Parts for Temp. Up to 650° |
| A 351 | Ferritic & Austenitic Stee H.T. Service |
| A 377 | Cast and Ductile Iron Pressure Pipe |
| A 389 | Ferritic Alloy Special Heat Treated |
| A 395 | Ductile Iron for Pressure Parts |

Structural Components

| | |
|---|---|
| A 36 | Carbon Steel for General Purposes |
| A 125 | Heat Treated Steel Helical Springs |

Structural Components (Cont'd)

| | |
|---|---|
| A 229 | Steel Wire for Mechanical Springs |
| A 242 | High Strength Low Alloy Steel |
| A 276 | Stainless & Heat Resisting Bars & Shapes |
| A 283 | Low & Intermediate T.S. Carbon Steel Plates |
| A 322 | Hot Rolled Alloy Steel Bars |
| A 479 | Stainless & Heat Resisting Steel Bars & Shapes for Use in Boilers and Other Pressure Vessels |
| A 575 | Merchant Quality Hot Rolled Carbon Steel Bars |
| A 576 | Special Quality Hot Rolled Carbon Steel Bars |

ASTM Standard Test Methods

| | |
|---|---|
| E 109 | Magnetic Particle Inspection |
| E 125 | Reference Photographs—Magnetic Particle Inspections—Ferrous Castings |
| E 142 | Controlling Quality of Radiographic Testing |
| E 165 | Liquid Penetrant Inspection |
| E 186 | Reference Radiographs—Steel Castings 2 to 4 1/2 in. |
| E 280 | Reference Radiographs—Steel Castings 4 1/2 to 12 in. |
| E 466 | Std. Reference Radiographs—Steel Castings Up to 2 in. in Thickness |

ASTM Nonferrous Material Specifications

Pipe—Seamless

| | |
|---|---|
| B 42 | Copper (Standard Sizes) |
| B 43 | Red Brass (Standard Sizes) |
| B 302 | Threadless Copper |

Pipe and Tube—Seamless & Welded

| | |
|---|---|
| B 161 | Nickel |

Pipe and Tube—Seamless & Welded (Cont'd)

| | |
|---|---|
| B 165 | Nickel-Copper Alloy |
| B 167 | Nickel-Chromium-Iron Alloy |
| B 241 | Aluminum Alloy Extruded |
| B 315 | Copper Silicon Alloy |
| B 423 | Nickel-Iron-Chromium-Molybdenum-Copper Alloy |
| B 466 | Copper-Nickel |
| B 547 | Aluminum-Alloy Formed & Arc Welded |

Pipe and Tube—Seamless & Welded (Cont'd)

| | |
|---|---|
| B 467 | Copper-Nickel |
| B 464 | Chromium-Nickel-Iron-Molybdenum-Copper-Columbium Alloy |
| B 468 | Chromium-Nickel-Iron-Molybdenum-Copper-Columbium Alloy |

Tube—Copper Seamless

| | |
|---|---|
| B 68 | Bright Annealed |
| B 75 | For Gen. Engr. Purpose |
| B 88 | For Gen. Plumbing Purpose |
| B 251 | General Requirements Wrought Copper & Copper Alloy |
| B 280 | For Air Conditioning & Refrigeration Field Service |

Tube—Aluminum Seamless

| | |
|---|---|
| B 210 | Drawn |
| B 234 | Drawn for Condensers & Heat Exchangers |

Welding Fittings—Factory

| | |
|---|---|
| B 361 | Wrought Aluminum & Aluminum Alloy |

Bars, Rods, Shapes, & Tubes

| | |
|---|---|
| B 150 | Aluminum Bronze |
| B 221 | Aluminum Alloy Extruded |

Bars, Rods, Shapes, & Tubes (Cont'd)

| | |
|---|---|
| B 425 | Nickel-Iron-Chromium-Molybdenum Copper Alloy |
| B 462 | Chromium-Nickel-Iron-Molybdenum-Copper-Columbium Alloy |
| B 473 | Chromium-Nickel-Iron-Molybdenum-Copper-Columbium Alloy |

Castings

| | |
|---|---|
| B 26 | Aluminum Alloy Sand |
| B 61 | Steam or Valve Bronze |
| B 62 | Composition Bronze or Ounce Metal |
| B 108 | Aluminum Bronze Sand |
| B 148 | Aluminum Alloy Permanent Mold |

Forgings

| | |
|---|---|
| B 247 | Aluminum Alloy Die & Hand Forgings |

Plates

| | |
|---|---|
| B 168 | Nickel-Chromium-Iron Alloy |
| B 209 | Aluminum Alloy |
| B 424 | Nickel-Iron Chromium-Molybdenum-Copper Alloy |
| B 402 | Copper-Nickel for Pressure Vessels |
| B 463 | Chromium-Nickel-Iron-Molybdenum-Copper-Columbium Alloy |

Solder

| | |
|---|---|
| B 32 | Tin-Lead and Silver-Lead Alloys |

**APPENDIX 6** *continued*

ASTM Nonmetallic Material Specifications

Asbestos Cement

C 296 Asbestos Cement For Pressure Pipe

ANSI Standards

| | |
|---|---|
| A21.1 | Computation—Strength & Thickness—C.I. Pipe |
| A21.6 | Centrifugally C.I. Pipe in Metal Molds for Water or Other Liquids |
| A21.8 | Centrifugally C.I. Pipe in Sand Lined Molds for Water or Other Liquids |
| A21.10 | Gray and Ductile Iron Fittings |
| A21.11 | Rubber Gasket Joint for C.I. Pipe and Fittings |
| A21.12 | 2 inch and $2^1/_4$ inch C.I. Pipe Centrifugally Cast for Water or Other Liquids |
| A21.50 | Thickness Design of Ductile Iron |
| A21.51 | Ductile-Iron Pipe, Centrifugally Cast in Metal Molds or Sand-Lined Molds for Water or Other Liquids |
| B1.1 | Unified Screw Threads |
| B2.1 | Pipe Threads (Except Dryseal) |
| B2.2 | Dryseal Pipe Threads |
| B16.1 | C.I. Pipe Flanges and Flanged Fittings—25, 125, 250 & 800 lbs. |
| B16.3 | M.I. Thd. Fittings—150 & 300 lbs. |
| B16.4 | C.I. Thd. Fittings—125 & 250 lbs. |
| B16.5 | Steel Pipe Flanges & Flanged Fittings |
| B16.9 | Wrought Steel Butt Welding Fittings |
| B16.10 | Dimensions of Ferrous Valves |
| B16.11 | Steel Fittings S.W. & Threaded |
| B16.14 | Ferrous Plugs, Bushings & Locknuts with Pipe Threads |
| B16.15 | Cast Bronze Thd. Fittings—150 & 300 lbs. |
| B16.18 | Cast Bronze Solder-Joints Pressure Fittings |
| B16.20 | Ring Joint Gaskets—Steel Flanges |
| B16.21 | Non-metallic Gaskets for Flanges |
| B16.22 | Wrought Copper & Bronze Solder Joints Pressure Fittings |
| B16.24 | Bronze Glanes & Fittings—150 & 300 lbs. |
| B16.25 | Butt Welding Ends—Pipe, Valves, Flanges & Fittings |
| B16.28 | Wrought Steel Butt Welding Short Radius Elbows and Returns |
| B16.34 | Steel Butt-Welding End Valves |
| B18.2.1 | Square and Hex Bolts and Screws |
| B18.2.2 | Square and Hex Nuts |
| B18.21.1 | Lock Washers |
| B18.22.1 | Plain Washers |
| B18.23.1 | Beveled Washers |
| B31.3 | Chemical Plant & Petroleum Refinery Piping |
| B31.4 | Oil Transportation Piping |
| B31.8 | Gas Transmission & Distribution Piping |
| B36.10 | Wrought Steel & Iron Pipe |
| B36.19 | Stainless Steel Pipe |
| B181.1 | Screw Threads and Gasket for Fire Hose Connections |

MSS Standard Practices

| | | | |
|---|---|---|---|
| SP-6 | Finishes-On Flanges, Valves & Fittings | SP-53 | Magnetic Particle Insp:—Steel Casting |
| SP-9 | Spot-Facing for Bronze, Iron & Steel Flanges | SP-55 | Visual Inspection—Steel Castings |
| SP-25 | Marking for Valves, Fittings, Flanges & Unions | SP-58 | Pipe Hangers & Supports, Materials & Design |
| SP-42 | Corrosion Resistant Cast Flanged Valve | SP-61 | Hydrostatic Testing Steel Valves |
| SP-43 | Wrought S.S. Butt Welding Fittings | SP-67 | Butterfly Valves |
| SP-44 | Steel Pipe Line Flanges | SP-69 | Pipe Hangers & Supports—Selection and Application |
| SP-45 | Bypass & Drain Connection | SP-75 | High Strength Wrought Welding Fittings |
| SP-51 | Corrosion Resistant Cast Flanges & Flange Fittings—150 lbs | SP-79 | Socket Welding Reducer Inserts |
| | | SP-80 | Bronze Gate, Globe, Angle & Check Valve |

API Specifications

| | | | |
|---|---|---|---|
| 5L | Line Pipe | Standard 605 | Large Diameter Carbon Steel Flanges |

ASME Codes

ASME Boiler and Pressure Vessel Code

AWS Filler Metal Specifications

| | | | |
|---|---|---|---|
| A5.1 | Mild Steel Covered Arc-Welding Electrodes | A5.11 | Nickel and Nickel-Alloy Covered Welding Electrodes |
| A5.2 | Iron and Steel Gas-Welding Rods | A5.12 | Tungsten Arc-Welding Electrodes |
| A5.4 | Corrosion-Resisting Chromium and Chromium-Nickel Steel Covered Welding Electrodes | A5.14 | Nickel and Nickel-Alloy Bare Welding Rods and Electrodes |
| A5.5 | Low-Alloy Steel Covered Arc-Welding Electrodes | A5.17 | Bare Mild Steel Electrodes and Fluxes for Submerged-Arc Welding |
| A5.6 | Copper and Copper-Alloy Arc-Welding Electrodes | | |
| A5.7 | Copper and Copper-Alloy Welding Rods | A5.18 | Mild Steel Electrodes for Gas Metal Arc Welding |
| A5.8 | Brazing Filler Metal | A5.20 | Mild Steel Electrodes for Flux Cored Arc Welding |
| A5.9 | Corrosion-Resisting Chromium and Chromium-Nickel Steel-Welding Rods & Bare Electrodes | A5.22 | Flux Cored Corrosion-Resisting Chromium & Chromium-Nickel-Steel Electrodes |
| A5.10 | Aluminum and Aluminum-Alloy Welding Rods & Bare Electrodes | A5.23 | Spec. for Bare Low-Alloy Steel Electrodes and Fluxes for Submerged Arc Welding |

**APPENDIX 6** *continued*

AWWA Standards

| | | | |
|---|---|---|---|
| C-101 | Computation of Strength Thickness of C.I. Pipe | C-208 | Dimensions for Steel Water Pipe Fittings |
| C-106 | C.I. Pipe Centrifugally Cast in Metal Molds | C-300 | Reinforced Concrete Water Pipe-Steel Cylinder Type not Prestressed |
| C-108 | C.I. Pipe Centrifugally Cast in Sand-Lined Molds | C-301 | Reinforced Concrete Water Pipe-Steel Cylinder Type Prestressed |
| C-110 | C.I. Fittings 2″ thru 48″ | C-302 | Reinforced Concrete Water Pipe-Non Cylinder Type—Not Prestressed |
| C-111 | Rubber Gasketed Joint for C.I. Pipe and Fittings | C-400 | Asbestos Cement Pipe |
| C-112 | 2″ and 2¼″ C.I. Pipe Centrifugally Cast | C-500 | Gate Valves for Ordinary Water Works Service |
| C-150 | Thickness Design of Ductile-Iron Pipe | C-600 | Installation of C.I. Water Mains |
| C-151 | Ductile Iron Pipe Centrifugally Cast in Metal or Sand-Lined Molds | C-603 | Installation of Asbestos Cement Pipe |
| C-200 | Fabricated Electrically Welded Steel Pipe | | |
| C-207 | Steel Pipe Flanges | | |

Federal Specifications

| | | | |
|---|---|---|---|
| SS-P-381 | Pipe: Concrete, (Pressure, Reinforced Pretensioned Reinforcement Steel Cylinder Type) | WW-P-421 | Pipe, Cast Gray and Ductile Iron, Pressure (for Water & other Liquids) |

PFI Standards

ES-16 Access Holes and Plugs for Radiographic Inspection of Pipe Welds

---

Specifications and Standards of the Following Organizations Appear in This List:

| | | | |
|---|---|---|---|
| API | American Petroleum Institute<br>1801 K. Street, N.W.<br>Washington, D.C. 20026 | AWWA | American Water Works Association<br>666 W. Quincy Avenue<br>Denver, CO 80235 |
| ANSI | American National Standards Institute<br>1430 Broadway<br>New York, NY 10018 | MSS | Manufacturers Standardization Society of the Valve & Fittings Industry<br>1815 N. Fort Myer Drive |

ASME American Society of Mechanical Engineers, Inc.
345 East 47th Street
New York, NY 10017

ASTM American Society of Testing & Materials
1916 Race Street
Philadelphia, PA 19103

AWS American Welding Society, Inc.
2501 N.W. Seventh Street
Miami, FL 33125
Arlington, VA 22209

PFI Pipe Fabrication Institute
1326 Freeport Road
Pittsburgh, PA 15238

Federal Specifications:
Superintendent of Documents
United States Government Printing Office
Washington, DC 20402

ABBREVIATIONS:

| | |
|---|---|
| B.W. | Butt Welding |
| C.I. | Cast iron |
| M.I. | Malleable iron |
| S.S. | Stainless steel |
| S.W. | Socket welding |
| H.T. | High temperature |
| L.T. | Low temperature |
| T.S. | Tensil strength |
| Thd. | Threaded |
| Serv. | Service(s) |
| Temp. | Temperature |
| Gen. | General |
| Engr. | Engineering |

# APPENDIX 7

## Corrosion resistance data

*(Data collected by Flowline Corp. from primary producers)*

The following table lists commonly known corrosive media and tabulates the theoretical corrosion resistance of Stainless Steels 20 $Cb_3$, 304, 304L, 309, 310, 316, 316L, 321, 347 and 348; Nickel 200, Monel 400, Inconel 600, Hastelloy B and C under various temperature conditions. The symbols A, B, C, D, and E represent appropximate corrosion ranges as defined in the accompanying table.

See footnotes for circumstances applicable under certain conditions.

| SYMBOL | | | | |
|---|---|---|---|---|
| A | B | C | D | E |
| DEFINITION | | | | |
| Fully Resistant | Satisfactorily Resistant | Fairly Resistant | Slightly Resistant | Not Resistant |
| ▼ | ▼ | ▼ | ▼ | ▼ |
| Less than 0.00035 inches penetration per month | 0.00035 to 0.0035 inches penetration per month | 0.0035 to 0.010 inches penetration per month | 0.010 to 0.035 inches penetration per month | Over 0.035 inches penetration per month |

This data is intended for general guidance only. Selection of a particular alloy for specific corrosion service should be based on actual tests and technical advice that is available from the basic metal producers.

1. Corrosion resistance data for stainless steels 309 and 310 is based on producer's report that corrosion data for type 316 is approximately representative of types 309 and 310.

† Pitting may sometimes occur under certain conditions, such as at the air line or when allowed to dry or when solutions are stagnant.

* May attack when sulphuric acid is present.

§ May attack when hydrochloric acid is present.

‡ Alkaline solutions

** Applies to low-carbon nickel

Hastelloy is a registered trademark of Union Carbide Corporation. 20 $Cb_3$ is a registered trademark of the Carpenter Steel Co.

Inconel and Monel are registered trademarks of the International Nickel Co.

ANSI/ASTM Specification B-334 for Hastelloy C materials has been discontinued since 1979.

| Substance and its Condition | Temp. F | 304 304L 347 348 | 316 316L 309(1) 310(1) | 321 | 20Cb$_{-3}$ | Nickel 200 | Monel 400 | Inconel 600 | | Hastelloy Alloy B | Hastelloy Alloy C |
|---|---|---|---|---|---|---|---|---|---|---|---|
| ANSI/ASTM | | A-312 | A-312 | A-312 | . . . | B-161 | B-165 | B-167 | . . . | B-333 | B-334 |
| A | | | | | | | | | | | |
| Acetic Acid | | | | | | | | | | | |
| 5% and 10% | 70 | A | A | A | A | A | A | A | . . | A | A |
| 20% | 70 | A | A | A | A | A | A | A | . . | A | A |
| 50% | 70 | A | A | A | A | B | A | A | . . | A | A |
| 50% | Boiling | C | B | B | A | B | A | B | . . | A | A |
| 80% | 70 | A | A | A | A | B | A | A | . . | A | A |
| 80% | Boiling | D | B | C | B | B | A | B | . . | A | A |
| 100% | 70 | A | A | A | A | A | A | A | . . | A | A |
| 100% | Boiling | C | B | D | B | C | B | B | . . | A | A |
| Acetic Anhydride | 70 | A | A | A | A | A | A | A | . . | A | A |
| | Boiling | A | A | B | B | A | A | A | . . | A | A |
| Acetic Vapors | | | | | | | | | | | |
| 30% | Hot | C | B | C | B | . . | . . | . . | . . | . . | . . |
| 100% | Hot | E | C | E | B | C | B | B | . . | . . | A |
| Acetone | 70 | A | A | A | A | A | A | A | . . | A | A |
| | Boiling | A | A | A | A | A | A | A | . . | A | A |
| Alcohol | | | | | | | | | | | |
| Ethyl 100% | 70 | A | A | A | A | A | A | A | . . | . . | . . |
| | Boiling | A | A | . . | A | A | A | A | . . | A | A |
| Methyl | 70 | A | A | . . | A | A | A | A | . . | A | A |
| | 150 | †C | B | . . | A | A | A | A | . . | . . | . . |
| Alum. (Chrome 5%) | 70 | A | A | . . | . . | C | C | A | . . | . . | . . |
| Aluminum Acetate | | | | | | | | | | | |
| Saturated | | A | A | A | . . | . . | . . | . . | . . | . . | . . |
| Aluminum Chloride | 70 | D | C | D | †A | B | B | C | . . | A | . . |
| Aluminum Chloride, | | | | | | | | | | | |
| Cold 100% | 85 | . . | . . | . . | . . | C | E | E | . . | A | . . |
| Aluminum Fluoride | 70 | D | C | D | . . | A | A | B | . . | . . | . . |
| Aluminum Hydroxide | | | | | | | | | | | |
| Saturated | | A | A | A | . . | A | A | A | . . | . . | . . |

## APPENDIX 7 *(continued)*

| Substance and its Condition | Temp. F | 304 304L 347 348 | 316 316L 309(1) 310(1) | 321 | 20Cb-3 | Nickel 200 | Monel 400 | Inconel 600 | | Hastelloy Alloy B | Hastelloy Alloy C |
|---|---|---|---|---|---|---|---|---|---|---|---|
| ANSI/ASTM | | A-312 | A-312 | A-312 | . . . | B-161 | B-165 | B-167 | . . . | B-333 | B-334 |
| Aluminum | Molten | E | E | E | . . | E | E | F | . . | . . | . . |
| Aluminum Potassium | | | | | | | | | | | |
| Sulphate 2% (alum.) | 70 | A | A | . . | A | A | A | A | . . | . . | . . |
| 10% | 70 | A | A | A | A | A | A | A | . . | . . | B |
| 10% | Boiling | B | A | B | A | B | A | B | . . | . . | . . |
| Saturated | Boiling | C | B | C | B | C | B | B | . . | . . | . . |
| Aluminum Sulfate | | | | | | | | | | | |
| 10% | 70 | A | A | †B | A | A | A | A | . . | . . | A |
| 10% | Boiling | B | A | B | A | B | A | B | . . | . . | B |
| Saturated | 70 | A | A | †B | A | A | A | A | . . | . . | . . |
| Saturated | Boiling | B | A | C | B | B | A | B | . . | . . | . . |
| Ammonia | | | | | | | | | | | |
| All Concentrations | 70 | A | A | A | A | A | A | A | . . | B | B |
| Gas | Hot | D | . . | . . | . . | A | A | A | . . | . . | . . |
| Ammonia Liquor | 70 | A | A | A | . . | C | C | A | . . | . . | . . |
| | Boiling | A | A | A | . . | C | C | A | . . | . . | . . |
| Ammonium | | | | | | | | | | | |
| 0-95% | 85 | . . | . . | . . | . . | E | E | A | . . | . . | . . |
| 100% | 600 | . . | . . | . . | . . | A | A | A | . . | . . | . . |
| Ammonium Bicarbonate | 70 | A | A | A | . . | A | A | A | . . | . . | . . |
| | Hot | A | A | A | . . | A | A | A | . . | . . | . . |
| 0-100% | 212 | . . | . . | A | . . | E | E | E | . . | . . | . . |
| Ammonium Carbonate | | | | | | | | | | | |
| 1 and 5% | 70 | A | A | A | A | A | A | A | . . | . . | . . |
| Ammonium Chloride | | | | | | | | | | | |
| 1% Solution | 70 | A | A | . . | †A | A | A | A | . . | . . | . . |
| 10% | Boiling | †A | †A | †B | †B | A | A | B | . . | A | A |
| 28% | Boiling | †B | †A | †B | †B | A | A | B | . . | A | A |
| 50% | Boiling | †B | †A | †C | †B | A | A | B | . . | A | A |
| Ammonium Nitrate | | | | | | | | | | | |
| 0-100% Agitated or Aerated | 70 | A | A | . . | A | C | C | A | . . | . . | . . |
| Saturated | Boiling | A | A | A | A | E | E | B | . . | . . | . . |

| | | | | | | | | | | | |
|---|---|---|---|---|---|---|---|---|---|---|---|
| Ammonium Oxalate | | | | | | | | | | | |
| 5% | 70 | A | A | A | . . | A | A | A | . . | . . | . . |
| Ammonium Persulphate | | | | | | | | | | | |
| 5% | 70 | A | A | A | . . | E | E | A | . . | . . | . . |
| Ammonium Phosphate | | | | | | | | | | | |
| 5% | 70 | A | A | . . | A | A | A | A | . . | . . | A |
| | 212 | . . | . . | . . | . . | C | C | C | . . | . . | . . |
| Ammonium Sulphate | | | | | | | | | | | |
| 1% and 5% Agitated | 70 | A | A | . . | B | A | A | A | . . | . . | B |
| 1% and 5% Aerated | 70 | A | A | . . | B | A | A | A | . . | . . | B |
| 10% | Boiling | †B | †A | †B | B | B | A | B | . . | | |
| Saturated | Boiling | †B | †A | . . | . . | B | A | B | . . | . . | . . |
| Ammonium Sulphite | Cold | A | A | A | . . | C | B | B | . . | . . | . . |
| | Boiling | A | A | B | . . | E | C | C | . . | . . | . . |
| Aniline | | | | | | | | | | | |
| 3% | 70 | A | A | A | A | A | A | A | . . | . . | B |
| Conc. Crude | 70 | A | A | A | A | A | A | A | . . | . . | B |
| 100% | 85 | . . | . . | . . | . . | C | C | C | . . | . . | B |
| Aniline Hypochloride | 70 | E | D | E | . . | B | B | C | . . | . . | . . |
| Antimony Trichloride | 70 | E | D | E | . . | A | A | A | . . | . . | . . |
| **B** | | | | | | | | | | | |
| Barium Carbonate | 70 | A | A | A | A | A | A | A | . . | B | B |
| Barium Chloride | | | | | | | | | | | |
| 5% | 70 | A | A | †B | A | A | A | A | . . | B | B |
| 0-40% | 212 | . . | . . | . . | A | C | C | C | . . | B | B |
| Saturated | 70 | A | B | †B | . . | A | A | A | . . | A | A |
| Aqueous Solution | Hot | †B | †A | . . | . . | A | A | B | . . | . . | . . |
| Barium Nitrate | | | | | | | | | | | |
| Aqueous Solution | Hot | A | A | A | A | C | C | B | . . | B | B |
| Barium Sulphate | | | | | | | | | | | |
| Barytes-BlancFixe | 70 | A | A | A | A | A | A | A | . . | . . | . . |
| Benzene | 70 | A | A | A | A | A | A | A | . . | B | B |
| 0-75% | 212 | . . | . . | A | A | A | A | A | . . | . . | . . |
| 100% | 212 | . . | . . | . . | A | A | A | A | . . | B | B |
| Benzoic, Acid | 70 | A | A | A | A | A | A | A | . . | . . | . . |
| Benzol | Hot | A | A | A | . . | A | A | A | . . | B | A |
| Boracic Acid | | | | | | | | | | | |
| 5% | Hot or Cold | A | A | A | A | A | A | A | . . | A | A |

## APPENDIX 7 *(continued)*

| Substance and its Condition | Temp. F | 304 304L 347 348 | 316 316L 309(1) 310(1) | 321 | 20Cb$_{-3}$ | Nickel 200 | Monel 400 | Inconel 600 | | Hastelloy Alloy B | Hastelloy Alloy C |
|---|---|---|---|---|---|---|---|---|---|---|---|
| ANSI/ASTM | | A-312 | A-312 | A-312 | . . . | B-161 | B-165 | B-167 | . . . | B-333 | B-334 |
| 0-20% (Air Free) | 212 | . . | . . | . . | A | C | C | C | . . | A | A |
| Borax—5% | Hot | A | A | A | . . | A | A | A | . . | . . | . . |
| Bromide—Dry | 70 | . . | . . | . . | A | A | A | A | . . | . . | . . |
| Bromine Water | 70 | E | D | . . | E | D | D | D | . . | . . | . . |
| Butyric Acid | | | | | | | | | | | |
| 5% | 70 | A | A | A | B | A | A | A | . . | . . | . . |
| 5% | 150 | A | A | A | B | A | A | A | . . | . . | . . |
| Aqueous Solution—Sp. gr. 0.964 | Boiling | A | A | . . | . . | . . | . . | A | . . | . . | . . |
| C | | | | | | | | | | | |
| Calcium Carbonate | 70 | A | A | A | B | A | A | A | . . | . . | . . |
| Calcium Chlorate | | | | | | | | | | | |
| Dilute Solution | 70 | A | A | A | B | A | A | A | . . | . . | . . |
| Dilute Solution | Hot | A | A | A | B | A | A | B | . . | . . | . . |
| Calcium Chloride | | | | | | | | | | | |
| Dilute Solution | 70 | ‡B | †A | **B | A | A | A | A | . . | . . | . . |
| Conc. Solution | 70 | ‡B | †A | **B | B | A | A | A | . . | . . | . . |
| Calcium Hydroxide | | | | | | | | | | | |
| 10% | Boiling | A | A | A | B | A | A | A | . . | B | A |
| 20% | Boiling | A | A | A | . . | A | A | A | . . | B | A |
| 50% | Boiling | A | A | A | . . | A | A | A | . . | B | A |
| Calcium Hypochlorite | | | | | | | | | | | |
| 2% | 70 | †B | †A | D | B | C | C | B | . . | . . | . . |
| Calcium Sulphate | | | | | | | | | | | |
| Saturated | 70 | A | A | A | B | A | A | A | . . | B | B |
| Carbolic Acid | | | | | | | | | | | |
| C.P. | Boiling | A | A | A | . . | A | A | A | . . | . . | . . |
| Crude | Boiling | A | A | A | B | A | A | A | . . | . . | . . |
| C.P. | 70 | A | A | A | . . | A | A | A | . . | . . | . . |

| | | | | | | | | | | | |
|---|---|---|---|---|---|---|---|---|---|---|---|
| Carbon Bisulphide | 70 | A | A | A | B | . . | . . | . . | . . | . . | . . |
| | 100 | . . | . . | . . | B | A | B | A | . . | . . | . . |
| Carbon Monoxide Gas | 1400 | A | A | . . | A | . . | . . | . . | . . | A | A |
| | 1600 | A | A | A | A | . . | . . | . . | . . | . . | . . |
| Carbon Tetrachloride | | | | | | | | | | | |
| Pure (Dry) | 70 | A | A | A | A | A | A | A | . . | A | A |
| Aqueous—5-10% | 70 | †C | †A | *C | B | A | A | A | . . | A | A |
| Dry | Boiling | A | . . | A | B | . . | . . | . . | . . | . . | . . |
| Chloracetic Acid | 70 | D | C | D | . . | A | A | A | . . | . . | . . |
| Chlorbenzol Conc.-Pure | 70 | A | A | A | . . | A | A | A | . . | . . | . . |
| Chlorine (Dry) (100% | Limit | . . | . . | . . | . . | . . | . . | . . | . . | . . | A |
| Gas—Dry | 70 | A | A | A | A | C | C | C | . . | . . | A |
| Gas—Moist | 70 | D | C | E | A | C | C | C | . . | . . | A |
| Gas | 212 | E | D | E | A | A | A | A | . . | . . | A |
| Chloroform | 70 | A | A | A | B | A | A | A | . . | B | B |
| | 212 | . . | . . | . . | A | A | A | A | . . | B | B |
| Dry | 70 | A | . . | . . | A | . . | . . | . . | . . | B | B |
| Chromic Acid | | | | | | | | | | | |
| 5% | 70 | A | A | . . | B | A | A | A | . . | . . | B |
| 10% C.P. | Boiling | C | B | . . | E | C | B | C | . . | . . | B |
| 10% C.P. (Free of $SO_3$) | 70 | B | . . | A | E | . . | . . | . . | . . | . . | B |
| 50% Com (Cont. $SO_3$) | Boiling | †D | C | E | E | . . | . . | . . | . . | . . | B |
| 50% C.P. (Free of $SO_3$) | 70 | A | . . | C | A | . . | . . | . . | . . | . . | B |
| | Boiling | C | . . | . . | E | . . | . . | . . | . . | . . | . . |
| Chromium Plating Bath | 70 | A | A | . . | . . | C | C | A | . . | . . | A |
| Citric Acid | | | | | | | | | | | |
| Air free 0-60% | 212 | . . | . . | . . | A | C | C | C | . . | A | A |
| 5% Still | 70 | A | A | . . | A | A | A | A | . . | A | A |
| | 150 | A | A | . . | A | A | A | A | . . | A | A |
| 15% | 70 | A | A | . . | A | A | A | A | . . | A | A |
| | Boiling | B | A | . . | A | A | A | A | . . | A | A |
| Concentrated | Boiling | C | B | . . | A | B | A | B | . . | A | A |
| 50% | 70 | A | . . | A | A | . . | . . | . . | . . | A | A |
| | Boiling | D | . . | . . | A | . . | . . | . . | . . | A | A |
| Copper Acetate | | | | | | | | | | | |
| Saturated Sol. | 70 | A | A | A | . . | C | C | A | . . | . . | . . |
| Copper Carbonate | | | | | | | | | | | |
| Sat. Sol. in 50% $NH_3OH$ | | A | A | A | . . | C | C | A | . . | . . | . . |
| Copper Chloride | | | | | | | | | | | |
| 1% Agitated | 70 | †B | †A | . . | . . | B | B | A | . . | . . | . . |

## APPENDIX 7 *(continued)*

| Substance and its Condition | Temp. F | 304 304L 347 348 | 316 316L 309(1) 310(1) | 321 | $20Cb_{-3}$ | Nickel 200 | Monel 400 | Inconel 600 | | Hastelloy Alloy B | Hastelloy Alloy C |
|---|---|---|---|---|---|---|---|---|---|---|---|
| ANSI/ASTM | | A-312 | A-312 | A-312 | . . . | B-161 | B-165 | B-167 | . . . | B-333 | B-334 |
| 1% Aerated | 70 | †B | †A | . . | . . | B | B | A | . . | . . | . . |
| 10% | Boiling | E | . . | E | . . | . . | . . | . . | . . | . . | . . |
| 5% Agitated | 70 | †B | †A | . . | . . | C | C | C | . . | . . | . . |
| 5% Aerated | 70 | †E | †D | . . | . . | D | D | C | . . | . . | . . |
| Copper Cyanide | | | | | | | | | | | |
| Saturated Sol. | Boiling | A | A | A | B | B | B | B | . . | B | A |
| Copper Nitrate | | | | | | | | | | | |
| 1% and 5% | 70 | A | A | A | A | C | C | A | . . | . . | B |
| 90% Aqueous | Hot | A | A | . . | B | E | E | C | . . | . . | . . |
| Copper Sulphate (Blue Vitrol) | | | | | | | | | | | |
| Saturated | 212 | A | A | A | A | . . | . . | . . | . . | . . | A |
| Copper Sulphate | | | | | | | | | | | |
| 5% | 70 | A | A | . . | A | B | B | A | . . | . . | . . |
| 0-30% | 212 | . . | . . | . . | A | E | E | E | . . | . . | A |
| Saturated Sol. | Boiling | A | A | . . | B | C | C | C | . . | . . | A |
| Creosote (Coal Tar) | Hot | A | A | A | . . | A | A | A | . . | . . | . . |
| Oil | Hot | A | A | A | . . | A | A | A | . . | . . | . . |
| Cyanogen Gas | 70 | A | A | A | . . | A | A | A | . . | . . | . . |
| D | | | | | | | | | | | |
| Dinitrochlorobenzol | | | | | | | | | | | |
| Melted and Solidified | 70 | A | A | A | . . | . . | . . | . . | . . | . . | . . |
| E | | | | | | | | | | | |
| Ether—100% | 70 | A | A | A | . . | A | A | A | . . | . . | . . |
| Ethyl Chloride | 70 | A | A | . . | . . | A | A | A | . . | . . | . . |

| | | | | | | | | | | | | |
|---|---|---|---|---|---|---|---|---|---|---|---|---|
| Ethylene Chloride .................... | 85 | A | A | A | . . | A | A | A | . . | . . | . . |
| F | | | | | | | | | | | | |
| Ferric Chloride | | | | | | | | | | | | |
| 1% Solution, Still .................. | 70 | §†B | †A | †C | †E | B | C | C | . . | . . | A |
| .................. | Boiling | §†D | †D | D | . . | E | E | E | . . | . . | A |
| 5% Solution, Still | 70 | §†C | †B | C | E | D | D | E | . . | . . | A |
| 5% Agitated ........................ | 70 | §†C | A | . . | E | D | D | C | . . | . . | A |
| 5% Aerated ......................... | 70 | §†C | †C | . . | E | D | D | C | . . | . . | A |
| Ferric Hydroxide | | | | | | | | | | | | |
| (Hydrated Iron Oxide) ............ | 70 | A | A | A | . . | A | A | A | . . | . . | . . |
| Ferric Nitrate 1% and 5% .......... | 70 | A | A | A | A | D | D | A | . . | . . | A |
| Ferric Sulphate | | | | | | | | | | | | |
| 1% and 5% ....................... | 70 | †A | A | . . | A | C | C | A | . . | . . | A |
| 10% ................................ | 70 | †A | . . | . . | A | . . | . . | . . | . . | . . | . . |
| | Boiling | A | . . | . . | A | . . | . . | . . | . . | . . | . . |
| 0-30% ............................... | 85 | . . | . . | . . | . . | E | C | E | . . | . . | A |
| Ferrous Sulphate | | | | | | | | | | | | |
| Dilute Sol. .......................... | 70 | A | A | †B | B | A | A | A | . . | B | B |
| Fluorine ............................... | 70 | E | E | E | . . | A | A | A | . . | B | B |
| Formic Acid | | | | | | | | | | | A |
| 5% Still ............................. | 70 | B | A | B | A | A | A | A | . . | A | |
| ............................. | 150 | B | A | B | B | A | A | A | . . | A | A |
| Air-Free—100% .................. | 212 | . . | . . | . . | B | C | C | C | . . | A | A |
| Fuel Oil ............................... | Hot | A | A | A | . . | B | B | A | . . | . . | . . |
| Containing $H_2SO_4$ ................ | | C | B | . . | . . | B | B | B | . . | . . | . . |
| G | | | | | | | | | | | | |
| Gallic Acid | | | | | | | | | | | | |
| 5% Solution ........................ | 70 | A | A | . . | B | A | A | A | . . | B | B |
| 5% Solution ........................ | 150 | A | A | . . | B | A | A | A | . . | B | B |
| 100% ................................ | 85 | . . | . . | . . | B | C | C | C | . . | B | B |
| Saturated ............................ | 212 | A | . . | A | B | . . | . . | . . | . . | B | B |
| Glue | | | | | | | | | | | | |
| Acid Free ........................... | Hot | A | . . | A | . . | . . | . . | . . | . . | . . | . . |
| Acid Sol ............................. | 85 | †B | A | . . | . . | A | A | A | . . | . . | . . |

## APPENDIX 7 *(continued)*

| Substance and its Condition | Temp. F | 304<br>304L<br>347<br>348 | 316<br>316L<br>309(1)<br>310(1) | 321 | 20Cb$_{-3}$ | Nickel<br>200 | Monel<br>400 | Inconel<br>600 | | Hastelloy Alloy<br>B | <br>C |
|---|---|---|---|---|---|---|---|---|---|---|---|
| ANSI/ASTM | | A-312 | A-312 | A-312 | . . . | B-161 | B-165 | B-167 | . . . | B-333 | B-334 |
| Glycerine | 70-85 | A | A | A | A | A | A | A | . . | . . | . . |
| H | | | | | | | | | | | |
| Hydrochloric Acid 1:85 | 70 | E | E | C | B | B | B | C | . . | C | A |
| | Boiling | E | . . | E | E | . . | . . | . . | . . | B | D |
| Hydrocyanic Acid | 70 | A | A | A | . . | A | A | A | | . . | . . |
| Vapors | 70 | D | . . | . . | . . | . . | . . | . . | . . | . . | . . |
| | 212 | D | . . | . . | . . | . . | . . | . . | . . | . . | . . |
| Hydrogen Peroxide | 70 | *A | A | A | A | A | A | A | . . | . . | . . |
| | | | | A | B | . . | . . | . . | . . | . . | . . |
| Acid Free | 85 | . . | . . | . . | A | C | C | C | . . | A | A |
| Hydrogen Sulfide | | . . | . . | . . | B | E | A | A | . . | . . | . . |
| Dry | | A | A | A | B | A | A | A | . . | . . | . . |
| Wet | | *B | *A | A | B | A | A | A | . . | . . | . . |
| Hydrosulphite Soda (Hypo) | | A | †A | †A | . . | A | A | A | . . | . . | . . |
| I | | | | | | | | | | | |
| Iodine | | E | D | D | B | D | D | D | . . | . . | B |
| Iodoform | | A | A | A | . . | . . | . . | . . | . . | . . | . . |
| L | | | | | | | | | | | |
| Lactic Acid | | | | | | | | | | | |
| 5% | 70 | A | A | A | B | A | A | A | . . | B | B |
| 5% | 150 | B | A | B | B | A | A | A | . . | B | B |
| 10% | Boiling | D | B | D | . . | C | C | B | . . | B | B |
| 10% | 150 | C | B | C | B | B | B | B | . . | B | B |

| | | | | | | | | | | | |
|---|---|---|---|---|---|---|---|---|---|---|---|
| Lead ........................ | Molten | B | B | C | . . | D | D | B | . . | . . | . . |
| Linseed Oil ........................ | 70-85 | A | A | A | . . | A | A | A | . . | . . | . . |
| M | | | | | | | | | | | |
| Magnesium Chloride | | | | | | | | | | | |
| 1% and 5%, Still ........................ | 70 | †A | A | †B | A | A | A | A | . . | A | A |
| | Hot | †C | †B | †C | A | A | A | A | . . | A | A |
| Magnesium Sulphate ........................ | Hot or Cold | A | A | †B | A | A | A | A | . . | A | A |
| 0-60% ........................ | 212 | . . | . . | . . | A | C | C | C | . . | . . | . . |
| Malic Acid ........................ | Hot or Cold | B | A | B | B | A | A | A | . . | . . | . . |
| Mayonnaise ........................ | 70 | †A | A | †A | . . | B | B | A | . . | . . | . . |
| Mercuric Chloride | | | | | | | | | | | |
| All Solutions ........................ | | †E | †D | E | . . | D | D | D | . . | . . | |
| Mercury ........................ | | A | A | A | . . | A | A | A | . . | . . | . . |
| Methanol (Methyl Alcohol) ........................ | 70 | A | A | A | . . | A | A | A | . . | . . | . . |
| Methyl Chloride 100% ........................ | 85 | . . | . . | A | A | C | C | C | . . | . . | . . |
| Mixed Acids | | | | | | | | | | | |
| 45% $HNO_3$ ........................ | Cold | A | A | . . | . . | D | D | A | . . | . . | . . |
| 50% $H_2SO_4$ + 50% | | | | | | | | | | | |
| $HNO_3$ ........................ | 140 | A | B | A | . . | . . | . . | . . | . . | . . | . . |
| ........................ | 200 | B | . . | B | . . | . . | . . | . . | . . | . . | . . |
| ........................ | 250 | C | . . | C | . . | . . | . . | . . | . . | . . | . . |
| 75% $H_2SO_4$ + 25% | | | | | | | | | | | |
| $HNO_3$ ........................ | 140 | A | B | A | . . | . . | . . | . . | . . | . . | . . |
| ........................ | 200 | B | . . | B | . . | . . | . . | . . | . . | . . | . . |
| ........................ | 315 | D | . . | D | . . | . . | . . | . . | . . | . . | . . |
| 70% $H_2SO_4$ + 10% $HNO_3$ | | | | | | | | | | | |
| + 20% Water ........................ | 140 | A | . . | A | . . | . . | . . | . . | . . | . . | . . |
| ........................ | 200 | B | . . | B | . . | . . | . . | . . | . . | . . | . . |
| ........................ | 335 | E | . . | E | . . | . . | . . | . . | . . | . . | . . |
| 30% $H_2SO_4$ + 75% | | | | | | | | | | | |
| $HNO_3$ + 65% Water ......... | 140 | A | B | A | . . | . . | . . | . . | . . | . . | . . |
| ......... | 200 | A | . . | A | . . | . . | . . | . . | . . | . . | . . |
| ......... | 230 | B | . . | B | . . | . . | . . | . . | . . | . . | . . |
| 15% $H_2SO_4$ + 5% | | | | | | | | | | | |
| $HNO_3$ + 80% Water ......... | 140 | A | B | A | . . | . . | . . | . . | . . | . . | . . |
| ......... | 200 | A | . . | A | . . | . . | . . | . . | . . | . . | . . |
| ......... | 220 | A | . . | A | . . | . . | . . | . . | . . | . . | . . |

## APPENDIX 7 *(continued)*

| Substance and its Condition | Temp. F | 304 304L 347 348 | 316 316L 309(1) 310(1) | 321 | $20Cb_{-3}$ | Nickel 200 | Monel 400 | Inconel 600 | | Hastelloy Alloy B | Hastelloy Alloy C |
|---|---|---|---|---|---|---|---|---|---|---|---|
| ANSI/ASTM | | A-312 | A-312 | A-312 | . . . | B-161 | B-165 | B-167 | . . . | B-333 | B-334 |
| Mixtures of Acids and Salts | | | | | | | | | | | |
| Fuming Nitric Acid (Sp.gr. 1.52) + 10% Potassium Nitrate | Boiling | B | . . | B | . . | . . | . . | . . | . . | . . | . . |
| Fuming Nitric Acid (Sp.gr. 1.52) + 10% Alum. Nitrate | Boiling | B | . . | . . | . . | . . | . . | . . | . . | . . | . . |
| 10% Sulphuric Acid + 10% Copper Sulphate | Boiling | A | . . | A | . . | . . | . . | . . | . . | . . | . . |
| 10% Sulphuric Acid + 10% Ferrous Sulphate | Boiling | B | . . | . . | . . | . . | . . | . . | . . | . . | . . |
| Molasses | | A | A | A | . . | A | A | A | . . | . . | . . |
| Muriatic Acid | 70 | E | E | . . | . . | B | B | C | . . | . . | . . |
| Mustard | 70 | †A | †A | †A | . . | A | B | B | . . | . . | . . |
| N | | | | | | | | | | | |
| Nickel Chloride Sol. | 70 | †A | †A | . . | . . | B | B | B | . . | A | A |
| Nickel Sulfate | Hot or Cold | A | A | . . | B | A | A | A | . . | . . | . . |
| Nitre Cake | Fused | B | A | B | . . | . . | . . | . . | . . | . . | . . |
| Nitric Acid | | | | | | | | | | | |
| 5% Sol. | 70 | A | A | A | A | E | E | B | . . | E | A |
| 20% Sol. | 70 | A | A | . . | A | E | E | A | . . | E | A |
| 50% Sol. | 70 | A | A | A | A | E | E | A | . . | E | A |
| | Boiling | A | A | A | B | E | E | C | . . | E | A |
| 65% Sol. | Boiling | B | B | . . | B | E | E | D | . . | E | C |
| Conc. | 70 | A | A | A | E | E | E | A | . . | E | A |
| Conc. | Boiling | D | D | B | C | E | E | D | . . | . . | . . |
| Nitrous Acid—5% Sol. | 70 | A | A | . . | B | D | D | A | . . | B | B |

| | | | | | | | | | | | |
|---|---|---|---|---|---|---|---|---|---|---|---|
| **O** | | | | | | | | | | | |
| Oils, Crude | | | | | | | | | | | |
| Asphalt Base | Cold or Hot | *A | *A | †A | . . | A | A | A | . . | . . | . . |
| Oils, Essential | 85 | . . | . . | . . | . . | A | A | A | . . | . . | . . |
| Oils, Veg. Mineral | 70-85 | A | A | A | . . | A | A | A | . . | . . | . . |
| Oleic Acid | 70-85 | †A | A | . . | B | A | A | A | . . | . . | . . |
| Oxalic Acid | | | | | | | | | | | |
| 5% | 70 | A | A | . . | B | A | A | A | . . | B | B |
| 10% | 70 | A | A | A | B | A | A | A | . . | B | B |
| 10% | Boiling | D | C | D | B | B | A | B | . . | B | B |
| 25% | Boiling | B | . . | D | B | . . | . . | . . | . . | B | B |
| 50% | Boiling | D | B | D | B | . . | . . | . . | . . | B | B |
| Air-free | 85 | . . | . . | . . | . . | E | C | C | . . | . . | . . |
| **P** | | | | | | | | | | | |
| Paraffin | Hot or Cold | A | A | A | . . | A | A | A | . . | . . | . . |
| Petroleum Ether | | A | A | A | . . | A | A | A | . . | . . | . . |
| Phenol | | A | A | . . | B | A | A | A | . . | . . | . . |
| Phosphoric Acid | | | | | | | | | | | |
| 1% | 70 | §A | §A | A | A | A | A | A | . . | A | A |
| | Boiling | A | . . | B | . . | . . | . . | . . | . . | A | A |
| 5% | 70 | A | A | . . | B | A | A | A | . . | A | A |
| 10%, Still | 70 | C | A | . . | B | A | A | A | . . | . . | . . |
| 10% Agitated | 70 | C | B | . . | . . | B | B | B | . . | . . | . . |
| 10% Aerated | 70 | C | B | . . | A | C | C | B | . . | . . | . . |
| 10% | Boiling | D | . . | D | . . | . . | . . | . . | . . | A | A |
| 45% | Boiling | D | . . | D | . . | . . | . . | . . | . . | A | A |
| 80% | 140 | C | . . | C | . . | . . | . . | . . | . . | A | A |
| | 230 | E | . . | E | . . | . . | . . | . . | . . | B | C |
| Picric Acid | 70 | A | A | A | B | . . | . . | . . | . . | . . | B |
| Potassium Bichromate | | | | | | | | | | | |
| 0-20% | 212 | . . | . . | A | . . | C | C | C | . . | . . | . . |
| | 70 | A | A | . . | . . | A | A | A | . . | . . | . . |
| Potassium Bromide | 70 | †B | †A | †B | B | A | A | A | . . | B | B |
| 0-30% Air-Free | 212 | . . | . . | . . | B | C | C | C | . . | . . | . . |
| Potassium Carbonate | 212 | . . | . . | . . | . . | C | C | C | . . | B | B |
| 1% | 70 | A | A | . . | B | A | A | A | . . | B | B |

## APPENDIX 7 *(continued)*

| Substance and its Condition | Temp. F | 304<br>304L<br>347<br>348 | 316<br>316L<br>309(1)<br>310(1) | 321 | 20Cb$_{-3}$ | Nickel<br>200 | Monel<br>400 | Inconel<br>600 | | Hastelloy Alloy<br>B | C |
|---|---|---|---|---|---|---|---|---|---|---|---|
| ANSI/ASTM | | A-312 | A-312 | A-312 | . . . | B-161 | B-165 | B-167 | . . . | B-333 | B-334 |
| 1% | Hot | A | A | A | B | A | A | A | . . | B | B |
| Potassium Chlorate | | A | A | A | B | A | A | A | . . | . . | . . |
| Potassium Chloride | | | | | | | | | | | |
| 1% and 5% | 70 | †A | †A | †A | †A | A | A | A | . . | . . | . . |
| 1% and 5% | Boiling | A | A | †A | †A | A | A | A | . . | . . | . . |
| Potassium Ferricyanide | | | | | | | | | | | |
| 5% | 70 | A | A | A | B | . . | . . | . . | . . | B | B |
| 25% | 70 | A | . . | . . | B | . . | . . | . . | . . | B | B |
| Potassium Ferrocyanide 5% | 70 | A | A | . . | B | . . | . . | . . | . . | B | B |
| Potassium Hydroxide | | | | | | | | | | | |
| 5% | 70 | A | A | . . | B | A | A | A | . . | B | . . |
| 27% | Boiling | A | A | A | B | A | A | A | . . | B | . . |
| 50% | Boiling | B | A | B | B | A | A | A | . . | B | . . |
| Melting | 675 | E | . . | E | . . | . . | . . | . . | . . | . . | . . |
| Potassium Nitrate | | | | | | | | | | | |
| Air-free | 212 | . . | . . | . . | . . | C | C | C | . . | . . | B |
| 1% and 5% | 70 | A | A | . . | B | A | A | A | . . | . . | B |
| 1% and 5% | Hot | A | A | . . | B | A | A | A | . . | . . | B |
| Potassium Nitrate (saltpeter) 50% | 70 | A | . . | A | B | . . | . . | . . | . . | . . | B |
| | Boiling | A | . . | A | B | . . | . . | . . | . . | . . | B |
| Potassium Oxalate | | A | A | A | B | A | A | A | . . | . . | . . |
| Potassium Permanganate | | | | | | | | | | | |
| 5% | 70 | A | A | A | B | A | A | A | . . | . . | B |
| Potassium Sulphate | | | | | | | | | | | |
| 1% and 5% | 70 | A | A | B | B | A | A | A | . . | D | B |
| | Hot | A | A | . . | B | A | A | A | . . | B | B |
| 10% | 85 | . . | . . | . . | B | C | C | C | . . | B | C |
| Potassium Sulphide (salt) | | A | A | A | . . | A | A | A | . . | . . | . . |

| | | | | | | | | | | | |
|---|---|---|---|---|---|---|---|---|---|---|---|
| Q | | | | | | | | | | | |
| Quinine Bisulphate—Dry | | B | A | B | . . | A | A | A | . . | . . | . . |
| Quinine Sulphate—Dry | | A | A | A | . . | A | A | A | . . | . . | . . |
| R | | | | | | | | | | | |
| Rosin 100% | Molten | A | A | A | B | A | A | A | . . | . . | . . |
| S | | | | | | | | | | | |
| Sea Water | | †A | †A | †B | A | A | A | A | . . | . . | A |
| | 212 | . . | . . | . . | . . | A | A | A | . . | . . | A |
| Sewage | | †A | †A | †A | . . | . . | A | A | . . | . . | A |
| Silver Bromide | | †B | †A | †C | . . | . . | . . | . . | . . | . . | . . |
| Silver Chloride | | E | E | . . | . . | . . | . . | . . | . . | . . | . . |
| Silver Nitrate—0-100% | Boiling | A | A | A | B | A | E | E | . . | . . | . . |
| Soap | 70 | A | A | A | . . | A | A | A | . . | . . | . . |
| Sodium Acetate | 85 | . . | . . | . . | B | C | C | C | . . | B | B |
| Moist | | †A | A | †A | . . | A | A | A | . . | . . | . . |
| Sodium Bicarbonate | 70 | A | A | A | . . | A | A | A | . . | . . | B |
| 5% Still | 150 | A | A | . . | A | A | A | A | . . | . . | . . |
| Sodium Carbonate 5% | 70 | A | A | A | A | A | A | A | . . | B | B |
| 5% | 150 | A | A | . . | A | A | A | A | . . | B | B |
| 5% | Boiling | A | . . | . . | A | . . | . . | . . | . . | B | B |
| 30% | 85 | . . | . . | . . | . . | C | C | C | . . | B | . . |
| 50% | Boiling | A | . . | A | . . | . . | . . | . . | . . | B | . . |
| Sodium Chloride | | | | | | | | | | | |
| 5%, Still | 70 | †A | A | . . | B | A | A | A | . . | A | A |
| | 150 | †A | A | . . | B | A | A | A | . . | A | A |
| 20%, Aerated | 70 | †A | A | . . | . . | A | A | A | . . | B | B |
| Saturated | 70 | †A | A | †A | . . | A | A | A | . . | . . | . . |
| | Boiling | †B | A | †B | . . | A | A | A | . . | . . | . . |
| Sodium Fluoride | | | | | | | | | | | |
| 5% Sol. | | †B | †A | †B | . . | A | A | A | . . | . . | . . |
| Sodium Hydroxide | 70 | A | A | . . | . . | A | A | A | . . | . . | . . |
| 20% | 230 | A | . . | A | B | A | . . | A | . . | A | A |
| 34% | 212 | A | . . | A | B | A | . . | A | . . | . . | . . |

## APPENDIX 7 *(continued)*

| Substance and its Condition | Temp. F | 304<br>304L<br>347<br>348 | 316<br>316L<br>309(1)<br>310(1) | 321 | 20Cb$_{-3}$ | Nickel<br>200 | Monel<br>400 | Inconel<br>600 | | Hastelloy Alloy<br>B | <br>C |
|---|---|---|---|---|---|---|---|---|---|---|---|
| ANSI/ASTM | | A-312 | A-312 | A-312 | . . . | B-161 | B-165 | B-167 | . . . | B-333 | B-334 |
| Melting | 610 | B | . . | B | . . | B | . . | C | . . | . . | . . |
| Sodium Hypochlorite—5% Still | | †B | A | †C | . . | C | C | C | . . | . . | . . |
| Sodium Hyposulphite | 70 | *A | A | †A | . . | A | A | A | . . | . . | . . |
| 0-50% | 212 | . . | . . | . . | . . | A | A | A | . . | . . | . . |
| 50%-75% | 212 | . . | . . | . . | . . | A | A | A | . . | . . | . . |
| 75-100% | 800 | . . | . . | . . | . . | **B | . . | **B | . . | . . | . . |
| Sodium Nitrate | Fused | C | B | . . | . . | A | B | A | . . | . . | . . |
| 0-50% | 212 | . . | . . | . . | A | C | C | A | . . | . . | . . |
| Sodium Sulphate | | | | | | | | | | | |
| 5%, Sill, All | 70 | A | A | B | A | A | A | A | . . | B | B |
| 0-30% | 212 | . . | . . | . . | B | C | C | C | . . | B | B |
| Sodium Sulphide | | | | | | | | | | | |
| Saturated | | †B | A | †A | B | A | A | A | . . | . . | A |
| Sodium Sulphite 5% | 70 | A | A | . . | A | A | A | A | . . | . . | B |
| 10% | 150 | A | A | . . | A | A | A | A | . . | . . | B |
| Stannic Chloride | | | | | | | | | | | |
| Sol. | 70 | D | D | D | . . | A | B | B | . . | . . | . . |
| Sp. Gr. 1.21 | Boiling | E | E | E | . . | B | B | C | . . | . . | . . |
| Stannous Chloride | | | | | | | | | | | |
| Saturated | | C | A | C | . . | A | B | B | . . | B | . . |
| Dry—100% | 85 | . . | . . | . . | . . | E | E | E | . . | . . | . . |
| Stearic Acid | | A | A | A | B | A | A | A | . . | A | A |
| Sulphate Black Liquor | | . . | . . | | B | . . | . . | A | . . | A | A |
| Green Liquor | | . . | . . | . . | B | . . | . . | C | . . | B | B |
| Sulphur | 265 | A | . . | . . | A | . . | . . | . . | . . | . . | A |
| Boiling | 830 | E | . . | . . | A | . . | . . | . . | . . | . . | A |
| Air Free | 400 | . . | . . | . . | . . | C | C | A | . . | . . | A |
| Sulphur Chloride | | E | D | . . | B | A | A | A | . . | . . | B |
| Sulphur Dioxide, Dry | Limit | . . | . . | . . | B | . . | . . | . . | . . | . . | B |
| Sulphur Dioxide Gas | | | | | | | | | | | |
| Moist | 70 | B | A | B | . . | D | C | C | . . | . . | A |
| Dry | 575 | A | A | A | . . | . . | . . | . . | . . | . . | A |

| | | | | | | | | | | | |
|---|---|---|---|---|---|---|---|---|---|---|---|
| Sulphur—Dry | Molten | A | A | . . | A | A | A | A | . . | . . | A |
| Wet | | †B | †A | . . | A | B | B | A | . . | . . | A |
| Sulphuric Acid | | | | | | | | | | | |
| 5% | 70 | C | B | B | A | A | A | A | . . | A | A |
| | Boiling | E | C | E | B | D | A | C | . | A | B |
| 10% | 70 | C | B | C | A | B | A | B | . . | A | A |
| | Boiling | E | D | E | B | C | A | C | . . | A | B |
| 50% | 70 | D | C | C | A | B | A | B | . . | A | A |
| | Boiling | E | D | E | C | E | E | E | . . | A | B |
| Conc. | 70 | A | A | A | A | B | B | B | . . | A | A |
| | Boiling | D | D | . . | . . | E | E | E | . . | E | D |
| | 300 | E | E | E | E | E | E | E | . . | . . | . . |
| Fuming | 70 | C | B | . . | . . | C | B | B | . . | E | A |
| Sulphurous Acid | | | | | | | | | | | |
| Saturated | 70 | C | B | . . | B | E | E | E | . . | . . | B |
| 60 psi | 250 | C | B | B | . . | E | E | E | . . | . . | B |
| 70/125 psi | 310 | C | B | E | . . | E | E | E | . . | . . | B |
| 150 psi | 375 | C | B | E | . . | E | E | E | . . | . . | B |
| T | | | | | | | | | | | |
| Tannic Acid | | | | | | | | | | | |
| 10% | 85 | A | A | A | B | C | C | C | . . | B | B |
| | Boiling | A | . . | A | B | . . | C | C | . . | B | B |
| 0-100% | 212 | . . | . . | . . | B | . . | . . | . . | . . | . . | . . |
| 50% | Boiling | A | . . | A | B | . . | . . | . . | . . | . . | . . |
| Tartaric Acid, Air-free | 70 | A | A | A | B | A | A | A | . . | . . | . . |
| | 150 | B | A | A | B | A | A | A | . . | . . | . . |
| 0-50% | 212 | . . | . . | B | B | C | C | C | . . | B | B |
| Tin | Molten | C | C | E | . . | E | E | E | . . | . . | . . |
| Trichloracetic Acid | 70 | E | E | E | . . | B | C | B | . . | . . | . . |
| V | | | | | | | | | | | |
| Varnish | 85 | A | A | A | . . | A | A | A | . . | . . | . . |
| | Hot | A | A | . . | . . | A | A | A | . . | . . | . . |
| Vinegar—Still, Agitated or Aerated | 70-85 | A | A | A | . . | A | A | A | . . | . . | . . |

## APPENDIX 7 *(continued)*

| Substance and its Condition | Temp. F | 304<br>304L<br>347<br>348 | 316<br>316L<br>309(1)<br>310(1) | 321 | 20Cb$_{-3}$ | Nickel<br>200 | Monel<br>400 | Inconel<br>600 | | Hastelloy Alloy<br>B | Hastelloy Alloy<br>C |
|---|---|---|---|---|---|---|---|---|---|---|---|
| ANSI/ASTM | | A-312 | A-312 | A-312 | . . . | B-161 | B-165 | B-167 | . . . | B-333 | B-334 |
| Z | | | | | | | | | | | |
| Zinc | Molten | E | E | E | . . | E | E | E | . . | . . | . . |
| Zinc Chloride | | | | | | | | | | | |
| 5% Still | 70 | †A | †A | . . | A | A | A | A | . . | B | . . |
| | Boiling | †B | †B | . . | B | B | B | B | . . | B | . . |
| Zinc Chloride Sol. | | | | | | | | | | | |
| Sp. gr. 2.05 | 100 | †A | . . | †C | . . | . . | . . | . . | . . | . . | . . |
| 1.09 | Boiling | †A | . . | D | . . | . . | . . | . . | . . | . . | . . |
| Zinc Nitrate Sol. | Hot | A | . . | A | . . | . . | . . | . . | . . | . . | . . |
| Zinc Sulphate | | | | | | | | | | | |
| 5% | 70 | A | A | . . | A | A | A | A | . . | B | B |
| Saturated | 70 | A | A | B | . . | A | A | A | . . | . . | . . |
| 25% | Boiling | A | A | B | B | A | A | A | . . | B | B |

# APPENDIX 8

# ANSI & API metallic piping standards

## STEEL PIPE

| | |
|---|---|
| ANSI B36.10 | Welded and seamless wrought-steel pipe |
| ANSI/ASTM A 53 | Welded and seamless steel pipe |
| ANSI/ASTM A 106 | Seamless carbon steel pipe for high-temperature service |
| ANSI/ASTM A 120 | Black and hot-dipped zinc-coated (galvanized) welded and seamless steel pipe for ordinary uses |
| ANSI/ASTM A 134 | Electric-fusion (arc)-welded steel-plate pipe (sizes 16 in. and over) |
| ANSI/ASTM A 135 | Electric-resistance-welded steel pipe |
| ANSI/ASTM A 139 | Electric-fusion (arc)-welded steel pipe (sizes 4 in. and over) |
| ANSI/ASTM A 155 | Electric-fusion-welded steel pipe for high-pressure service |
| ANSI/ASTM A 211 | Spiral-welded steel or iron pipe |
| ANSI/ASTM A 381 | Metal-arc-welded steel pipe for high-pressure transmission systems |
| ANSI/ASTM A 524 | Seamless carbon steel pipe for process |
| ANSI/ASTM A 530 | General requirements for specialized carbon and alloy steel pipe |
| ANSI/ASTM A 587 | Electric-welded low-carbon steel pipe for the chemical industry |
| ANSI/ASTM A 589 | Seamless and welded carbon steel water-well pipe |
| ANSI/ASTM A 660 | Centrifugally cast carbon steel pipe for high-temperature service |
| ANSI/ASTM A 672 | Electric-fusion-welded steel pipe for high-pressure service at moderate temperatures |
| API 5 L | Line pipe |
| API 5 LS | Spiral-weld line pipe |
| API 5 LU | Ultrahigh test heat-treated line pipe |
| API 5 LX | High-test line pipe |

## CAST-IRON AND DUCTILE-IRON PIPE

| | |
|---|---|
| ANSI A 21.52 | Ductile-rion pipe, centrifugally cast, in metal molds or sand-lined molds for gas |

(Suffixes indicating latest year of issue have not been included in specification number)

| | |
|---|---|
| ANSI A 40.5 | Threated cast-iron pipe for drainage, vent, and waste services |
| ANSI/ASTM A 74 | Cast-iron soil pipe and fittings |
| ANSI/ASTM A 142 | Cast-iron culvert pipe |
| ANSI/ASTM A 716 | Ductile-iron culvert pipe |
| ANSI/AWWA C 101 | Thickness design of cast-iron pipe |
| ANSI/AWWA C 150 | Thickness design of ductile-iron pipe |
| ANSI/AWWA C 106 | Cast-iron pipe centrifugally cast in metal molds for water or other liquids |
| ANSI/AWWA C 108 | Cast-iron pipe centrifugally cast in sand-lined molds for water or other liquids |
| ANSI/AWWA C 112 | 2-in. and 2¹/₄-in. cast-iron pipe, centrifugally cast for water or other liquids |
| ANSI/AWWA C 115 | Flanged cast-iron and ductile-iron pipe with threaded flanges |
| ANSI/AWWA C 151 | Ductile-iron pipe, centrifugally cast in metal molds or sand-lined molds for water or other liquids |

## STEEL PIPE FOR LOW-TEMPERATURE SERVICE

| | |
|---|---|
| ANSI/ASTM A 333 | Seamless and welded steel pipe for low-temperature service |
| ANSI/ASTM A 671 | Electric-fusion welded pipe for atmospheric and lower temperatures |

## FERRITIC-ALLOY PIPING AND TUBING

| | |
|---|---|
| ANSI/ASTM A 268 | Seamless and welded ferritic stainless-steel tubing |
| ANSI/ASTM A 335 | Seamless ferritic-alloy steel pipe for high-temperature service |
| ANSI/ASTM A 369 | Ferritic-alloy steel-forged and bored pipe for high-temperature service |
| ANSI/ASTM A 405 | Seamless ferritic-alloy steel pipe specially heat-treated for high-temperature service |
| ANSI/ASTM A 426 | Centrifugally cast ferritic-alloy steel pipe for high-temperature service |
| ANSI/ASTM A 669 | Seamless ferritic-austenitic alloy steel tubes |

## NICKEL AND NICKEL ALLOY PIPE AND TUBING

| | |
|---|---|
| ANSI/ASTM B 161 | Nickel seamless pipe and tube |
| ANSI/ASTM B 165 | Nickel-copper alloy seamless pipe and tube |
| ANSI/ASTM B 167 | Nickel-chromium-iron alloy seamless pipe and tube |

| | |
|---|---|
| ANSI/ASTM B 407 | Nickel-iron-chromium alloy seamless pipe and tube |
| ANSI/ASTM B 423 | Nickel-iron chromium-molybdenum-copper alloy seamless pipe and tube |
| ANSI/ASTM B 444 | Nickel-chromium-molybdenum-columbium alloy seamless pipe and tube |
| ANSI/ASTM B 445 | Nickel-chromium-iron-columbium-molybdenum-tungsten alloy seamless pipe and tube |
| ANSI/ASTM B 513 | Supplementary requirements for nickel-alloy seamless pipe and tube for nuclear applications |
| ANSI/ASTM B 514 | Welded nickel-iron-chromium alloy pipe |
| ANSI/ASTM B 517 | Welded nickel-chromium-iron alloy pipe |

## ALUMINUM AND ALUMINUM ALLOY PIPE AND TUBING

| | |
|---|---|
| ANSI/ASTM B 210 | Aluminum-alloy drawn seamless tubes |
| ANSI/ASTM B 221 | Aluminum-alloy extruded bars, rods, wire shapes, and tubes |
| ANSI/ASTM B 241 | Aluminum-alloy seamless pipe and seamless extruded tube |
| ANSI/ASTM B 313 | Aluminum-alloy round welded tubes |
| ANSI/ASTM B 345 | Aluminum-alloy seamless extruded tube and seamless pipe for gas and oil transmission and distribution piping systems |
| ANSI/ASTM B 483 | Aluminum-alloy drawn tubes for general purpose applications |
| ANSI/ASTM B 547 | Aluminum-alloy formed and arc-welded round tube |

## COPPER AND COPPER ALLOY PIPE AND TUBING

| | |
|---|---|
| ANSI/ASTM B 42 | Seamless copper pipe, standard sizes, hard drawn |
| ANSI/ASTM B 43 | Seamless red brass pipe, standard sizes, annealed |
| ANSI/ASTM B 68 | Seamless copper tube, bright annealed |
| ANSI/ASTM B 75 | Seamless copper tube, annealed or drawn |
| ANSI/ASTM B 88 | Seamless copper water tube |
| ANSI/ASTM B 135 | Seamless brass tube |
| ANSI/ASTM B 165 | Nickel-copper-alloy seamless pipe and tube |
| ANSI/ASTM B 251 | General requirements for wrought seamless copper and copper-alloy tube |
| ANSI/ASTM B 280 | Seamless copper tube for refrigeration field service |
| ANSI/ASTM B 302 | Threadless copper pipe |
| ANSI/ASTM B 306 | Copper drainage tube (DWV) |

| | |
|---|---|
| ANSI/ASTM B 315 | Copper-silicon-alloy seamless pipe and tube |
| ANSI/ASTM B 447 | Welded copper tube |
| ANSI/ASTM B 466 | Seamless copper-nickel pipe and tube |
| ANSI/ASTM B 467 | Welded copper-nickel pipe and tube |
| ANSI/ASTM B 469 | Copper-iron alloy tubes for pressure applications |
| ANSI/ASTM B 543 | Welded copper and copper-alloy tube |
| ANSI/ASTM B 552 | Seamless and welded copper-nickel tubes for water desalting plants |
| ANSI/ASTM B 587 | Welded brass tube |
| ANSI/ASTM B 608 | Welded copper-alloy pipe |

## AUSTENITIC STAINLESS-STEEL PIPE AND TUBING

| | |
|---|---|
| ANSI B36.19 | Stainless-steel pipe |
| ANSI/ASTM A 269 | Seamless and welded austenitic stainless-steel tubing for general services |
| ANSI/ASTM A 270 | Seamless and welded austenitic stainless-steel sanitary tubing |
| ANSI/ASTM A 271 | Seamless austenitic-steel still tubes for refinery service |
| ANSI/ASTM A 312 | Seamless and welded austenitic stainless-steel pipe |
| ANSI/ASTM A 358 | Electric-fusion-welded austenitic chromium-nickel-alloy steel pipe for high-temperature service |
| ANSI/ASTM A 376 | Seamless austenitic-steel pipe for high-temperature central-station service |
| ANSI/ASTM A 409 | Welded large-diameter light-wall austenitic chromium-nickel alloy steel pipe for corrosive or high-temperature service |
| ANSI/ASTM A 430 | Austenitic-steel forged and bored pipe for high-temperature service |
| ANSI/ASTM A 450 | General requirements for carbon, ferritic, and austenitic tubes |
| ANSI/ASTM A 451 | Centrifugally cast austenitic-steel pipe for high-temperature service |
| ANSI/ASTM A 452 | Centrifugally cast austenitic cold-wrought pipe for high-temperature service |
| ANSI/ASTM A 530 | General requirements for specialized carbon and alloy steel pipe |
| ANSI/ASTM A 632 | Seamless and welded austenitic stainless-steel tubing |
| ANSI/ASTM A 651 | Stainless-steel water-DWV tubes |
| ANSI/ASTM A 669 | Seamless ferritic-austenitic-alloy steel tubes |

# APPENDIX 9

# Fitting and Flange Dimensions

*(Reprinted by permission of Tube Turns Division, Chemetron Corp.)*

| NOM PIPE SIZE | OD | WALL THICKNESS T | | | | 90° ELBOWS | | 180° RETURNS | | 45° ELBOWS | TEES | CAPS | CROSSES | STUB ENDS | |
|---|---|---|---|---|---|---|---|---|---|---|---|---|---|---|---|
| | | ST | XS | 160 | XX | LONG R A | SHORT R A | LONG R K | SHORT R K | B | C | E | C | F | G |
| 1/2 | .840 | .109 | .147 | — | — | 1 1/2 | — | 1 15/16 | — | 5/8 | 1 | 1 | — | 3 | 1 3/8 |
| 3/4 | 1.050 | .113 | .154 | — | .308 | 1 1/8 | — | 1 11/16 | — | 7/16 | 1 1/8 | 1 1/4 | — | 3 | 1 11/16 |
| 1 | 1.315 | .133 | .179 | .250 | .358 | 1 1/2 | 1 | 2 3/16 | 1 5/8 | 7/8 | 1 1/2 | 1 1/2 | — | 4 | 2 |
| 1 1/4 | 1.660 | .140 | .191 | .250 | .382 | 1 7/8 | 1 1/4 | 2 3/4 | 2 1/16 | 1 | 1 7/8 | 1 1/2 | 1 7/8 | 4 | 2 1/2 |
| 1 1/2 | 1.900 | .145 | .200 | .281 | .400 | 2 1/4 | 1 1/2 | 3 1/4 | 2 7/16 | 1 1/8 | 2 1/4 | 1 1/2 | 2 1/4 | 4 | 2 7/8 |
| 2 | 2.375 | .154 | .218 | .334 | .436 | 3 | 2 | 4 3/16 | 3 3/16 | 1 3/8 | 2 1/2 | 1 1/2* | 2 1/2 | 6 | 3 5/8 |
| 2 1/2 | 2.875 | .203 | .276 | .375 | .552 | 3 3/4 | 2 1/2 | 5 3/16 | 3 15/16 | 1 3/4 | 3 | 1 1/2* | 3 | 6 | 4 1/8 |
| 3 | 3.500 | .216 | .300 | .438 | .600 | 4 1/2 | 3 | 6 1/4 | 4 3/4 | 2 | 3 3/8 | 2* | 3 3/8 | 6 | 5 |
| 3 1/2 | 4.000 | .226 | .318 | — | .636 | 5 1/4 | 3 1/2 | 7 1/4 | 5 1/2 | 2 1/4 | 3 3/4 | 2 1/2* | 3 3/4 | 6 | 5 1/2 |
| 4 | 4.500 | .237 | .337 | .531 | .674 | 6 | 4 | 8 1/4 | 6 1/4 | 2 1/2 | 4 1/8 | 2 1/2* | 4 1/8 | 6 | 6 3/16 |
| 5 | 5.563 | .258 | .375 | .625 | .750 | 7 1/2 | 5 | 10 5/16 | 7 3/4 | 3 1/8 | 4 7/8 | 3* | 4 7/8 | 8 | 7 5/16 |
| 6 | 6.625 | .280 | .432 | .719 | .864 | 9 | 6 | 12 5/16 | 9 5/16 | 3 3/4 | 5 5/8 | 3 1/2* | 5 5/8 | 8 | 8 1/2 |

**APPENDIX 9** *(continued)*

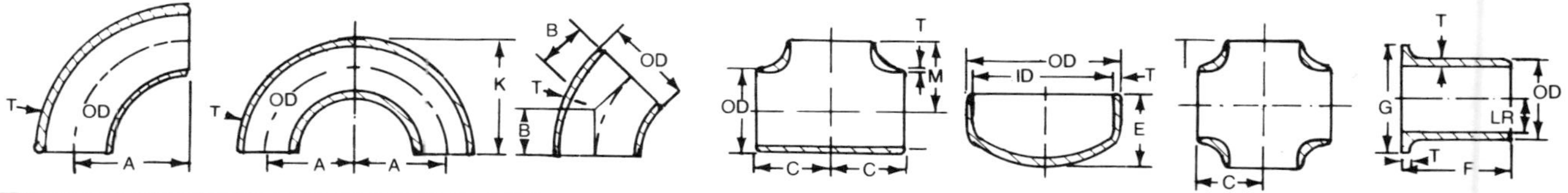

| NOM PIPE SIZE | OD | WALL THICKNESS T | | | | 90° ELBOWS | | 180° RETURNS | | 45° ELBOWS | TEES | CAPS | CROSSES | STUB ENDS | |
|---|---|---|---|---|---|---|---|---|---|---|---|---|---|---|---|
| | | ST | XS | 160 | XX | LONG R A | SHORT R A | LONG R K | SHORT R K | B | C | E | C | F | G |
| 8 | 8.625 | .322 | .500 | .906 | .875 | 12 | 8 | 16 5/16 | 12 5/16 | 5 | 7 | 4* | 7 | 8 | 10 5/8 |
| 10 | 10.750 | .365 | .500 | 1.125 | 1.000 | 15 | 10 | 20 3/8 | 15 3/8 | 6 1/4 | 8 1/2 | 5* | 8 1/2 | 10 | 12 3/4 |
| 12 | 12.750 | .375 | .500 | 1.312 | 1.000 | 18 | 12 | 24 3/8 | 18 3/8 | 7 1/2 | 10 | 6* | 10 | 10 | 15 |
| 14 | 14.000 | .375 | .500 | — | — | 21 | 14 | 28 | 21 | 8 3/4 | 11 | 6 1/2* | 11 | 12 | 16 1/4 |
| 16 | 16.000 | .375 | .500 | — | — | 14 | 16 | 32 | 24 | 10 | 12 | 7* | 12 | 12 | 18 1/2 |
| 18 | 18.000 | .375 | .500 | — | — | 17 | 18 | 36 | 27 | 11 1/4 | 13 1/2 | 8* | 13 1/2 | 12 | 21 |
| 20 | 20.000 | .375 | .500 | — | — | 30 | 20 | 40 | 30 | 12 1/2 | 15 | 9* | 15 | 12 | 23 |
| 22 | 22.000 | .375 | .500 | — | — | 33 | — | 44 | — | 13 1/2 | 16 1/2 | 10 | 16 1/2 | — | — |
| 24 | 24.000 | .375 | .500 | — | — | 36 | 24 | 48 | 36 | 15 | 17 | 10 1/2 | 17 | 12 | 27 1/4 |
| 26 | 26.000 | .375 | .500 | — | — | 39 | — | 52 | — | 16 | 19 1/2 | 10 1/2 | 19 1/2 | — | — |
| 30 | 30.000 | .375 | .500 | — | — | 45 | 30 | 60 | 45 | 18 1/2 | 22 | 10 1/2 | 22 | — | — |
| 34 | 34.000 | .375 | .500 | — | — | 51 | — | — | — | 21 | 25 | 10 1/2 | 25 | — | — |
| 36 | 36.000 | .375 | .500 | — | — | 54 | 36 | — | 54 | 22 1/4 | 26 1/2 | 10 1/2 | — | — | — |
| 42 | 42.000 | .375 | .500 | — | — | 63 | 48 | — | — | 26 | — | 12 | — | — | — |

| NOM PIPE SIZE | | CONCENTRIC AND ECCENTRIC REDUCERS | REDUCING OUTLET TEES | |
|---|---|---|---|---|
| | | H | C | M |
| 1/2 × | 1/4 | — | 1 | 1 |
| | 3/8 | — | 1 | 1 |
| 3/4 × | 3/8 | 1 1/2 | 1 1/2 | 1 1/8 |
| | 1/2 | 1 1/2 | 1 1/2 | 1 1/8 |
| 1 × | 3/8 | 2 | 1 1/2 | 1 1/2 |
| | 1/2 | 2 | 1 1/2 | 1 1/2 |
| | 3/4 | 2 | 1 1/2 | 1 1/2 |
| 1 1/4 × | 1/2 | 2 | 1 7/8 | 1 7/8 |
| | 3/4 | 2 | 1 7/8 | 1 7/8 |
| | 1 | 2 | 1 7/8 | 1 7/8 |
| 1 1/2 × | 1/2 | 2 1/2 | 2 1/4 | 2 1/4 |
| | 3/4 | 2 1/2 | 2 1/4 | 2 1/4 |
| | 1 | 2 1/2 | 2 1/4 | 2 1/4 |
| | 1 1/4 | 2 1/2 | 2 1/4 | 2 1/4 |
| 2 × | 3/4 | 3 | 2 1/2 | 1 3/4 |
| | 1 | 3 | 2 1/2 | 2 |
| | 1 1/4 | 3 | 2 1/2 | 2 1/4 |
| | 1 1/2 | 3 | 2 1/2 | 2 3/8 |
| 2 1/2 × | 1 | 3 1/2 | 3 | 2 1/4 |
| | 1 1/4 | 3 1/2 | 3 | 2 1/2 |
| | 1 1/2 | 3 1/2 | 3 | 2 3/8 |
| | 2 | 3 1/2 | 3 | 2 3/4 |
| 3 × | 1 | — | 3 3/8 | 2 3/8 |
| | 1 1/4 | 3 1/2 | 3 3/8 | 2 3/4 |
| | 1 1/2 | 3 1/2 | 3 3/8 | 2 7/8 |
| | 2 | 3 1/2 | 3 3/8 | 3 |
| | 2 1/2 | 3 1/2 | 3 3/8 | 3 1/4 |
| 3 1/2 × | 1 1/4 | 4 | — | — |
| | 1 1/2 | 4 | 3 3/4 | 3 1/8 |
| | 2 | 4 | 3 3/4 | 3 1/4 |
| | 2 1/2 | 4 | 3 3/4 | 3 1/2 |
| | 3 | 4 | 3 3/4 | 3 5/8 |
| 4 × | 1 1/2 | 4 | 4 1/8 | 3 3/8 |
| | 2 | 4 | 4 1/8 | 3 1/2 |
| | 2 1/2 | 4 | 4 1/8 | 3 3/4 |
| | 3 | 4 | 4 1/8 | 3 7/8 |
| | 3 1/2 | 4 | 4 1/8 | 4 |
| 5 × | 2 | 5 | 4 7/8 | 4 1/8 |
| | 2 1/2 | 5 | 4 7/8 | 4 1/4 |
| | 3 | 5 | 4 7/8 | 4 3/8 |
| | 3 1/2 | 5 | 4 7/8 | 4 1/2 |
| | 4 | 5 | 4 7/8 | 4 5/8 |

*Dimensions apply to ST and XS only.

| NOM PIPE SIZE | | CONCENTRIC AND ECCENTRIC REDUCERS | REDUCING OUTLET TEES | |
|---|---|---|---|---|
| | | H | C | M |
| 6 × | $2^1/_2$ | $5^1/_2$ | $5^5/_8$ | $4^3/_4$ |
| | 3 | $5^1/_2$ | $5^5/_8$ | $4^7/_8$ |
| | $3^1/_2$ | $5^1/_2$ | $5^5/_8$ | 5 |
| | 4 | $5^1/_2$ | $5^5/_8$ | $5^1/_8$ |
| | 5 | $5^1/_2$ | $5^5/_8$ | $5^1/_8$ |
| 8 × | 3 | — | 7 | 6 |
| | $3^1/_2$ | 6 | 7 | 6 |
| | 4 | 6 | 7 | $6^1/_8$ |
| | 5 | 6 | 7 | $6^3/_8$ |
| | 6 | 6 | 7 | $6^5/_8$ |
| 10 × | 4 | 7 | $8^1/_2$ | $7^1/_4$ |
| | 5 | 7 | $8^1/_2$ | $7^1/_2$ |
| | 6 | 7 | $8^1/_2$ | $7^5/_8$ |
| | 8 | 7 | $8^1/_2$ | 8 |
| 12 × | 5 | 8 | 10 | $8^1/_2$ |
| | 6 | 8 | 10 | $8^5/_8$ |
| | 8 | 8 | 10 | 9 |
| | 10 | 8 | 10 | $9^1/_2$ |
| 14 × | 6 | 13 | 11 | $9^3/_8$ |
| | 8 | 13 | 11 | $9^1/_4$ |
| | 10 | 13 | 11 | $10^1/_8$ |
| | 12 | 13 | 11 | $10^5/_8$ |
| 16 × | 6 | — | 12 | $10^3/_8$ |
| | 8 | 14 | 12 | $10^3/_4$ |
| | 10 | 14 | 12 | $11^1/_8$ |
| | 12 | 14 | 12 | $11^5/_8$ |
| | 14 | 14 | | 12 |
| | | | | 12 |
| 18 × | 8 | — | $13^1/_2$ | $11^3/_4$ |
| | 10 | 15 | $13^1/_2$ | $12^1/_8$ |
| | 12 | 15 | $13^1/_2$ | $12^5/_8$ |
| | 14 | 15 | $13^1/_2$ | 13 |
| | 16 | 15 | $13^1/_2$ | 13 |
| 20 × | 8 | — | 15 | $12^3/_4$ |
| | 10 | — | 15 | $13^1/_8$ |
| | 12 | 20 | 15 | $13^5/_8$ |
| | 14 | 20 | 15 | 14 |
| | 16 | 20 | 15 | 14 |
| | 18 | 20 | 15 | $14^1/_2$ |
| 22 × | 10 | — | $16^1/_2$ | $14^1/_8$ |
| | 12 | — | $16^1/_2$ | $14^5/_8$ |
| | 14 | 20 | $16^1/_2$ | 15 |
| | 16 | 20 | $16^1/_2$ | 15 |
| | 18 | 20 | $16^1/_2$ | $15^1/_2$ |
| | 20 | 20 | $16^1/_2$ | 16 |

| NOM PIPE SIZE | CONCENTRIC AND ECCENTRIC REDUCERS | REDUCING OUTLET TEES | |
|---|---|---|---|
| | H | C | M |
| 24 × 10 | — | 17 | $15\frac{1}{8}$ |
| 12 | — | 17 | $15\frac{5}{8}$ |
| 14 | — | 17 | 16 |
| 16 | 20 | 17 | 16 |
| 18 | 20 | 17 | $16\frac{1}{2}$ |
| 20 | 20 | 17 | 17 |
| 26 × 12 | — | $19\frac{1}{2}$ | $16\frac{5}{8}$ |
| 14 | — | $19\frac{1}{2}$ | 17 |
| 16 | — | $19\frac{1}{2}$ | 17 |
| 18 | 24 | $19\frac{1}{2}$ | $17\frac{1}{2}$ |
| 20 | 24 | $19\frac{1}{2}$ | 18 |
| 22 | 24 | $19\frac{1}{2}$ | $18\frac{1}{2}$ |
| 24 | 24 | $19\frac{1}{2}$ | 19 |
| 30 × 14 | — | 22 | 19 |
| 16 | — | 22 | 19 |
| 18 | — | 22 | $19\frac{1}{2}$ |
| 20 | 24 | 22 | 20 |
| 22 | — | 22 | $20\frac{1}{2}$ |
| 24 | 24 | 22 | 21 |
| 26 | 24 | 22 | $21\frac{1}{2}$ |
| 28 | 24 | 22 | $21\frac{1}{2}$ |

| NOM PIPE SIZE | CONCENTRIC AND ECCENTRIC REDUCERS | REDUCING OUTLET TEES | |
|---|---|---|---|
| | H | C | M |
| 34 × 16 | — | 25 | 21 |
| 18 | — | 25 | $21\frac{1}{2}$ |
| 20 | — | 25 | 22 |
| 22 | — | 25 | $22\frac{1}{2}$ |
| 24 | 24 | 25 | 23 |
| 26 | 24 | 25 | $23\frac{1}{2}$ |
| 28 | — | 25 | $23\frac{1}{2}$ |
| 30 | 24 | 25 | 24 |
| 32 | 24 | 25 | $24\frac{1}{2}$ |
| 36 × 16 | — | $26\frac{1}{2}$ | 22 |
| 18 | — | $26\frac{1}{2}$ | $22\frac{1}{2}$ |
| 20 | — | $26\frac{1}{2}$ | 23 |
| 22 | — | $26\frac{1}{2}$ | $23\frac{1}{2}$ |
| 24 | 24 | $26\frac{1}{2}$ | 24 |
| 26 | 24 | $26\frac{1}{2}$ | $24\frac{1}{2}$ |
| 28 | — | $26\frac{1}{2}$ | $24\frac{1}{2}$ |
| 30 | 24 | $26\frac{1}{2}$ | 25 |
| 32 | 24 | $26\frac{1}{2}$ | $25\frac{1}{2}$ |
| 34 | 24 | $26\frac{1}{2}$ | 26 |
| 42 × 24 | 24 | — | — |
| 26 | 24 | — | — |
| 30 | 24 | — | — |
| 32 | 24 | — | — |
| 34 | 24 | — | — |
| 36 | 24 | — | — |

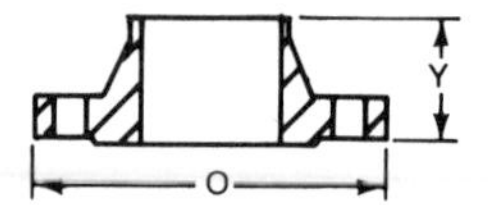


## WELDING NECK FLANGES

| Nom Pipe Size | 150 lb. | | 300 Lb. | | 400 Lb. | | 600 Lb. | |
|---|---|---|---|---|---|---|---|---|
| | Outside Diam Flange O | Length Thru Hub Y(1) | Outside Diam of Flange O | Length Thru Hub Y(1) | Outside Diam of Flange O | Length Thru Hub Y(2) | Outside Diam of Flange O | Length Thru Hub Y(2) |
| 1/2 | 3 1/2 | 1 7/8 | 3 3/4 | 2 1/16 | For sizes 3 1/2 and smaller use 600 Lb. Standard | | 3 3/4 | 2 1/16 |
| 3/4 | 3 7/8 | 2 1/16 | 4 5/8 | 2 1/4 | | | 4 5/8 | 2 1/4 |
| 1 | 4 1/4 | 2 3/16 | 4 7/8 | 2 7/8 | | | 4 7/8 | 2 7/16 |
| 1 1/4 | 4 5/8 | 2 1/4 | 5 1/4 | 2 9/16 | | | 5 1/4 | 2 5/8 |
| 1 1/2 | 5 | 2 7/16 | 6 1/8 | 2 11/16 | | | 6 1/8 | 2 3/4 |
| 2 | 6 | 2 1/2 | 6 1/2 | 2 3/4 | | | 6 1/2 | 2 7/8 |
| 2 1/2 | 7 | 2 3/4 | 7 1/2 | 3 | | | 7 1/2 | 3 1/8 |
| 3 | 7 1/2 | 2 3/3 | 8 1/4 | 3 1/8 | | | 8 1/4 | 3 1/4 |
| 3 1/2 | 8 1/2 | 2 13/16 | 9 | 3 3/16 | | | 9 | 3 3/8 |
| 4 | 9 | 3 | 10 | 3 3/8 | 10 | 3 1/2 | 10 3/4 | 4 |
| 5 | 10 | 3 1/2 | 11 | 3 7/8 | 11 | 4 | 13 | 4 1/2 |
| 6 | 11 | 3 1/2 | 12 1/2 | 3 7/8 | 12 1/2 | 4 1/16 | 14 | 4 5/8 |
| 8 | 13 1/2 | 4 | 15 | 4 3/8 | 15 | 4 5/8 | 16 1/2 | 5 1/4 |
| 10 | 16 | 4 | 17 1/2 | 4 5/8 | 17 1/2 | 4 7/8 | 20 | 6 |
| 12 | 19 | 4 1/2 | 20 1/2 | 5 1/8 | 20 1/2 | 5 3/8 | 22 | 6 1/8 |
| 14 | 21 | 5 | 23 | 5 5/8 | 23 | 5 7/8 | 23 3/4 | 6 1/2 |
| 16 | 23 1/2 | 5 | 25 1/2 | 5 3/4 | 25 1/2 | 6 | 27 | 7 |
| 18 | 25 | 5 1/2 | 28 | 6 1/4 | 28 | 6 1/2 | 29 1/4 | 7 1/4 |
| 20 | 27 1/2 | 5 11/16 | 30 1/2 | 6 3/8 | 30 1/2 | 6 5/8 | 32 | 7 1/2 |
| 22 | 29 1/2 | 5 7/8 | 33 | 6 1/2 | 33 | 6 3/4 | 34 1/2 | 7 3/4 |
| 24 | 32 | 6 | 36 | 6 5/8 | 36 | 6 7/8 | 37 | 8 |
| 26 | 34 1/4 | 5 | 38 1/4 | 7 1/4 | 38 1/4 | 7 5/8 | 40 | 8 3/4 |
| 30 | 38 3/4 | 5 1/8 | 43 | 8 1/4 | 43 | 8 2/8 | 44 1/2 | 9 3/4 |
| 34 | 43 3/4 | 5 5/16 | 47 1/2 | 9 1/8 | 47 1/2 | 9 1/2 | 49 | 10 5/8 |
| 36 | 46 | 5 3/8 | 50 | 9 1/2 | 50 | 9 7/8 | 51 3/4 | 11 1/8 |
| 42 | 53 | 5 5/8 | 57 | 10 7/8 | 57 | 11 3/8 | 58 3/4 | 12 3/4 |

(1) The 1/16" raised face *is* included in "Length thru Hub 'Y'."
(2) The 1/4" raised face *is not* included in "Length thru Hub 'Y'."

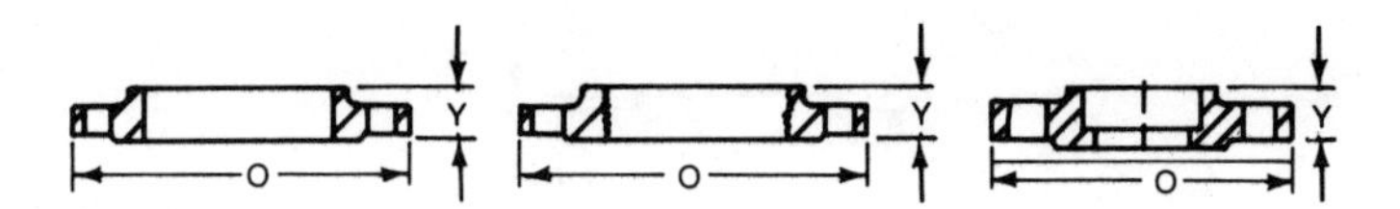


## SLIP-ON, THREADED, AND SOCKET-TYPE FLANGES

| Nom Pipe Size | 150 Lb. | | 300 Lb. | | 400 Lb. | | 600 Lb. | |
|---|---|---|---|---|---|---|---|---|
| | Outside Diam O | Length Thru Hub Y(1) | Outside Diam O | Length Thru Hub Y(1) | Outside Diam O | Length Thru Hub Y(2) | Outside Diam O | Length Thru Hub Y(2) |
| 1/4 | 3 1/2 | 5/8*† | 3 3/4 | 7/8*† | For sizes 3 1/2 and smaller use 600 Lb. Standard | | 3 3/4 | 7/8*† |
| 3/8 | 3 1/2 | 5/8*† | 3 3/4 | 7/8*† | | | 3 3/4 | 7/8*† |
| 1/2 | 3 1/2 | 5/8 | 3 3/4 | 7/8 | | | 3 3/4 | 7/8 |
| 3/4 | 3 7/8 | 5/8 | 4 5/8 | 1 | | | 4 5/8 | 1 |
| 1 | 4 1/4 | 11/16 | 4 7/8 | 1 1/16 | | | 4 7/8 | 1 1/16 |
| 1 1/4 | 4 5/8 | 13/16 | 5 1/4 | 1 1/16 | | | 5 1/4 | 1 1/8 |
| 1 1/2 | 5 | 7/8 | 6 1/8 | 1 3/16 | | | 6 1/8 | 1 1/4 |
| 2 | 6 | 1 | 6 1/2 | 1 5/16 | | | 6 1/2 | 1 7/16 |
| 2 1/2 | 7 | 1 1/8 | 7 1/2 | 1 1/2 | | | 7 1/2 | 1 5/8 |
| 3 | 7 1/2 | 1 3/16 | 8 1/4 | 1 11/16 | | | 8 1/4 | 1 13/16 |
| 3 1/2 | 8 1/2 | 1 1/4 | 9 | 1 3/4 | | | 9 | 1 15/16 |
| 4 | 9 | 1 5/16 | 10 | 1 7/8 | 10 | 2‡ | 10 3/4 | 2 1/8‡ |
| 5 | 10 | 1 7/16 | 11 | 2‡ | 11 | 2 1/8‡ | 13 | 2 3/8‡ |
| 6 | 11 | 1 9/16 | 12 1/2 | 2 1/16‡ | 12 1/2 | 2 1/4‡ | 14 | 2 5/8‡ |
| 8 | 13 1/2 | 1 3/4 | 15 | 2 7/16‡ | 15 | 2 11/16‡ | 16 1/2 | 3‡ |
| 10 | 16 | 1 15/16 | 17 1/2 | 2 5/8‡ | 17 1/2 | 2 7/8‡ | 20 | 3 3/8‡ |
| 12 | 19 | 2 3/16 | 20 1/2 | 2 7/8‡ | 20 1/2 | 3 1/8‡ | 22 | 3 5/8‡ |
| 14 | 21 | 2 1/4 | 23 | 3‡ | 23 | 3 5/16‡ | 23 3/4 | 3 11/16‡ |
| 16 | 23 1/2 | 2 1/2 | 25 1/2 | 3 1/4‡ | 25 1/2 | 11/16‡ | 27 | 4 3/16‡ |
| 18 | 25 | 2 11/16 | 28 | 3 1/2‡ | 28 | 3 7/8‡ | 29 1/4 | 4 5/8‡ |
| 20 | 27 1/2 | 2 7/8 | 30 1/2 | 3 3/4‡ | 30 1/2 | 4‡ | 32 | 5‡ |
| 22 | 29 1/2 | 3 1/8†‡ | 33 | 4†‡ | 33 | 4 1/4†‡ | 34 1/4 | 5 1/4†‡ |
| 24 | 32 | 3 1/4 | 36 | 4 3/16‡ | 36 | 4 1/2‡ | 37 | 5 1/2‡ |
| 26 | 34 1/4 | 3 3/8†‡ | 38 1/4 | 7 1/4†‡ | 38 1/4 | 7 5/8†‡ | 40 | 8 3/4†‡ |
| 30 | 38 3/4 | 3 1/2†‡ | 43 | 8 1/4†‡ | 43 | 8 5/8†‡ | 44 1/2 | 9 3/4†‡ |
| 34 | 43 3/4 | 3 11/16†‡ | 47 1/2 | 9 1/8†‡ | 47 1/2 | 9 1/2†† | 49 | 10 5/8†‡ |
| 36 | 46 | 3 3/4†‡ | 50 | 9 1/2†‡ | 50 | 9 7/8†‡ | 51 3/4 | 11 1/8†‡ |
| 42 | 53 | 4†‡ | 57 | 10 7/8†‡ | 57 | 11 3/8†‡ | 58 3/4 | 12 3/4†‡ |

*Not available in slip-on type.
†Not available in threaded type.
‡Not available in socket type.

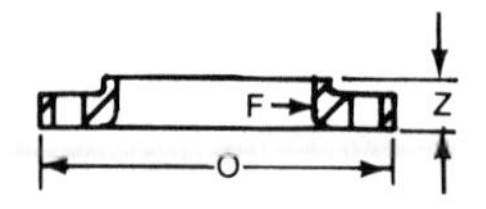


# LAP JOINT FLANGES

| Nom Pipe Size | 150 Lb. Outside Diam of Flange O | 150 Lb. Length Thru Hub Z(1) | 300 Lb. Outside Diam of Flange O | 300 Lb. Length Thru Hub Z(1) | 400 Lb. Outside Diam of Flange O | 400 Lb. Length Thru Hub Z(2) | 600 Lb. Outside Diam of Flange O | 600 Lb. Length Thru Hub Z(2) |
|---|---|---|---|---|---|---|---|---|
| 1/2 | 3 1/2 | 5/8 | 3 3/4 | 7/8 | For sizes 3 1/2 and smaller use 600 Lb. Standard | | 3 3/4 | 7/8 |
| 3/4 | 3 7/8 | 5/8 | 4 5/8 | 1 | | | 4 5/8 | 1 |
| 1 | 4 1/4 | 11/16 | 4 7/8 | 1 1/16 | | | 4 7/8 | 1 1/16 |
| 1 1/4 | 4 5/8 | 13/16 | 5 1/4 | 1 1/16 | | | 5 1/4 | 1 1/8 |
| 1 1/2 | 5 | 7/8 | 6 1/8 | 1 3/16 | | | 6 1/8 | 1 1/4 |
| 2 | 6 | 1 | 6 1/2 | 1 5/16 | | | 6 1/2 | 1 7/16 |
| 2 1/2 | 7 | 1 1/8 | 7 1/2 | 1 1/2 | | | 7 1/2 | 1 5/8 |
| 3 | 7 1/2 | 1 3/16 | 8 1/4 | 1 11/16 | | | 8 1/4 | 1 13/16 |
| 3 1/2 | 8 1/2 | 1 1/4 | 9 | 1 3/4 | | | 9 | 1 15/16 |
| 4 | 9 | 1 5/16 | 10 | 1 7/8 | 10 | 2 | 10 3/4 | 2 1/8 |
| 5 | 10 | 1 7/16 | 11 | 2 | 11 | 2 1/8 | 13 | 2 3/8 |
| 6 | 11 | 1 9/16 | 12 1/2 | 2 1/16 | 12 1/2 | 2 1/4 | 14 | 2 5/8 |
| 8 | 13 1/2 | 1 3/4 | 15 | 2 7/16 | 15 | 2 11/16 | 16 1/2 | 3 |
| 10 | 16 | 1 15/16 | 17 1/2 | 3 3/4 | 17 1/2 | 4 | 20 | 4 3/8 |
| 12 | 19 | 2 3/16 | 20 1/2 | 4 | 20 1/2 | 4 1/4 | 22 | 4 5/8 |
| 14 | 21 | 3 1/8 | 23 | 4 3/8 | 23 | 4 5/8 | 23 3/4 | 5 |
| 15 | 23 1/2 | 3 7/16 | 25 1/2 | 4 3/4 | 25 1/2 | 5 | 27 | 5 1/2 |
| 18 | 25 | 3 13/16 | 28 | 5 1/8 | 28 | 5 3/8 | 29 1/4 | 6 |
| 20 | 27 1/2 | 4 1/16 | 30 1/2 | 5 1/2 | 30 1/2 | 5 3/4 | 32 | 6 1/2 |
| 24 | 32 | 4 3/8 | 36 | 6 | 36 | 6 1/4 | 37 | 7 1/2 |

(1) The 1/16″ raised face *is* included in "Length thru Hub 'Z' " and "Thickness 'Q'."
(2) The 1/4″ raised face *is not* included in "Length thru Hub 'Z' " and "Thickness 'Q'."

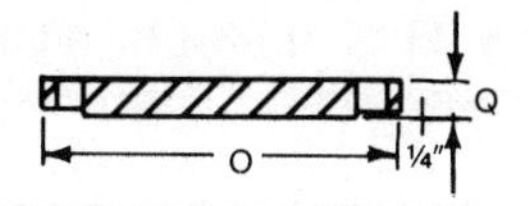


## BLIND FLANGES

| Nom Pipe Size | 150 Lb. | | 300 Lb. | | 400 Lb. | | 600 Lb. | |
|---|---|---|---|---|---|---|---|---|
| | Outside Diam of Flange O | Thick-ness $Q^{(1)}$ | Outside Diam of Flange O | Thick-ness $Q^{(1)}$ | Outside Diam of Flange O | Thick-ness $Q^{(2)}$ | Outside Diam of Flange O | Thick-ness $Q^{(2)}$ |
| 1/2 | 3 1/2 | 7/16 | 3 3/4 | 9/16 | For sizes 3 1/2 and smaller use 600 Lb. Standard | | 3 3/4 | 9/16 |
| 3/4 | 3 7/8 | 1/2 | 4 5/8 | 5/8 | | | 4 5/8 | 5/8 |
| 1 | 4 1/4 | 9/16 | 4 7/8 | 11/16 | | | 4 7/8 | 11/16 |
| 1 1/4 | 4 5/8 | 5/8 | 5 1/4 | 3/4 | | | 5 1/4 | 13/16 |
| 1 1/2 | 5 | 11/16 | 6 1/8 | 13/16 | | | 6 1/8 | 7/8 |
| 2 | 6 | 3/4 | 6 1/2 | 7/8 | | | 6 1/2 | 1 |
| 2 1/2 | 7 | 7/8 | 7 1/2 | 1 | | | 7 1/2 | 1 1/8 |
| 3 | 7 1/2 | 15/16 | 8 1/4 | 1 1/8 | | | 8 1/4 | 1 1/4 |
| 3 1/2 | 8 1/2 | 15/16 | 9 | 1 3/8 | | | 9 | 1 3/8 |
| 4 | 9 | 15/16 | 10 | 1 1/4 | 10 | 1 3/8 | 10 3/4 | 1 1/2 |
| 5 | 10 | 15/16 | 11 | 1 3/8 | 11 | 1 1/2 | 13 | 1 3/4 |
| 6 | 11 | 1 | 12 1/2 | 1 7/16 | 12 1/2 | 1 5/8 | 14 | 1 7/8 |
| 8 | 13 1/2 | 1 1/8 | 15 | 1 5/8 | 15 | 1 7/8 | 16 1/2 | 2 3/16 |
| 10 | 16 | 1 3/16 | 17 1/2 | 1 7/8 | 17 1/2 | 2 1/8 | 20 | 2 1/2 |
| 12 | 19 | 1 1/4 | 20 1/2 | 2 | 20 1/2 | 2 1/4 | 22 | 2 5/8 |
| 14 | 21 | 1 3/8 | 23 | 2 1/8 | 23 | 2 3/8 | 23 3/4 | 2 3/4 |
| 16 | 23 1/2 | 1 7/16 | 25 1/2 | 2 1/4 | 25 1/2 | 2 1/2 | 27 | 3 |
| 18 | 25 | 1 9/16 | 28 | 2 3/8 | 28 | 2 5/8 | 29 1/4 | 3 1/4 |
| 20 | 27 1/2 | 1 11/16 | 30 1/2 | 2 1/2 | 30 1/2 | 2 3/4 | 32 | 3 1/2 |
| 22 | 29 1/2 | 1 13/16 | 33 | 2 5/8 | 33 | 2 7/8 | 34 1/4 | 3 3/4 |
| 24 | 32 | 1 7/8 | 36 | 2 3/4 | 36 | 3 | 37 | 4 |
| 26 | 34 1/4 | 2 | 38 1/4 | 3 1/8 | 38 1/4 | 3 1/2 | 40 | 4 1/4 |
| 30 | 38 3/4 | 2 1/8 | 43 | 3 5/8 | 43 | 4 | 44 1/2 | 4 1/2 |
| 34 | 43 3/4 | 2 5/16 | 47 1/2 | 4 | 47 1/2 | 4 3/4 | 49 | 4 3/4 |
| 36 | 46 | 2 3/8 | 50 | 4 1/8 | 50 | 4 1/2 | 51 3/4 | 4 7/8 |
| 42 | 53 | 2 5/8 | 57 | 4 5/8 | 57 | 5 1/8 | 58 3/4 | 5 1/2 |

# BOLTING DIMENSIONS FOR 150-LB. FLANGES

| Nom Pipe Size | 150 Lb. Steel Flanges | | | | |
|---|---|---|---|---|---|
| | Diam of Bolt Circle | Diam of Bolts | No. of Bolts | Length of Studs $^1/_{16}$" Raised Face | Bolt Length for 125 Lb. Cast Iron Flanges |
| $^1/_2$ | $2^3/_8$ | $^1/_2$ | 4 | $2^1/_4$ | |
| $^3/_4$ | $2^3/_4$ | $^1/_2$ | 4 | $2^1/_4$ | |
| 1 | $3^1/_8$ | $^1/_2$ | 4 | $2^1/_2$ | $1^3/_4$ |
| $1^1/_4$ | $3^1/_2$ | $^1/_2$ | 4 | $2^1/_2$ | 2 |
| $1^1/_2$ | $3^7/_8$ | $^1/_2$ | 4 | $2^3/_4$ | 2 |
| 2 | $4^3/_4$ | $^5/_8$ | 4 | 3 | $2^1/_4$ |
| $2^1/_2$ | $5^1/_2$ | $^5/_8$ | 4 | $3^1/_4$ | $2^1/_2$ |
| 3 | 6 | $^5/_8$ | 4 | $3^1/_2$ | $2^1/_2$ |
| $3^1/_2$ | 7 | $^5/_8$ | 8 | $3^1/_2$ | $2^3/_4$ |
| 4 | $7^1/_2$ | $^5/_8$ | 8 | $3^1/_2$ | 3 |
| 5 | $8^1/_2$ | $^3/_4$ | 8 | $3^3/_4$ | 3 |
| 6 | $9^1/_2$ | $^3/_4$ | 8 | $3^3/_4$ | $3^1/_4$ |
| 8 | $11^3/_4$ | $^3/_4$ | 8 | 4 | $3^1/_2$ |
| 10 | $14^1/_4$ | $^7/_8$ | 12 | $4^1/_2$ | $3^3/_4$ |
| 12 | 17 | $^7/_8$ | 12 | $4^1/_2$ | $3^3/_4$ |
| 14 | $18^3/_4$ | 1 | 12 | 5 | $4^1/_4$ |
| 16 | $21^1/_4$ | 1 | 16 | $5^1/_4$ | $4^1/_2$ |
| 18 | $22^3/_4$ | $1^1/_8$ | 16 | $5^3/_4$ | $4^3/_4$ |
| 20 | 25 | $1^1/_8$ | 20 | 6 | 5 |
| 22 | $27^1/_4$ | $1^1/_4$ | 20 | $6^1/_2$ | |
| 24 | $29^1/_2$ | $1^1/_4$ | 20 | $6^3/_4$ | $5^1/_2$ |
| 26 | $31^3/_4$ | $1^1/_4$ | 24 | 7 | |
| 30 | 36 | $1^1/_4$ | 28 | $7^1/_4$ | $6^1/_4$ |
| 34 | $40^1/_2$ | $1^1/_2$ | 32 | 8 | |
| 36 | $42^3/_4$ | $1^1/_2$ | 32 | $8^1/_4$ | 7 |
| 42 | $49^1/_2$ | $1^1/_2$ | 36 | $8^3/_4$ | $7^1/_2$ |

Stud lengths for lap joint flanges are equal to lengths shown plus the thickness of two laps of the stub ends.

The size and number of bolts, and the bolt circle diameter of 125 lb. cast iron flanges are the same as shown for 150 lb. steel flanges.

## BOLTING DIMENSIONS FOR 300 LB. FLANGES

| Nom Pipe Size | 300 Lb. Steel Flanges | | | | |
|---|---|---|---|---|---|
| | Diam of Bolt Circle | Diam of Bolts | No. of Bolts | Length of Studs 1 1/16" Raised Face | Bolt Length for 250 Lb. Cast Iron Flanges |
| 1/2 | 2 5/8 | 1/2 | 4 | 2 1/2 | |
| 3/4 | 3 1/4 | 5/8 | 4 | 2 3/4 | |
| 1 | 3 1/2 | 5/8 | 4 | 3 | 2 1/2 |
| 1 1/4 | 3 7/8 | 5/8 | 4 | 3 | 2 1/2 |
| 1 1/2 | 4 1/2 | 3/4 | 4 | 3 1/2 | 2 3/4 |
| 2 | 5 | 5/8 | 8 | 3 1/4 | 2 3/4 |
| 2 1/2 | 5 7/8 | 3/4 | 8 | 3 3/4 | 3 1/4 |
| 3 | 6 5/8 | 3/4 | 8 | 4 | 3 1/2 |
| 3 1/2 | 7 1/4 | 3/4 | 8 | 4 1/4 | 3 1/2 |
| 4 | 7 7/8 | 3/4 | 8 | 4 1/4 | 3 3/4 |
| 5 | 9 1/4 | 3/4 | 8 | 4 1/2 | 4 |
| 6 | 10 5/8 | 3/4 | 12 | 4 3/4 | 4 |
| 8 | 13 | 7/8 | 12 | 5 1/4 | 4 1/2 |
| 10 | 15 1/4 | 1 | 16 | 6 | 5 1/4 |
| 12 | 17 3/4 | 1 1/8 | 16 | 6 1/2 | 5 1/2 |
| 14 | 20 1/4 | 1 1/8 | 20 | 6 3/4 | 6 |
| 16 | 22 1/2 | 1 1/4 | 20 | 7 1/4 | 6 1/4 |
| 18 | 24 3/4 | 1 1/4 | 24 | 7 1/2 | 6 1/2 |
| 20 | 27 | 1 1/4 | 24 | 8 | 6 3/4 |
| 22 | 29 1/4 | 1 1/2 | 24 | 8 3/4 | |
| 24 | 32 | 1 1/2 | 24 | 9 | 7 3/4 |
| 26 | 34 1/2 | 1 5/8 | 28 | 10 | |
| 30 | 39 1/4 | 1 3/4 | 28 | 11 1/4 | 8 1/2 |
| 34 | 43 1/2 | 1 7/8 | 28 | 12 1/4 | |
| 36 | 46 | 2 | 32 | 12 3/4 | 9 1/2 |
| 42 | 52 3/4 | 2 | 36 | 13 3/4 | 10 1/4 |

The size and number of bolts, and the bolt circle diameter of 250 lb. cast iron flanges are the same as shown for 300 lb. steel flanges.

# BOLTING DIMENSIONS FOR 400 AND 600-LB FLANGES

| Nom Pipe Size | 400 Lb. Steel Flanges | | | | 600 Lb. Steel Flanges | | | |
|---|---|---|---|---|---|---|---|---|
| | Diam of Bolt Circle | Diam of Bolts | No. of Bolts | Length of Studs 1/4" Raised Face | Diam of Bolt Circle | Diam of Bolts | No. of Bolts | Length of Studs 1/4" Raised Face |
| 1/2 | 2 5/8 | 1/2 | 4 | 3 | 2 5/8 | 1/2 | 4 | 3 |
| 3/4 | 3 1/4 | 5/8 | 4 | 3 1/4 | 3 1/4 | 5/8 | 4 | 3 1/4 |
| 1 | 3 1/2 | 5/8 | 4 | 3 1/2 | 3 1/2 | 5/8 | 4 | 3 1/2 |
| 1 1/4 | 3 7/8 | 5/8 | 4 | 3 3/4 | 3 7/8 | 5/8 | 4 | 3 3/4 |
| 1 1/2 | 4 1/2 | 3/4 | 4 | 4 | 4 1/2 | 3/4 | 4 | 4 |
| 2 | 5 | 5/8 | 8 | 4 | 5 | 5/8 | 8 | 4 |
| 2 1/2 | 5 7/8 | 3/4 | 8 | 4 1/2 | 5 7/8 | 3/4 | 8 | 4 1/2 |
| 3 | 6 5/8 | 3/4 | 8 | 4 3/4 | 6 5/8 | 3/4 | 8 | 4 3/4 |
| 3 1/2 | 7 1/4 | 7/8 | 8 | 5 1/4 | 7 1/2 | 7/8 | 8 | 5 1/4 |
| 4 | 7 7/8 | 7/8 | 8 | 5 1/4 | 1/2 | 7/8 | 8 | 5 1/2 |
| 5 | 9 1/4 | 7/8 | 8 | 6 1/2 | 10 1/2 | 1 | 8 | 6 1/4 |
| 6 | 10 5/8 | 7/8 | 12 | 5 3/4 | 11 1/2 | 1 | 12 | 6 1/2 |
| 8 | 13 | 1 | 12 | 6 1/2 | 13 3/4 | 1 1/8 | 12 | 7 1/2 |
| 10 | 15 1/4 | 1 1/8 | 16 | 7 1/4 | 17 | 1 1/4 | 16 | 8 1/4 |
| 12 | 17 3/4 | 1 1/4 | 16 | 7 3/4 | 19 1/4 | 1 1/4 | 20 | 8 1/2 |
| 14 | 20 1/4 | 1 1/4 | 20 | 8 | 20 3/4 | 1 3/8 | 20 | 9 |
| 16 | 22 1/2 | 1 3/8 | 20 | 8 1/2 | 23 3/4 | 1 1/2 | 20 | 9 3/4 |
| 18 | 24 3/4 | 1 3/8 | 24 | 8 3/4 | 25 3/4 | 1 5/8 | 20 | 10 1/2 |
| 20 | 27 | 1 1/2 | 24 | 9 1/2 | 28 1/2 | 1 5/8 | 24 | 11 1/4 |
| 22 | 29 1/4 | 1 5/8 | 24 | 10 | 30 5/8 | 1 3/4 | 24 | 12 |
| 24 | 32 | 1 3/4 | 24 | 10 1/2 | 33 | 1 7/8 | 24 | 12 3/4 |
| 26 | 34 1/2 | 1 3/4 | 28 | 11 1/2 | 36 | 1 7/8 | 28 | 13 1/4 |
| 30 | 39 1/4 | 2 | 28 | 13 | 40 1/4 | 2 | 28 | 14 |
| 34 | 43 1/2 | 2 | 28 | 13 3/4 | 44 1/2 | 2 1/4 | 28 | 15 |
| 36 | 46 | 2 | 32 | 14 | 47 | 2 1/2 | 28 | 15 3/4 |
| 42 | 52 3/4 | 2 1/2 | 32 | 16 1/4 | 53 3/4 | 2 3/4 | 28 | 17 1/2 |

Stud lengths for lap joint flanges are equal to lengths shown minus 1/2" plus the thickness of two laps of the stub ends.

# LARGE O.D. STEEL FLANGE DIMENSIONS MSS SP-44

*(Reprinted by permission of Manufacturers Standardization Society of the Valve and Fittings Industry)*

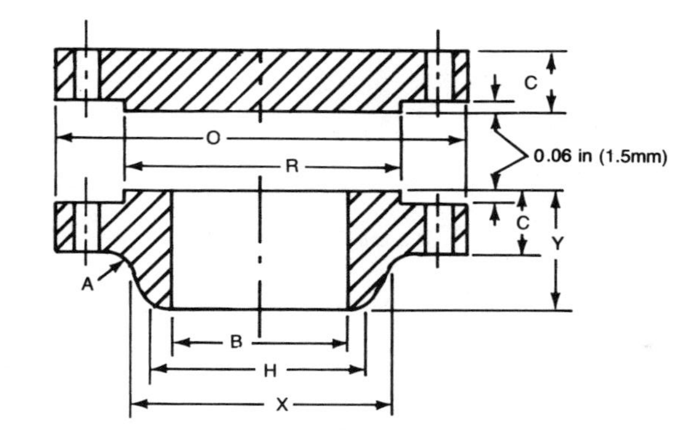


Raised Face

Class 150.285 psi (19.5 Bars) at Atmospheric Temperature Raised Face

Dimensions in inches

| | Flange Dimensions | | | Hub Dimensions | Drilling | | | | |
|---|---|---|---|---|---|---|---|---|---|
| | OD of Flange O | Thick. of Flange C (MIN) | Length Thru Hub Y | OD Large End Hub X | NO. of Bolt Holes | Dia. of Bolt Holes | Dia. of Bolt Circle | Face Dia. R | Fillet Radius (MIN) A |
| 12 | 19.00 | 1.25 | 4.50 | 14.38 | 12 | 1.00 | 17.00 | 15.00 | 0.38 |
| 14 | 21.00 | 1.38 | 5.00 | 15.75 | 12 | 1.12 | 18.75 | 16.25 | 0.38 |
| 16 | 23.50 | 1.44 | 5.00 | 18.00 | 16 | 1.12 | 21.25 | 18.50 | 0.38 |

## LARGE O.D. STEEL FLANGE DIMENSIONS MSS SP-44 *(continued)*

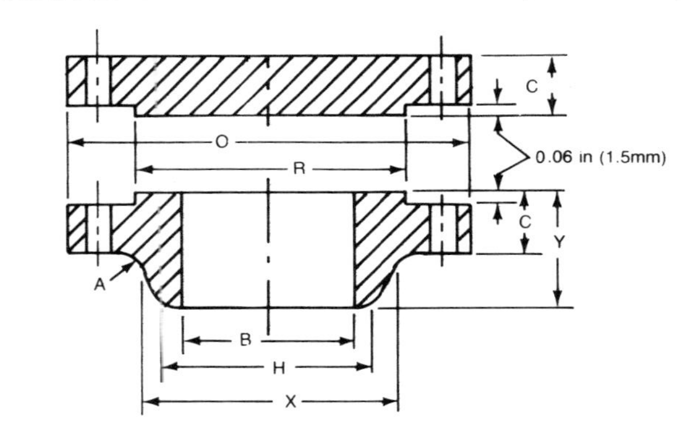


Raised Face

Class 150.285 psi (19.5 Bars) at Atmospheric Temperature Raised Face

Dimensions in inches

| | Flange Dimensions | | | Hub Dimensions | Drilling | | | | |
|---|---|---|---|---|---|---|---|---|---|
| | OD of Flange O | Thick. of Flange C (MIN) | Length Thru Hub Y | OD Large End Hub X | NO. of Bolt Holes | Dia. of Bolt Holes | Dia. of Bolt Circle | Face Dia. R | Fillet Radius (MIN) A |
| 18 | 25.00 | 1.56 | 5.50 | 19.88 | 16 | 1.25 | 22.75 | 21.00 | 0.38 |
| 20 | 27.50 | 1.69 | 5.69 | 22.00 | 20 | 1.25 | 25.00 | 23.00 | 0.38 |
| 22 | 29.50 | 1.81 | 5.88 | 24.00 | 20 | 1.38 | 27.25 | 25.25 | 0.38 |

| | | | | | | | | | |
|---|---|---|---|---|---|---|---|---|---|
| 24 | 32.00 | 1.88 | 6.00 | 26.12 | 20 | 1.38 | 29.50 | 27.25 | 0.38 |
| 26 | 34.25 | 2.69 | 4.75 | 26.62 | 24 | 1.38 | 31.75 | 29.50 | 0.38 |
| 28 | 36.50 | 2.81 | 4.94 | 28.62 | 28 | 1.38 | 34.00 | 31.50 | 0.44 |
| 30 | 38.75 | 2.94 | 5.38 | 30.75 | 28 | 1.38 | 36.00 | 33.75 | 0.44 |
| 32 | 41.75 | 3.18 | 5.69 | 32.75 | 28 | 1.62 | 38.50 | 36.00 | 0.44 |
| 34 | 43.75 | 3.25 | 5.88 | 34.75 | 32 | 1.62 | 40.50 | 38.00 | 0.50 |
| 36 | 46.00 | 3.56 | 6.18 | 36.75 | 32 | 1.62 | 42.75 | 40.25 | 0.50 |
| 38 | 48.75 | 3.44 | 6.19 | 39.00 | 32 | 1.62 | 45.25 | 42.25 | 0.50 |
| 40 | 50.75 | 3.56 | 6.44 | 41.00 | 36 | 1.62 | 47.25 | 44.25 | 0.50 |
| 42 | 53.00 | 3.81 | 6.75 | 43.00 | 36 | 1.62 | 49.50 | 47.00 | 0.50 |
| 44 | 55.25 | 4.00 | 7.00 | 45.00 | 40 | 1.62 | 51.75 | 49.00 | 0.50 |
| 46 | 57.25 | 4.06 | 7.31 | 47.12 | 40 | 1.62 | 53.75 | 51.00 | 0.50 |
| 48 | 59.50 | 4.25 | 7.56 | 49.12 | 44 | 1.62 | 56.00 | 53.50 | 0.50 |
| 50 | 61.75 | 4.38 | 8.00 | 51.25 | 44 | 1.88 | 58.25 | 55.50 | 0.50 |
| 52 | 64.00 | 4.56 | 8.25 | 53.25 | 44 | 1.88 | 60.50 | 57.50 | 0.50 |
| 54 | 66.25 | 4.75 | 8.50 | 55.25 | 44 | 1.88 | 62.75 | 59.50 | 0.50 |
| 56 | 68.75 | 4.88 | 9.00 | 57.38 | 48 | 1.88 | 65.00 | 62.00 | 0.50 |
| 58 | 71.00 | 5.06 | 9.25 | 59.38 | 48 | 1.88 | 67.25 | 64.00 | 0.50 |
| 60 | 73.00 | 5.19 | 9.44 | 61.38 | 52 | 1.88 | 69.25 | 66.00 | 0.50 |

# LARGE O.D. STEEL FLANGE DIMENSIONS MSS SP-44 *(continued)*

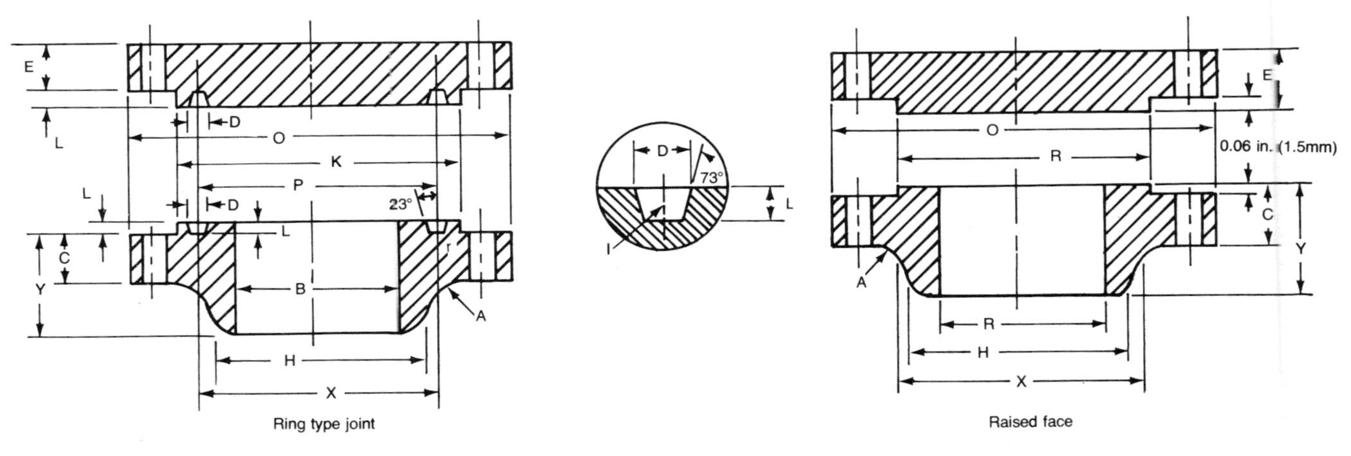


Ring type joint

Raised face

Class 300, 740 PSI (51 Bars) at Atmospheric Temperature Raised Face and Ring-Type Joints

Dimensions in inches

| Pipe Size | FLANGE DIMENSIONS | | | | HUB DIM. | DRILLING | | | FACING DIMENSIONS | | | | | | Fillet Radius (min) | Groove Fillet Radius |
|---|---|---|---|---|---|---|---|---|---|---|---|---|---|---|---|---|
| | OD of Flange | Thick of Flange: Weld-Neck | Thick of Flange: Bld. Flg. | Length Thru Hub | OD Large End Hub | No. of Bolt Holes | Dia. of Bolt Holes | Dia. of Bolt Circle | Raised Face Dia. | Ring-Type Joint: Facing Dia. | Ring-Type Joint: Depth of Groove | Ring-Type Joint: Pitch Dia. | Ring-Type Joint: Width of Groove | Ring-Type Joint: Ring No. | | |
| | O | C | E | Y | X | | | | R | K | L | P | D | | A | r |
| 12 | 20.50 | 2.00 | 2.00 | 5.12 | 14.75 | 16 | 1.25 | 17.75 | 15.00 | 16.50 | 0.312 | 15.000 | 0.469 | R57 | 0.38 | 0.03 |
| 14 | 23.00 | 2.12 | 2.12 | 5.62 | 16.75 | 20 | 1.25 | 20.25 | 16.25 | 18.00 | 0.312 | 16.500 | 0.469 | R61 | 0.38 | 0.03 |
| 16 | 25.50 | 2.25 | 2.25 | 5.75 | 19.00 | 20 | 1.38 | 22.50 | 18.50 | 20.00 | 0.312 | 18.500 | 0.469 | R65 | 0.38 | 0.03 |
| 18 | 28.00 | 2.38 | 2.38 | 6.25 | 21.00 | 24 | 1.38 | 24.75 | 21.00 | 22.62 | 0.312 | 21.000 | 0.469 | R69 | 0.38 | 0.03 |

| | | | | | | | | | | | | | | | | |
|---|---|---|---|---|---|---|---|---|---|---|---|---|---|---|---|---|
| 20 | 30.50 | 2.50 | 2.50 | 6.38 | 23.12 | 24 | 1.38 | 27.00 | 23.00 | 25.00 | 0.375 | 23.000 | 0.531 | R73 | 0.38 | 0.06 |
| 22 | 33.00 | 2.62 | 2.62 | 6.50 | 25.25 | 24 | 1.62 | 29.25 | 25.25 | 27.00 | 0.438 | 25.000 | 0.594 | R81 | 0.38 | 0.06 |
| 24 | 36.00 | 2.75 | 2.75 | 6.62 | 27.62 | 24 | 1.62 | 32.00 | 27.25 | 29.50 | 0.438 | 27.250 | 0.656 | R77 | 0.38 | 0.06 |
| 26 | 38.25 | 3.12 | 3.31 | 7.25 | 28.38 | 28 | 1.75 | 34.50 | 29.50 | 31.88 | 0.500 | 29.500 | 0.781 | R93 | 0.38 | 0.06 |
| 28 | 40.75 | 3.38 | 3.56 | 7.75 | 30.50 | 28 | 1.75 | 37.00 | 31.50 | 33.88 | 0.500 | 31.500 | 0.781 | R94 | 0.44 | 0.06 |
| 30 | 43.00 | 3.62 | 3.75 | 8.25 | 32.56 | 28 | 1.88 | 39.25 | 33.75 | 36.12 | 0.500 | 33.750 | 0.781 | R95 | 0.44 | 0.06 |
| 32 | 45.25 | 3.88 | 3.94 | 8.75 | 34.69 | 28 | 2.00 | 41.50 | 36.00 | 38.75 | 0.562 | 36.000 | 0.906 | R96 | 0.44 | 0.06 |
| 34 | 47.50 | 4.00 | 4.12 | 9.12 | 36.88 | 28 | 2.00 | 43.50 | 38.00 | 40.75 | 0.562 | 38.000 | 0.906 | R97 | 0.50 | 0.06 |
| 36 | 50.00 | 4.12 | 4.38 | 9.50 | 39.00 | 32 | 2.12 | 46.00 | 40.25 | 43.00 | 0.562 | 40.250 | 0.906 | R98 | 0.50 | 0.06 |
| 38 | 46.00 | 4.25 | 4.25 | 7.12 | 39.12 | 32 | 1.62 | 43.00 | 40.50 | — | — | — | — | — | 0.50 | — |
| 40 | 48.75 | 4.50 | 4.50 | 7.62 | 41.25 | 32 | 1.75 | 45.50 | 42.75 | — | — | — | — | — | 0.50 | — |
| 42 | 50.75 | 4.69 | 4.69 | 7.88 | 43.25 | 32 | 1.75 | 47.50 | 44.75 | — | — | — | — | — | 0.50 | — |
| 44 | 53.25 | 4.88 | 4.88 | 8.12 | 45.25 | 32 | 1.88 | 49.75 | 47.00 | — | — | — | — | — | 0.50 | — |
| 46 | 55.75 | 5.06 | 5.06 | 8.50 | 47.38 | 28 | 2.00 | 52.00 | 49.00 | — | — | — | — | — | 0.50 | — |
| 48 | 57.75 | 5.25 | 5.25 | 8.82 | 49.38 | 32 | 2.00 | 54.00 | 51.25 | — | — | — | — | — | 0.50 | — |
| 50 | 60.25 | 5.50 | 5.50 | 9.12 | 51.38 | 32 | 2.12 | 56.25 | 53.50 | — | — | — | — | — | 0.50 | — |
| 52 | 62.25 | 5.69 | 5.69 | 9.38 | 53.38 | 32 | 2.12 | 58.25 | 55.50 | — | — | — | — | — | 0.50 | — |
| 54 | 65.25 | 6.00 | 6.00 | 9.94 | 55.50 | 28 | 2.38 | 61.00 | 57.75 | — | — | — | — | — | 0.50 | — |
| 56 | 67.25 | 6.06 | 6.06 | 10.25 | 57.62 | 28 | 2.38 | 63.00 | 59.75 | — | — | — | — | — | 0.50 | — |
| 58 | 69.25 | 6.25 | 6.25 | 10.50 | 59.62 | 32 | 2.38 | 65.00 | 62.00 | — | — | — | — | — | 0.50 | — |
| 60 | 71.25 | 6.44 | 6.44 | 10.75 | 61.62 | 32 | 2.38 | 67.00 | 64.00 | — | — | — | — | — | 0.50 | — |

## LARGE O.D. STEEL FLANGE DIMENSIONS MSS SP-44 *(continued)*

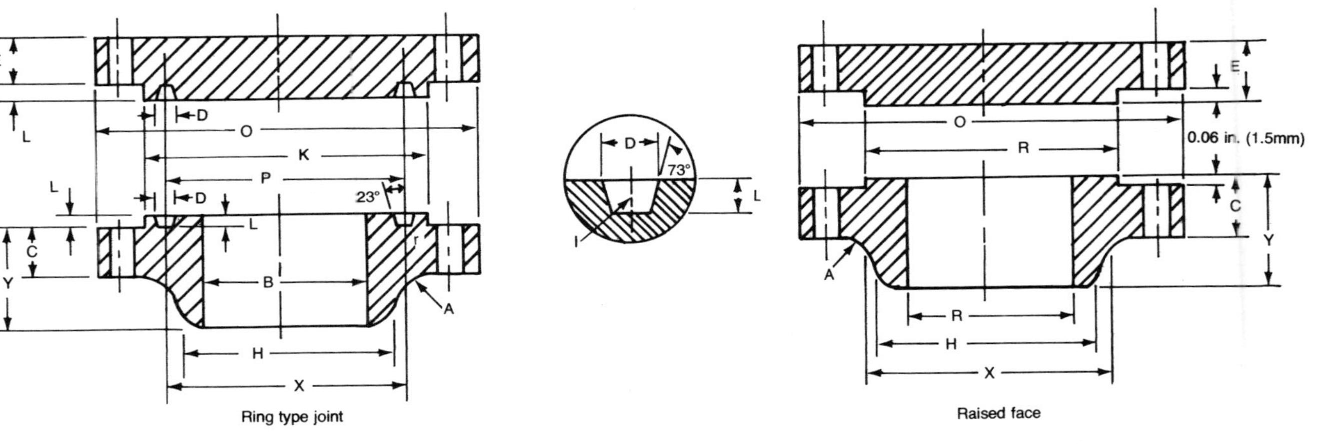


Class 400, 985 PSI (68 Bars) at Atmospheric Temperature Raised Face and Ring-Type Joints

Dimensions in inches

| Pipe Size | FLANGE DIMENSIONS | | | | HUB DIM. | DRILLING | | | FACING DIMENSIONS | | | | | | Fillet Radius (min) | Groove Fillet Radius |
|---|---|---|---|---|---|---|---|---|---|---|---|---|---|---|---|---|
| | OD of Flange | Thick of Flange | | Length Thru Hub | OD Large End Hub | No. of Bolt Holes | Dia. of Bolt Holes | Dia. of Bolt Circle | Raised Face Dia. | Ring-Type Joint | | | | | | |
| | | Weld-Neck | Bld. Flg. | | | | | | | Facing Dia. | Depth of Groove | Pitch Dia. | Width of Groove | Ring No. | | |
| | O | C | E | Y | X | | | | R | K | L | P | D | | A | r |
| 12 | 20.50 | 2.25 | 2.25 | 5.38 | 14.75 | 16 | 1.38 | 17.75 | 15.00 | 16.25 | 0.312 | 15.000 | 0.469 | R57 | 0.44 | 0.03 |
| 14 | 23.00 | 2.38 | 2.28 | 5.88 | 16.75 | 20 | 1.38 | 20.25 | 16.25 | 18.00 | 0.312 | 16.500 | 0.469 | R61 | 0.44 | 0.03 |
| 16 | 25.50 | 2.50 | 2.50 | 6.00 | 19.00 | 20 | 1.50 | 22.50 | 18.12 | 20.00 | 0.312 | 18.500 | 0.469 | R65 | 0.44 | 0.03 |
| 18 | 28.00 | 2.62 | 2.62 | 6.50 | 21.00 | 24 | 1.50 | 24.75 | 21.00 | 22.62 | 0.312 | 21.000 | 0.469 | R69 | 0.44 | 0.03 |

| | | | | | | | | | | | | | | | | |
|---|---|---|---|---|---|---|---|---|---|---|---|---|---|---|---|---|
| 20 | 30.50 | 2.75 | 2.75 | 6.62 | 23.12 | 24 | 1.62 | 27.00 | 23.00 | 25.00 | 0.375 | 23.000 | 0.531 | R73 | 0.44 | 0.06 |
| 22 | 33.00 | 2.88 | 2.88 | 6.75 | 25.25 | 24 | 1.75 | 29.25 | 25.25 | 27.00 | 0.438 | 25.000 | 0.594 | R81 | 0.44 | 0.06 |
| 24 | 36.00 | 3.00 | 3.00 | 6.88 | 27.62 | 24 | 1.88 | 32.00 | 27.25 | 29.50 | 0.438 | 27.250 | 0.656 | R77 | 0.44 | 0.06 |
| 26 | 38.25 | 3.50 | 3.88 | 7.62 | 28.62 | 28 | 1.88 | 34.50 | 29.50 | 31.88 | 0.500 | 29.500 | 0.781 | R93 | 0.44 | 0.06 |
| 28 | 40.75 | 3.75 | 4.12 | 8.12 | 30.81 | 28 | 2.00 | 37.00 | 31.50 | 33.88 | 0.500 | 31.500 | 0.781 | R94 | 0.50 | 0.06 |
| 30 | 43.00 | 4.00 | 4.38 | 8.62 | 32.94 | 28 | 2.12 | 39.25 | 33.75 | 36.12 | 0.500 | 33.750 | 0.781 | R95 | 0.50 | 0.06 |
| 32 | 45.25 | 4.25 | 4.56 | 9.12 | 35.00 | 28 | 2.12 | 41.50 | 36.00 | 38.75 | 0.562 | 36.000 | 0.906 | R96 | 0.50 | 0.06 |
| 34 | 47.50 | 4.38 | 4.81 | 9.50 | 37.19 | 28 | 2.12 | 43.50 | 38.00 | 40.75 | 0.562 | 38.000 | 0.906 | R97 | 0.56 | 0.06 |
| 36 | 50.00 | 4.50 | 5.06 | 9.88 | 19.38 | 32 | 2.12 | 46.00 | 40.25 | 43.00 | 0.562 | 40.250 | 0.906 | R98 | 0.56 | 0.06 |
| 38 | 47.50 | 4.88 | 4.88 | 8.12 | 39.50 | 32 | 1.88 | 44.00 | 40.75 | — | — | — | — | — | 0.56 | — |
| 40 | 50.00 | 5.12 | 5.12 | 8.50 | 41.50 | 32 | 2.00 | 46.25 | 43.00 | — | — | — | — | — | 0.56 | — |
| 42 | 52.00 | 5.25 | 5.25 | 8.81 | 43.62 | 32 | 2.00 | 48.25 | 45.00 | — | — | — | — | — | 0.56 | — |
| 44 | 54.50 | 5.50 | 5.50 | 9.18 | 45.62 | 32 | 2.12 | 50.50 | 47.25 | — | — | — | — | — | 0.56 | — |
| 46 | 56.75 | 5.75 | 5.75 | 9.62 | 47.75 | 36 | 2.12 | 52.75 | 49.50 | — | — | — | — | — | 0.56 | — |
| 48 | 59.50 | 6.00 | 6.00 | 10.12 | 49.88 | 28 | 2.38 | 55.25 | 51.50 | — | — | — | — | — | 0.56 | — |
| 50 | 61.75 | 6.19 | 6.25 | 10.56 | 52.00 | 32 | 2.38 | 57.50 | 53.62 | — | — | — | — | — | 0.56 | — |
| 52 | 63.75 | 6.38 | 6.44 | 10.88 | 54.00 | 32 | 2.38 | 59.50 | 55.62 | — | — | — | — | — | 0.56 | — |
| 54 | 67.00 | 6.69 | 6.75 | 11.38 | 56.12 | 28 | 2.62 | 62.25 | 57.88 | — | — | — | — | — | 0.56 | — |
| 56 | 69.00 | 6.88 | 6.94 | 11.75 | 58.25 | 32 | 2.62 | 64.25 | 60.12 | — | — | — | — | — | 0.56 | — |
| 58 | 71.00 | 7.00 | 7.12 | 12.06 | 60.25 | 32 | 2.62 | 66.25 | 62.12 | — | — | — | — | — | 0.56 | — |
| 60 | 74.25 | 7.31 | 7.44 | 12.56 | 62.38 | 32 | 2.88 | 69.00 | 64.38 | — | — | — | — | — | 0.56 | — |

## LARGE O.D. STEEL FLANGE DIMENSIONS MSS SP-44 *(continued)*

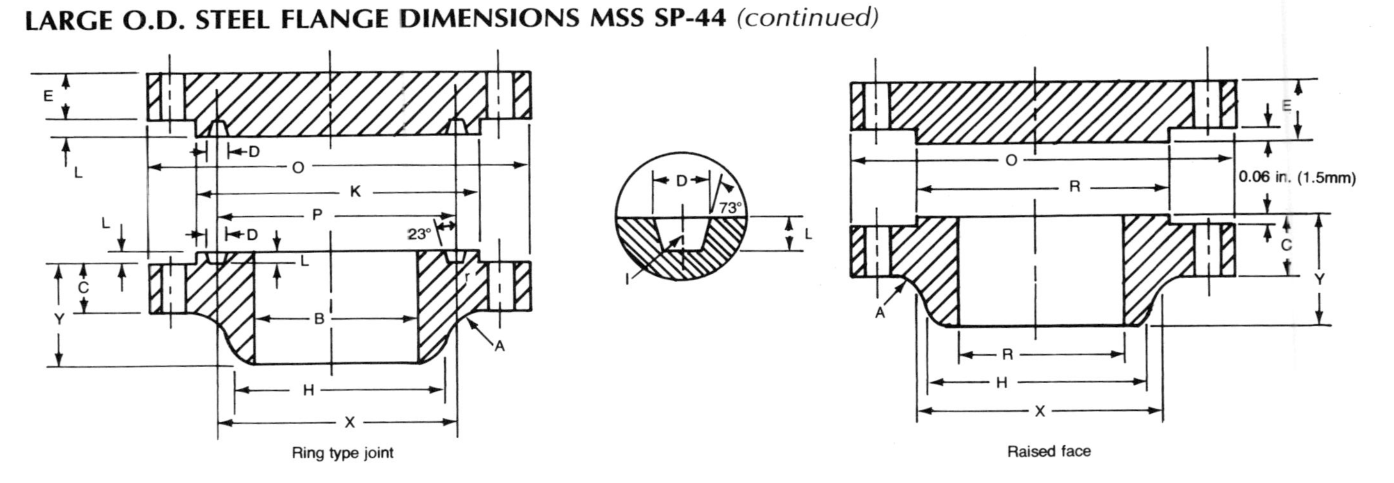


Class 600, 1480 PSI (102.1 Bars) at Atmospheric Temperature Raised Act and Ring-Type Joints

Dimensions in inches

| Pipe Size | FLANGE DIMENSIONS | | | | HUB DIM. | DRILLING | | | | FACING DIMENSIONS | | | | | | |
|---|---|---|---|---|---|---|---|---|---|---|---|---|---|---|---|---|
| | | Thick of Flange | | | | | | | | Ring-Type Joint | | | | | | |
| | OD of Flange | Weld-Neck | Bld. Flg. | Length Thru Hub | OD Large End Hub | No. of Bolt Holes | Dia. of Bolt Holes | Dia. of Bolt Circle | Raised Face Dia. | Facing Dia. | Depth of Groove | Pitch Dia. | Width of Groove | Ring No. | Fillet Radius (min) | Groove Fillet Radius |
| | O | C | E | Y | X | | | | R | K | L | P | D | | A | r |
| 12 | 22.00 | 2.62 | 2.62 | 6.12 | 15.75 | 20 | 1.28 | 19.25 | 15.00 | 16.25 | 0.312 | 15.000 | 0.469 | R57 | 0.44 | 0.03 |
| 14 | 23.75 | 2.75 | 2.75 | 6.50 | 17.00 | 20 | 1.50 | 20.75 | 16.25 | 18.00 | 0.312 | 16.500 | 0.469 | R61 | 0.44 | 0.03 |
| 16 | 27.00 | 3.00 | 3.00 | 7.00 | 19.50 | 20 | 1.62 | 23.75 | 18.50 | 20.00 | 0.312 | 18.500 | 0.469 | R65 | 0.44 | 0.03 |
| 18 | 29.25 | 3.25 | 3.25 | 7.25 | 21.50 | 20 | 1.75 | 25.75 | 21.00 | 22.62 | 0.312 | 21.000 | 0.469 | R69 | 0.44 | 0.03 |

| | | | | | | | | | | | | | | | |
|---|---|---|---|---|---|---|---|---|---|---|---|---|---|---|---|
| 22 | 34.25 | 3.75 | 3.75 | 7.75 | 26.25 | 24 | 1.88 | 30.62 | 25.25 | 27.00 | 0.438 | 25.000 | 0.594 | R81 | 0.44 | 0.06 |
| 24 | 37.00 | 4.00 | 4.00 | 8.00 | 28.25 | 24 | 2.00 | 33.00 | 27.25 | 29.50 | 0.438 | 27.250 | 0.656 | R77 | 0.44 | 0.06 |
| 26 | 40.00 | 4.25 | 4.94 | 8.75 | 29.44 | 28 | 2.00 | 36.00 | 29.50 | 31.88 | 0.500 | 29.500 | 0.781 | R93 | 0.50 | 0.06 |
| 28 | 42.25 | 4.38 | 5.19 | 9.25 | 31.62 | 28 | 2.12 | 38.00 | 31.50 | 33.88 | 0.500 | 31.500 | 0.781 | R94 | 0.50 | 0.06 |
| 30 | 44.50 | 4.50 | 5.50 | 9.75 | 33.94 | 28 | 2.12 | 40.25 | 33.75 | 36.12 | 0.500 | 33.750 | 0.781 | R95 | 0.50 | 0.06 |
| 32 | 47.00 | 4.62 | 5.81 | 10.25 | 36.12 | 28 | 2.38 | 42.50 | 36.00 | 38.75 | 0.562 | 36.000 | 0.906 | R96 | 0.50 | 0.06 |
| 34 | 49.00 | 4.75 | 6.06 | 10.62 | 38.19 | 28 | 2.38 | 44.50 | 38.00 | 40.75 | 0.562 | 38.000 | 0.906 | R97 | 0.56 | 0.06 |
| 36 | 51.75 | 4.88 | 6.38 | 11.12 | 40.62 | 28 | 2.62 | 47.00 | 40.25 | 43.00 | 0.562 | 40.250 | 0.906 | R98 | 0.56 | 0.06 |
| 38 | 50.00 | 6.00 | 6.12 | 10.00 | 40.25 | 28 | 2.38 | 45.75 | 41.50 | — | — | — | — | — | 0.56 | — |
| 40 | 52.00 | 6.25 | 6.38 | 10.38 | 42.25 | 32 | 2.38 | 47.75 | 43.75 | — | — | — | — | — | 0.56 | — |
| 42 | 55.25 | 6.62 | 6.75 | 11.00 | 44.38 | 28 | 2.62 | 50.50 | 46.00 | — | — | — | — | — | 0.56 | — |
| 44 | 57.25 | 6.81 | 7.00 | 11.38 | 46.50 | 32 | 2.62 | 52.50 | 48.25 | — | — | — | — | — | 0.56 | — |
| 46 | 59.50 | 7.06 | 7.31 | 11.81 | 48.62 | 32 | 2.62 | 54.75 | 50.25 | — | — | — | — | — | 0.56 | — |
| 48 | 62.75 | 7.44 | 7.69 | 12.44 | 50.75 | 32 | 2.88 | 57.50 | 52.50 | — | — | — | — | — | 0.56 | — |
| 50 | 65.75 | 7.75 | 8.00 | 12.94 | 52.88 | 28 | 3.12 | 60.00 | 54.50 | — | — | — | — | — | 0.56 | — |
| 52 | 67.75 | 8.00 | 8.25 | 13.25 | 54.88 | 32 | 3.12 | 62.00 | 56.50 | — | — | — | — | — | 0.56 | — |
| 54 | 70.00 | 8.25 | 8.56 | 13.75 | 57.00 | 32 | 3.12 | 64.25 | 58.75 | — | — | — | — | — | 0.56 | — |
| 56 | 73.00 | 8.56 | 8.88 | 14.25 | 59.12 | 32 | 3.38 | 66.75 | 60.75 | — | — | — | — | — | 0.62 | — |
| 58 | 75.00 | 8.75 | 9.12 | 14.56 | 61.12 | 32 | 3.38 | 68.75 | 63.00 | — | — | — | — | — | 0.62 | — |
| 60 | 78.50 | 9.19 | 9.56 | 15.31 | 63.38 | 28 | 3.62 | 71.75 | 65.25 | — | — | — | — | — | 0.69 | — |

# STANDARD CAST IRON COMPANION FLANGES AND BOLTS

*(For working pressures up to 125 psi steam, 175 psi WOG)*

| 0 SIZE IN INCHES | DIAM OF FLANGE, IN INCHES | BOLT CIRCLE, IN INCHES | NO. OF BOLTS | SIZE OF BOLTS, IN INCHES | LENGTH OF BOLTS, IN INCHES |
|---|---|---|---|---|---|
| 3/4 | 3 1/2 | 2 1/2 | 4 | 3/8 | 1 3/8 |
| 1 | 4 1/4 | 3 1/8 | 4 | 1/2 | 1 1/2 |
| 1 1/4 | 4 5/8 | 3 1/2 | 4 | 1/2 | 1 1/2 |
| 1 1/2 | 5 | 3 7/8 | 4 | 1/2 | 1 3/4 |
| 2 | 6 | 4 3/4 | 4 | 5/8 | 2 |
| 2 1/2 | 7 | 5 1/2 | 4 | 5/8 | 2 1/4 |
| 3 | 7 1/2 | 6 | 4 | 5/8 | 2 1/2 |
| 3 1/2 | 8 1/2 | 7 | 8 | 5/8 | 2 1/2 |
| 4 | 9 | 7 1/2 | 8 | 5/8 | 2 3/4 |
| 5 | 10 | 8 1/2 | 8 | 3/4 | 3 |
| 6 | 11 | 9 1/2 | 8 | 3/4 | 3 |
| 8 | 13 1/2 | 11 3/4 | 8 | 3/4 | 3 1/4 |
| 10 | 16 | 14 1/4 | 12 | 7/8 | 3 1/2 |
| 12 | 19 | 17 | 12 | 7/8 | 3 3/4 |
| 14 | 21 | 18 3/4 | 12 | 1 | 4 1/4 |
| 16 | 23 1/2 | 21 1/4 | 16 | 1 | 4 1/4 |

# EXTRA-HEAVY CAST IRON COMPANION FLANGES AND BOLTS

*(For working pressures up to 250 psi steam, 400 psi WOG)*

| PIPE SIZES INCHES | DIAM OF FLANGES | DIAM OF BOLT CIRCLE | NO. OF BOLTS | DIAM OF BOLTS | LENGTH OF BOLTS |
|---|---|---|---|---|---|
| 1 | 4 7/8 | 3 1/2 | 4 | 5/8 | 2 1/4 |
| 1 1/4 | 5 1/4 | 3 7/8 | 4 | 5/8 | 2 1/2 |
| 1 1/2 | 6 1/8 | 4 1/2 | 4 | 3/4 | 2 1/2 |
| 2 | 6 1/2 | 5 | 8 | 5/8 | 2 1/2 |
| 2 1/2 | 7 1/2 | 5 7/8 | 8 | 3/4 | 3 |
| 3 | 8 1/4 | 6 5/8 | 8 | 3/4 | 3 1/4 |
| 3 1/2 | 9 | 7 1/4 | 8 | 3/4 | 3 1/4 |
| 4 | 10 | 7 7/8 | 8 | 3/4 | 3 1/2 |
| 5 | 11 | 9 1/4 | 8 | 3/4 | 3 3/4 |
| 6 | 12 1/2 | 10 5/8 | 12 | 3/4 | 3 3/4 |
| 8 | 15 | 13 | 12 | 7/8 | 4 1/4 |
| 10 | 17 1/2 | 15 1/4 | 16 | 1 | 5 |
| 12 | 20 1/2 | 17 3/4 | 16 | 1 1/8 | 5 1/2 |
| 14 O.D. | 23 | 20 1/4 | 20 | 1 1/8 | 5 3/4 |
| 16 O.D. | 25 1/2 | 22 1/2 | 20 | 1 1/4 | 6 |
| 18 O.D. | 28 | 24 3/4 | 24 | 1 1/4 | 6 1/4 |
| 20 O.D. | 30 1/2 | 27 | 24 | 1 1/4 | 6 3/4 |
| 24 O.D. | 36 | 32 | 24 | 1 1/2 | 7 1/2 |
| 30 O.D. | 43 | 39 1/4 | 28 | 1 3/4 | 8 1/2 |
| 36 O.D. | 50 | 46 | 32 | 2 | 9 1/2 |
| 42 O.D. | 57 | 52 3/4 | 36 | 2 | 10 |
| 48 O.D. | 65 | 60 3/4 | 40 | 2 | 11 |

# DIMENSIONS OF CLASS 125 CAST-IRON FLANGED FITTINGS

*(Reprinted from ANSI B16.1-1-1975 with permission of American Society of Mechanical Engineers)*

Dimensions in Inches

| Nominal Pipe Size | Flanges | | General | | Straight Fittings | | | | | | Reducing Fittings (Short Body Patterns) | | |
|---|---|---|---|---|---|---|---|---|---|---|---|---|---|
| | | | | | | | | | | | Tees and Crosses | | |
| | Dia. of Flange | Thickness of Flange (Min) | Inside Dia. of Fittings | Wall Thickness | Center to Face 90 deg Elbow Tees, Crosses True "Y" and Double Branch Elbow A | Center to Face 90 deg Long Radius Elbow B | Center to Face 45 deg Elbow C | Center to Face Lateral D | Short Center to Face True "Y" and Lateral E | Face to Face Reducer F | NPS Size of Outlet and Smaller | Center to Face Run H | Center to Face Outlet or Side Outlet J |
| 1 | 4.25 | 0.44 | 1.00 | 0.31 | 3.50 | 5.00 | 1.75 | 5.75 | 1.75 | — | All reducing tees and crosses, sizes 16 in. and smaller, shall have same center to face dimensions as straight size fittings, corresponding to the size of the largest opening. | | |
| $1^1/_4$ | 5.62 | 0.50 | 1.25 | 0.31 | 3.75 | 5.50 | 2.00 | 6.25 | 1.75 | — | | | |
| $1^1/_2$ | 5.00 | 0.56 | 1.50 | 0.31 | 4.00 | 6.00 | 2.25 | 7.00 | 2.00 | — | | | |
| 2 | 6.00 | 0.62 | 2.00 | 0.31 | 4.50 | 6.50 | 2.50 | 8.00 | 2.50 | 5.0 | | | |
| $2^1/_2$ | 7.00 | 0.69 | 2.50 | 0.31 | 5.00 | 7.00 | 3.00 | 9.50 | 2.50 | 5.5 | | | |
| 3 | 7.50 | 0.75 | 3.00 | 0.38 | 5.50 | 7.75 | 3.00 | 10.00 | 3.00 | 6.0 | | | |
| $3^1/_2$ | 8.50 | 0.81 | 3.50 | 0.44 | 6.00 | 8.50 | 3.50 | 11.50 | 3.00 | 6.5 | | | |
| 4 | 9.00 | 0.94 | 4.00 | 0.50 | 6.50 | 9.00 | 4.00 | 12.00 | 3.00 | 7.0 | | | |
| 5 | 10.00 | 0.94 | 5.00 | 0.50 | 7.50 | 10.25 | 4.50 | 13.50 | 3.50 | 8.0 | | | |
| 6 | 11.00 | 1.00 | 6.00 | 0.56 | 8.00 | 11.50 | 5.00 | 14.50 | 3.50 | 9.0 | | | |
| 8 | 13.50 | 1.12 | 8.00 | 0.62 | 9.00 | 14.00 | 5.50 | 17.50 | 4.50 | 11.0 | | | |
| 10 | 16.00 | 1.19 | 10.00 | 0.75 | 11.00 | 16.50 | 6.50 | 20.50 | 5.00 | 12.0 | | | |
| 12 | 19.00 | 1.25 | 12.00 | 0.81 | 12.00 | 19.00 | 7.50 | 24.50 | 5.50 | 14.0 | | | |
| 14 | 21.00 | 1.38 | 14.00 | 0.88 | 14.00 | 21.50 | 7.50 | 27.00 | 6.60 | 16.0 | | | |
| 16 | 23.50 | 1.44 | 16.00 | 1.00 | 15.00 | 24.00 | 8.00 | 30.00 | 6.50 | 18.0 | | | |
| 18 | 25.00 | 1.56 | 18.00 | 1.06 | 16.50 | 26.50 | 8.50 | 32.00 | 7.00 | 19.0 | 12.0 | 13.0 | 15.5 |
| 20 | 27.50 | 1.69 | 20.00 | 1.12 | 18.00 | 29.00 | 9.50 | 35.00 | 8.00 | 20.0 | 14.0 | 14.0 | 17.0 |
| 24 | 32.00 | 1.88 | 24.00 | 1.25 | 22.00 | 34.00 | 11.00 | 40.50 | 9.00 | 24.0 | 16.0 | 15.0 | 19.0 |
| 30 | 38.75 | 2.12 | 30.00 | 1.44 | 25.00 | 41.50 | 15.00 | 49.00 | 10.00 | 30.0 | 20.0 | 18.0 | 23.0 |
| 36 | 46.00 | 2.38 | 36.00 | 1.62 | 28.00 | 49.00 | 18.00 | — | — | 36.0 | 24.0 | 20.0 | 26.0 |
| 42 | 53.00 | 2.62 | 42.00 | 1.81 | 31.00 | 56.50 | 21.00 | — | — | 42.0 | 24.0 | 23.0 | 30.0 |
| 48 | 59.50 | 2.75 | 48.00 | 2.00 | 34.00 | 64.00 | 24.00 | — | — | 48.0 | 30.0 | 26.0 | 34.0 |

Dimensions in inches

| Nominal Pipe Size | Reducing Fittings (Short Body Patterns) — Laterals: NPS Size of Branch and Smaller | Center to Face Run M | Center to Face Run N | Center to Face Branch P | Base Elbows and Tees: Center to Base R | Dia. of Round Base or Width of Square Base S | Thickness of Base T | Thickness of Ribs U | NPS Size of Supporting Pipe for Base | Base Drilling: Bolt Circle or Bolt Spacing W | Dia of Holes |
|---|---|---|---|---|---|---|---|---|---|---|---|
| 1 | All reducing lateral sizes 16 in. and smaller, shall have same center-to-face dimensions as straight size fittings corresponding to size of the largest opening. | | | | 3.50 | 3.50 | 0.44 | 0.38 | 3/4 | 2.75 | 0.62 |
| 1 1/4 | | | | | 3.62 | 3.50 | 0.44 | 0.38 | 3/4 | 2.75 | 0.62 |
| 1 1/2 | | | | | 3.75 | 4.25 | 0.44 | 0.50 | 1 | 3.12 | 0.62 |
| 2 | | | | | 4.12 | 4.62 | 0.50 | 0.50 | 1 1/4 | 3.50 | 0.62 |
| 2 1/2 | | | | | 4.50 | 4.62 | 0.50 | 0.50 | 1 1/4 | 3.50 | 0.62 |
| 3 | | | | | 4.88 | 5.00 | 0.56 | 0.50 | 1 1/2 | 3.88 | 0.62 |
| 3 1/2 | | | | | 5.25 | 5.00 | 0.56 | 0.50 | 1 1/2 | 3.88 | 0.62 |
| 4 | | | | | 5.50 | 6.00 | 0.62 | 0.50 | 2 | 4.25 | 0.75 |
| 5 | | | | | 6.25 | 7.00 | 0.69 | 0.62 | 2 1/2 | 5.50 | 0.75 |
| 6 | | | | | 7.00 | 7.00 | 0.69 | 0.62 | 2 1/2 | 5.50 | 0.75 |
| 8 | | | | | 8.75 | 9.00 | 0.94 | 0.88 | 4 | 7.50 | 0.75 |
| 10 | | | | | 9.75 | 9.00 | 0.94 | 0.88 | 4 | 7.50 | 0.75 |
| 12 | | | | | 11.25 | 11.00 | 1.00 | 1.00 | 6 | 9.50 | 0.88 |
| 14 | | | | | 12.50 | 11.00 | 1.00 | 1.00 | 6 | 9.50 | 0.88 |
| 16 | | | | | 13.75 | 11.00 | 1.00 | 1.00 | 6 | 9.50 | 0.88 |
| 18 | 8 | 25.0 | 1.0 | 27.5 | 15.00 | 13.50 | 1.12 | 1.12 | 8 | 11.75 | 0.88 |
| 20 | 10 | 27.0 | 1.0 | 29.5 | 16.00 | 13.50 | 1.12 | 1.12 | 8 | 11.75 | 0.88 |
| 24 | 12 | 31.5 | 0.5 | 34.5 | 18.50 | 13.50 | 1.12 | 1.12 | 8 | 11.75 | 0.88 |
| 20 | 14 | 39.0 | 0 | 42.0 | | | | | | | |

## CLASS 125 FLANGED FITTINGS *(Courtesy ASME)*

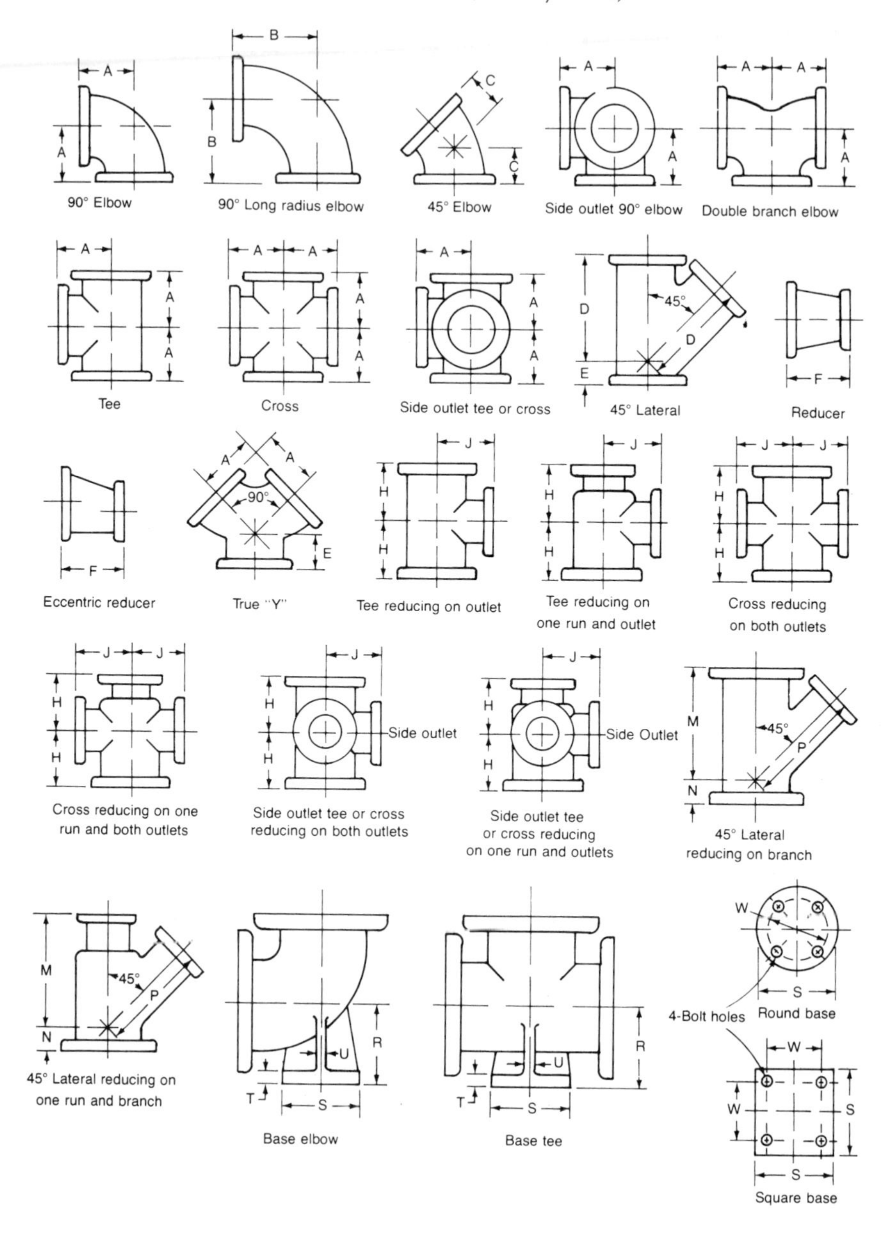

# CAST-IRON PIPE FLANGES AND FLANGE FITTINGS

*(Reprinted from ANSI B16.1-1975 with permission of American Society of Mechanical Engineers)*

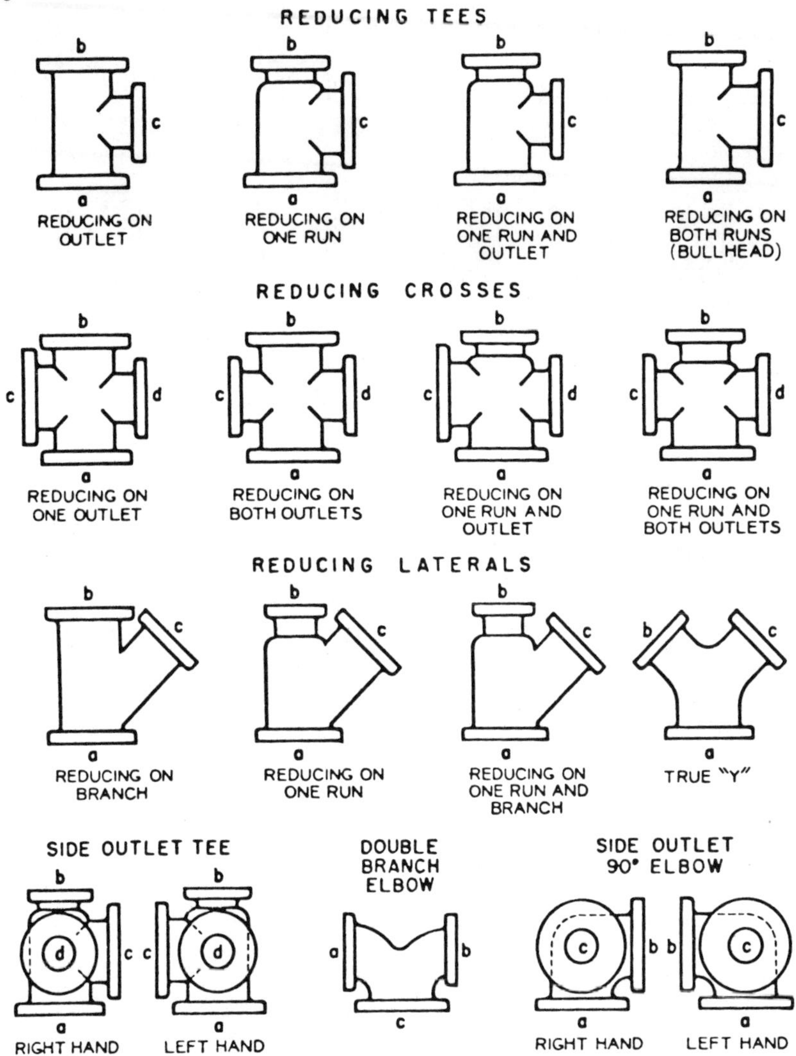


NOTE: The largest opening establishes the basic size of a reducing fitting. The largest opening is named first, except for bull head tees which are reducing on both runs and for double branch elbows where both branches are reducing; the outlet is the largest opening and named last in both cases.

In designating the openings of reducing fittings, they should be read in the order indicated by the sequence of the letters a, b, c, and d. In designating the outlets of side outlet reducing fittings the side outlet is named last and in the case of the cross which is not shown the side outlet is designated by the letter e.

Method of Designating Outlets of Reducing Fittings in Specifications

*(Reprinted from ANSI B16.1-1975 with permission of American Society of Mechanical Engineers)*

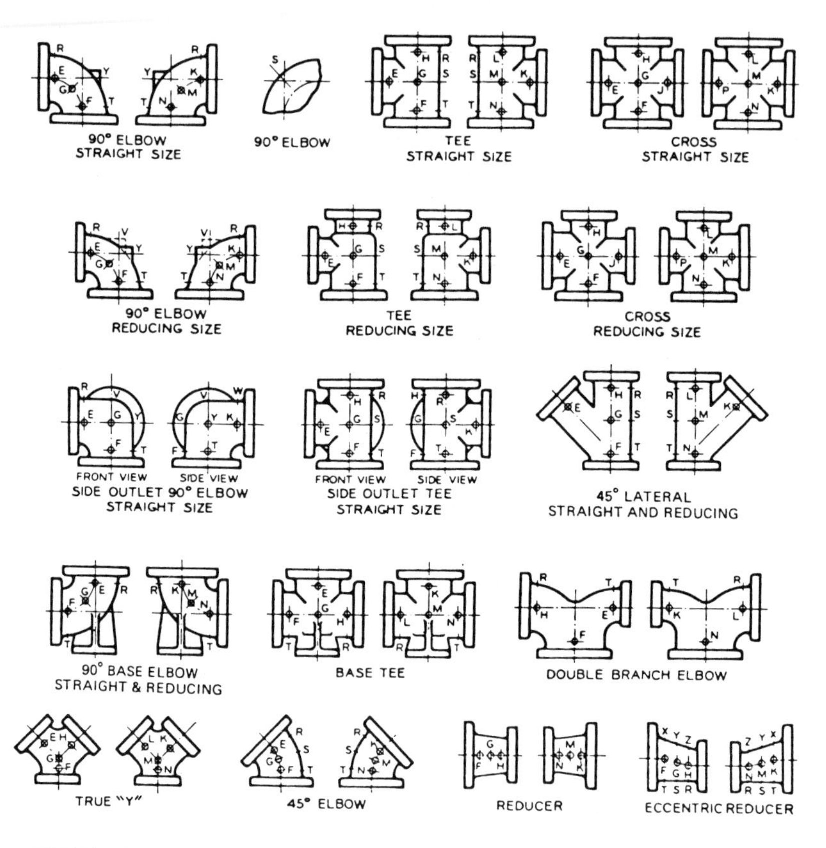


NOTE: The above sketches show two views of the same fitting and represent fittings with symmetrical shapes, with the exception of the side outlet elbow and the side outlet tee (straight sizes).

# APPENDIX 10
# Hangers and support data

*(Reprinted from MSS SP-58 with permission of Manufacturers Standardization Society of the Valve and Fittings Industry)*

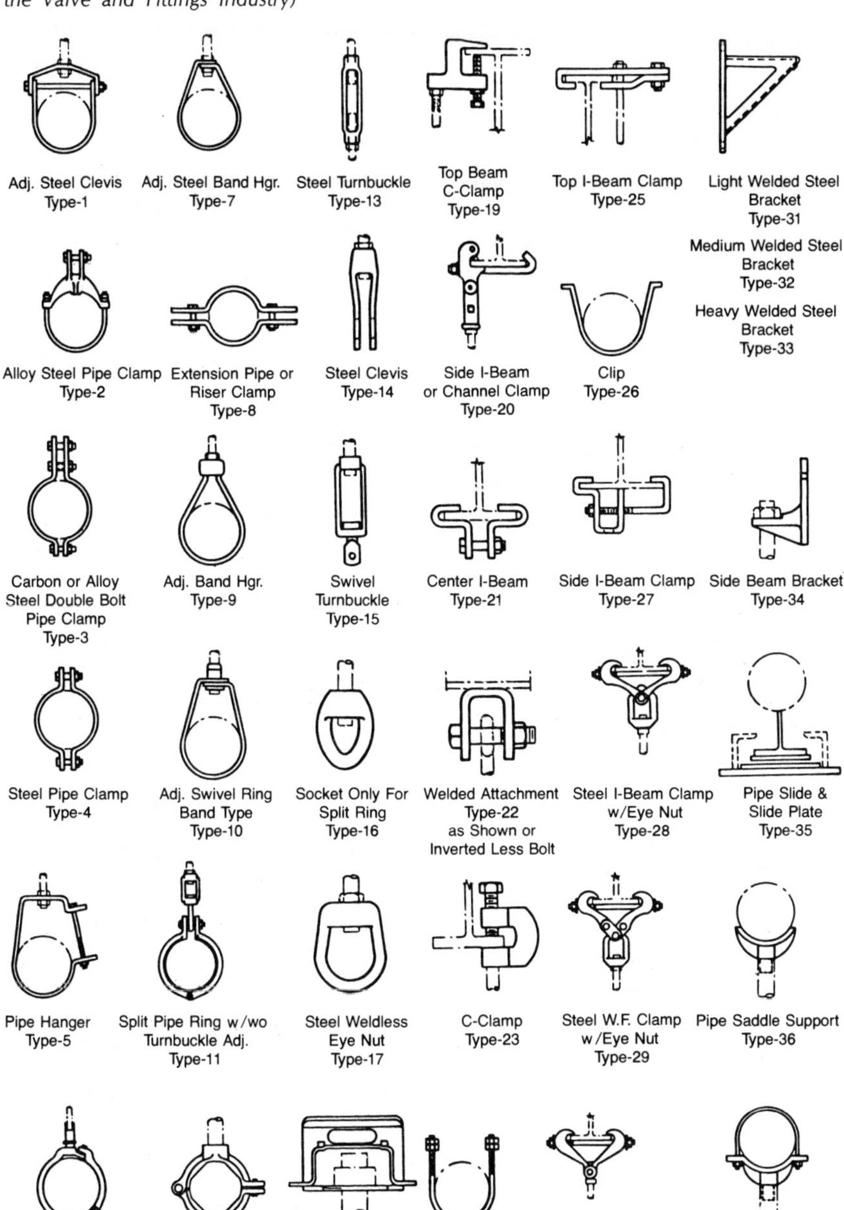

## APPENDIX 10 *continued*

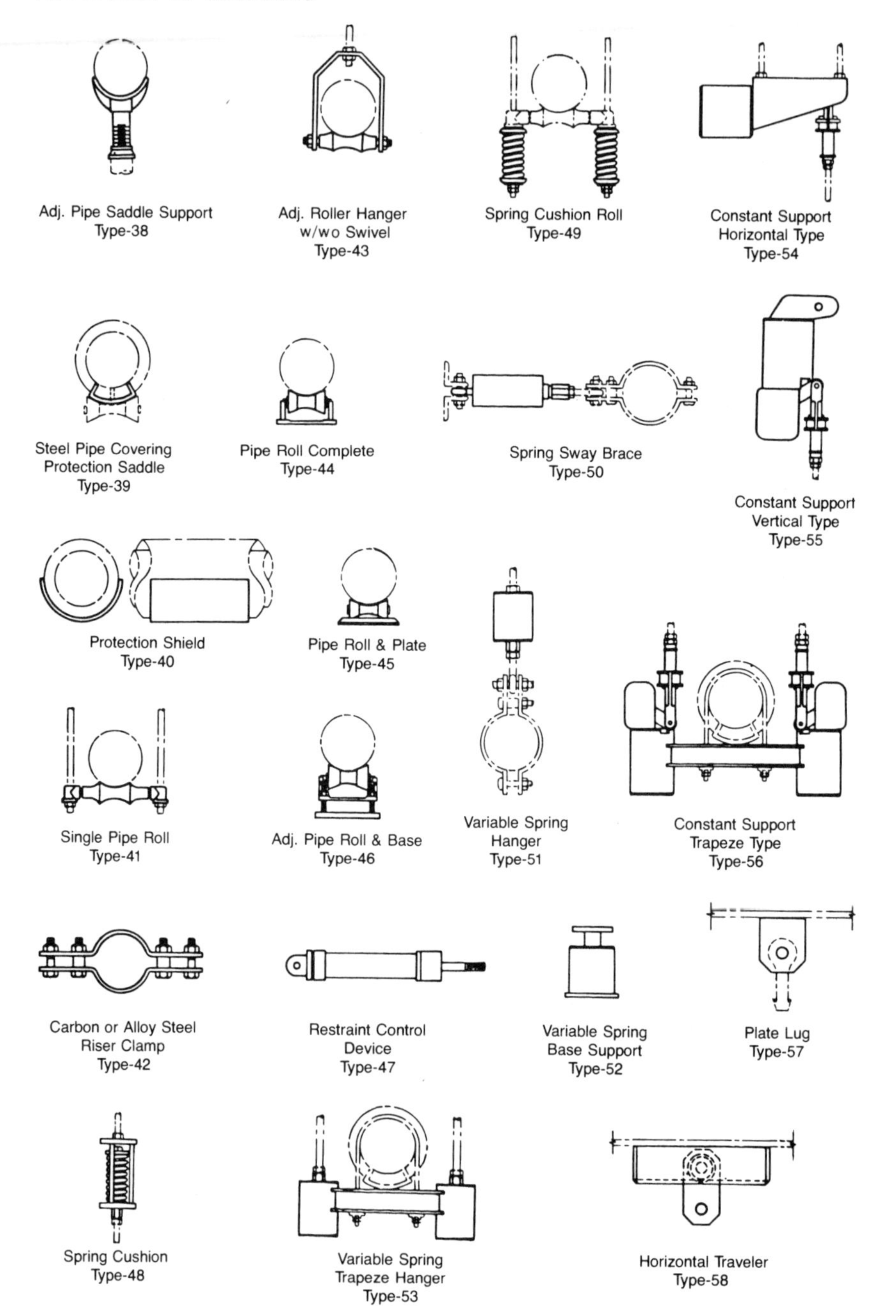

# SPRING SUPPORT SELECTION

*(Reprinted from MSS SP-69 with permission of Manufacturers Standardization Society of the Valve and Fittings Industry*

| VERTICAL EXPANSION | ALLOWABLE VARIABILITY OR DEVIATION | SINGLE ROD HANGER | DOUBLE ROD HANGER | BASE SUPPORT |
|---|---|---|---|---|
| | NOTE (1) | NOTE (2) AND NOTE (3) | | |
| MAX. 1/4 INCH (6.3mm) | 25%<br>6% | 48,51SS<br>51SS | 48,49,51SS,53SS<br>51S, 53S | 52SS<br>52S |
| MAX. 1 INCH (25.4mm) | 25%<br>6% | 51S<br>54,55 | 51S, 53S<br>54,55,56 | 52S<br>54,55 |
| MAX. 2 INCH (50.8mm) | 25%<br>6% | 51LS<br>54,55 | 51LS,53LS<br>54,55,56 | 52LS<br>54,55 |
| MAX. 3 INCH (76.2mm) | 25%<br>6% | 51LS<br>54.55 | 51LS,53LS<br>54,55,56 | 52LS<br>54,55 |
| OVER 3 INCH (76.2mm) | 25%<br>6% | 54,55<br>54,55 | 54,55,56<br>54,55,56 | 54,55<br>54,55 |

Note (1) Variable-spring hangers

$$\text{Variability factor} - \frac{\text{pipe travel, in.} \times \text{spring rate, lbs/in.}}{\text{operating load, lbs}}$$

Constant-support hangers

$$\text{Deviation} = \frac{\text{max. reading moving down} - \text{min. reading moving up}}{\text{max. reading moving down} + \text{min. reading moving up}}$$

(2) Numbers in columns are type numbers

(3) Variable-spring types 51, 52 and 53, i.e., standard spring, short spring, and long spring models, are identified as S, SS, and LS respectively

# PROTECTION SHIELDS FOR VAPOR BARRIER INSULATED PIPE AND TUBING

| NOM. PIPE SIZE | SHIELD LENGTH | | U.S. STANDARD GAGE | SPACING | |
|---|---|---|---|---|---|
| | in | mm | | ft | m |
| 1/2– 3 1/2 | 12 | 305 | 18 | 10 | 3.0 |
| 4 – | 12 | 305 | 16 | 10 | 3.0 |
| 5 – 6 | 18 | 457 | 16 | 10 | 3.0 |
| 8 –14 | 24 | 610 | 14 | 10 | 3.0 |
| 16 –24 | 24 | 610 | 12 | 10 | 3.0 |
| **NOM. TUBING SIZE** | **SHIELD LENGTH** | | **U.S. STANDARD GAGE** | **SPACING** | |
| | in | mm | | ft | –m |
| 1/4–1 | 12 | 305 | 18 | 5* | 1.5 |
| 1 1/4–2 1/2 | 12 | 305 | 18 | 8* | 2.4 |
| 3 –3 1/2 | 12 | 305 | 18 | 10 | 3.0 |
| 4 | 12 | 305 | 16 | 10 | 3.0 |
| 5 –6 | 18 | 457 | 16 | 10 | 3.0 |
| 8 | 24 | 610 | 14 | 10 | 3.0 |

Note: The listed spans and shield lengths are based on insulation with a compressive strength of 15 psi.

# MINIMUM ROD DIAMETER FOR SINGLE ROD HANGERS

| | COLUMNS 1, 2, 6, 7, 9 | | COLUMNS 3, 4, 8, 10, 11 | |
|---|---|---|---|---|
| NOMINAL PIPE OR TUBING DIA. | NOMINAL ROD DIA. | | NOMINAL ROD DIA. | |
| in | in | mm | in | mm |
| 1/4 | 3/8 | 9.6 | 3/8 | 9.6 |
| 3/8 | 3/8 | 9.6 | 3/8 | 99.6 |
| 1/2 | 3/8 | 9.6 | 3/8 | 9.6 |
| 3/4 | 3/8 | 9.6 | 3/8 | 9.6 |
| 1 | 3/8 | 9.6 | 3/8 | 9.8 |
| 1 1/4 | 3/8 | 9.6 | 3/8 | 9.6 |
| 1 1/2 | 3/8 | 9.6 | 3/8 | 9.6 |
| 2 | 3/8 | 9.6 | 3/8 | 12.7 |
| 2 1/2 | 1/2 | 12.7 | 1/2 | 12.7 |
| 3 | 1/2 | 12.7 | 1/2 | 12.7 |
| 3 1/2 | 1/2 | 12.7 | 1/2 | 12.7 |
| 4 | 5/8 | 15.8 | 1/2 | 12.7 |
| 5 | 5/8 | 15.8 | 1/2 | 15.8 |
| 6 | 3/4 | 19.1 | 5/8 | 19.1 |
| 8 | 7/8 | 22.2 | 3/4 | 19.1 |
| 10 | 7/8 | 22.2 | 3/4 | 19.1 |
| 12 | 7/8 | 22.2 | 3/4 | 22.2 |
| 14 | 1 | 25.4 | 7/8 | |
| 16 | 1 | 25.4 | | |
| 18 | 1 | 25.4 | | |
| 20 | 1 1/4 | 31.8 | | |
| 24 | 1 1/4 | 31.8 | | |

Note: Columns noted refer to table on next page.

## MAXIMUM HORIZONTAL PIPE HANGER AND SUPPORT SPACING

*(Reprinted from MSS SP-69 by permission of Manufacturers Standardization Society of the Valve and Fittings Industry)*

| | 1 | | 2 | | 3 | | 4 | | 5 | 6 | 7 | 8 | 9 | 10 | 11 |
|---|---|---|---|---|---|---|---|---|---|---|---|---|---|---|---|
| NOMINAL PIPE OR TUBE DIAM. | STD WT STEEL PIPE | | | | COPPER TUBE | | | | FIRE PRO-TECTION | CAST IRON PRES. | CAST IRON SOIL | ASBESTOS CEMENT | GLASS | PLASTIC | FIBERGLASS REIN-FORCED |
| | WATER SERVICE | | VAPOR SERVICE | | WATER SERVICE | | VAPOR SERVICE | | | | | | | | |
| | ft | m | ft | m | ft | m | ft | m | | | | | | | |
| 1/4 | 7 | 2.1 | 8 | 2.4 | 5 | 1.5 | 5 | 1.5 | FOLLOW REQUIREMENTS OF THE NATIONAL FIRE PROTECTION ASSOCIATION. | 12 ft (3.7m) MAX SPACING MIN OF ONE (1) HANGER PER PIPE SECTION CLOSE TO JOINT ON THE BARREL. ALSO AT CHANGE OF DIRECTION AND BRANCH CONNECTIONS. | 10 ft (3.0m) MAX SPACING MIN OF ONE (1) HANGER PER PIPE SECTION CLOSE TO JOINT ON THE BARREL. ALSO AT CHANGE OF DIRECTION AND BRANCH CONNECTIONS. | FOLLOW PIPE MANUFACTURER'S RECOMMENDATIONS. | 8 ft (2.4m) MAX SPACING, FOLLOW MANUFACTURER'S RECOMMENDATIONS. | FOLLOW PIPE MANUFACTURER'S RECOMMENDATIONS FOR MATERIAL AND SERVICE TEMPERATURE. | FOLLOW PIPE MANUFACTURER'S RECOMMENDATIONS FOR SPACING AND SERVICE CONDITIONS. |
| 3/8 | 7 | 2.1 | 8 | 2.4 | 5 | 1.5 | 6 | 1.8 | | | | | | | |
| 1/2 | 7 | 2.1 | 8 | 2.4 | 5 | 1.5 | 6 | 1.8 | | | | | | | |
| 3/4 | 7 | 2.1 | 9 | 2.7 | 5 | 1.5 | 7 | 2.1 | | | | | | | |
| 1 | 7 | 2.1 | 9 | 2.7 | 6 | 1.8 | 8 | 2.4 | | | | | | | |
| 1 1/4 | 7 | 2.1 | 9 | 2.7 | 7 | 2.1 | 9 | 2.7 | | | | | | | |
| 1 1/2 | 9 | 2.7 | 12 | 3.7 | 8 | 2.4 | 10 | 3.0 | | | | | | | |
| 2 | 10 | 3.0 | 13 | 4.0 | 8 | 2.4 | 11 | 3.4 | | | | | | | |
| 2 1/2 | 11 | 3.4 | 14 | 4.3 | 9 | 2.7 | 13 | 4.0 | | | | | | | |
| 3 | 12 | 3.7 | 15 | 4.6 | 10 | 3.0 | 14 | 4.3 | | | | | | | |
| 3 1/2 | 13 | 4.0 | 16 | 4.9 | 11 | 3.4 | 15 | 4.6 | | | | | | | |

| | | | | | | | | |
|---|---|---|---|---|---|---|---|---|
| 4 | 14 | 4.3 | 17 | 5.2 | 12 | 3.7 | 16 | 4.9 |
| 5 | 16 | 4.9 | 19 | 5.8 | 13 | 4.0 | 18 | 5.5 |
| 6 | 17 | 5.2 | 21 | 6.4 | 14 | 4.3 | 20 | 6.1 |
| 8 | 19 | 5.8 | 24 | 7.3 | 16 | 4.9 | 23 | 7.0 |
| 10 | 20 | 6.1 | 26 | 7.9 | 18 | 5.5 | 25 | 7.6 |
| 12 | 23 | 7.0 | 30 | 9.1 | 19 | 5.8 | 28 | 8.5 |
| 14 | 25 | 7.6 | 32 | 9.8 | | | | |
| 16 | 27 | 8.2 | 35 | 10.7 | | | | |
| 18 | 28 | 8.5 | 37 | 11.3 | | | | |
| 20 | 30 | 9.1 | 39 | 11.9 | | | | |
| 24 | 32 | 9.8 | 42 | 12.8 | | | | |
| 30 | 33 | 10.1 | 44 | 13.4 | | | | |

NOTE: (1) FOR SPACING SUPPORTS INCORPORATING TYPE 40 SHIELDS, SEE TABLE XXXIII

(2) DOES NOT APPLY WHERE SPAN CALCULATIONS ARE MADE OR WHERE THERE ARE CONCENTRATED LOADS BETWEEN SUPPORTS SUCH AS FLANGES, VALVES, SPECIALITIES, ETC., OR CHANGES IN DIRECTION REQUIRING ADDITIONAL SUPPORTS.

# APPENDIX 11 Dimensions and weights of seamless and welded wrought steel pipe ANSI B36.10

| Nominal | | Pipe Size | | | | | | Schedule Number | | | | | |
|---|---|---|---|---|---|---|---|---|---|---|---|---|---|
| Pipe Size | Outside Diameter | STD | | X.S. | | X.X.S. | | 10 | | 20 | | 30 | |
| | | Wall | Wt. | Wall | Wt. | Wall | Wt. | Wall | Wt. | Wall | Wt. | Wall | Wt. |
| 1/8 | 0.405 | 0.068 | 0.24 | 0.095 | 0.31 | | | | | | | | |
| 1/4 | 0.540 | 0.088 | 0.42 | 0.119 | 0.54 | | | | | | | | |
| 3/8 | 0.675 | 0.091 | 0.57 | 0.126 | 0.74 | | | | | | | | |
| 1/2 | 0.840 | 0.109 | 0.85 | 0.147 | 1.09 | 0.294 | 1.71 | | | | | | |
| 3/4 | 1.050 | 0.113 | 1.13 | 0.154 | 1.47 | 0.308 | 2.44 | | | | | | |
| 1 | 1.315 | 0.133 | 1.68 | 0.179 | 2.17 | 0.358 | 3.66 | | | | | | |
| 1 1/4 | 1.660 | 0.140 | 2.27 | 0.191 | 3.00 | 0.382 | 5.21 | | | | | | |
| 1 1/2 | 1.900 | 0.145 | 2.72 | 0.200 | 3.63 | 0.400 | 6.41 | | | | | | |
| 2 | 2.375 | 0.154 | 3.65 | 0.218 | 5.02 | 0.436 | 9.03 | | | | | | |
| 2 1/2 | 2.875 | 0.203 | 5.79 | 0.276 | 7.66 | 0.552 | 13.70 | | | | | | |
| 3 | 3.500 | 0.216 | 7.58 | 0.300 | 10.25 | 0.600 | 18.58 | | | | | | |
| 3 1/2 | 4.000 | 0.226 | 9.11 | 0.318 | 12.51 | | | | | | | | |
| 4 | 4.500 | 0.237 | 10.79 | 0.337 | 14.98 | 0.674 | 27.54 | | | | | | |
| 5 | 5.563 | 0.258 | 14.62 | 0.375 | 20.78 | 0.750 | 38.55 | | | | | | |
| 6 | 6.625 | 0.280 | 18.97 | 0.432 | 28.57 | 0.864 | 53.16 | | | | | | |
| 8 | 8.625 | 0.322 | 28.55 | 0.500 | 43.39 | 0.875 | 72.42 | | | 0.250 | 22.36 | 0.277 | 24.70 |
| 10 | 10.750 | 0.365 | 40.48 | 0.500 | 54.74 | 1.000 | 104.13 | | | 0.250 | 28.04 | 0.307 | 34.24 |
| 12 | 12.750 | 0.375 | 49.56 | 0.500 | 65.42 | 1.000 | 125.49 | | | 0.250 | 33.38 | 0.330 | 43.77 |
| 14 | 14.000 | 0.375 | 54.57 | 0.500 | 72.09 | | | 0.250 | 36.71 | 0.312 | 45.61 | 0.375 | 54.57 |
| 16 | 16.000 | 0.375 | 62.58 | 0.500 | 82.77 | | | 0.250 | 42.05 | 0.312 | 52.27 | 0.375 | 62.58 |
| 18 | 18.000 | 0.375 | 70.59 | 0.500 | 93.45 | | | 0.250 | 47.39 | 0.312 | 58.94 | 0.438 | 82.15 |
| 20 | 20.000 | 0.375 | 78.60 | 0.500 | 104.13 | | | 0.250 | 52.73 | 0.375 | 78.60 | 0.500 | 104.13 |
| 22 | 22.000 | 0.375 | 86.61 | 0.500 | 114.81 | | | 0.250 | 58.07 | 0.375 | 86.81 | 0.500 | 114.81 |
| 24 | 24.000 | 0.375 | 94.62 | 0.500 | 125.49 | | | 0.250 | 63.41 | 0.375 | 94.62 | 0.562 | 140.68 |
| 26 | 26.000 | 0.375 | 102.63 | 0.500 | 136.17 | | | 0.312 | 85.60 | 0.500 | 136.17 | | |
| 28 | 28.000 | 0.375 | 110.64 | 0.500 | 146.85 | | | 0.312 | 92.26 | 0.500 | 146.85 | 0.625 | 182.73 |
| 30 | 30.000 | 0.375 | 118.65 | 0.500 | 157.53 | | | 0.312 | 98.93 | 0.500 | 157.53 | 0.625 | 196.08 |
| 32 | 32.000 | 0.375 | 126.66 | 0.500 | 168.21 | | | 0.312 | 105.59 | 0.500 | 168.21 | 0.625 | 209.43 |
| 34 | 34.000 | 0.375 | 134.67 | 0.500 | 178.89 | | | 0.312 | 112.25 | 0.500 | 178.89 | 0.625 | 222.78 |
| 36 | 36.000 | 0.375 | 142.68 | 0.500 | 189.57 | | | 0.312 | 118.92 | 0.500 | 189.57 | 0.625 | 236.13 |
| 38 | 38.000 | 0.375 | 150.69 | 0.500 | 200.25 | | | | | | | | |
| 40 | 40.000 | 0.375 | 158.70 | 0.500 | 210.93 | | | | | | | | |
| 42 | 42.000 | 0.375 | 166.71 | 0.500 | 221.61 | | | | | | | | |
| 44 | 44.000 | 0.375 | 174.72 | 0.500 | 232.29 | | | | | | | | |
| 46 | 46.000 | 0.375 | 182.73 | 0.500 | 242.97 | | | | | | | | |
| 48 | 48.000 | 0.375 | 190.74 | 0.500 | 253.65 | | | | | | | | |

## APPENDIX 11 *continued*

| Schedule Number | | | | | | | | | | | | | |
|---|---|---|---|---|---|---|---|---|---|---|---|---|---|
| 40 | | 60 | | 80 | | 100 | | 120 | | 140 | | 160 | |
| Wall | Wt. | Wall | Wt. | Wall | Wt. | Wall | Wt. | Wall | Wt. | Wall | Wt. | Wall | Wt. |
| 0.068 | 0.24 | | | 0.095 | 0.31 | | | | | | | | |
| 0.088 | 0.42 | | | 0.119 | 0.54 | | | | | | | | |
| 0.091 | 0.57 | | | 0.126 | 0.74 | | | | | | | | |
| 0.109 | 0.85 | | | 0.147 | 1.09 | | | | | | | 0.188 | 1.31 |
| 0.113 | 1.13 | | | 0.154 | 1.47 | | | | | | | 0.219 | 1.94 |
| 0.133 | 1.68 | | | 0.179 | 2.17 | | | | | | | 0.250 | 2.84 |
| 0.140 | 2.27 | | | 0.191 | 3.00 | | | | | | | 0.250 | 3.76 |
| 0.145 | 2.72 | | | 0.200 | 3.63 | | | | | | | 0.281 | 4.86 |
| 0.154 | 3.65 | | | 0.218 | 5.02 | | | | | | | 0.344 | 7.46 |
| 0.203 | 5.79 | | | 0.276 | 7.66 | | | | | | | 0.375 | 10.01 |
| 0.216 | 7.58 | | | 0.300 | 10.25 | | | | | | | 0.438 | 14.31 |
| 0.226 | 9.11 | | | 0.318 | 12.51 | | | | | | | | |
| 0.237 | 10.79 | | | 0.337 | 14.98 | | | 0.438 | 18.98 | | | 0.531 | 22.52 |
| 0.258 | 14.62 | | | 0.375 | 20.78 | | | 0.500 | 27.04 | | | 0.625 | 32.96 |
| 0.280 | 18.97 | | | 0.432 | 28.57 | | | 0.562 | 36.39 | | | 0.719 | 45.35 |
| 0.322 | 28.55 | 0.406 | 35.64 | 0.500 | 43.39 | 0.594 | 50.95 | 0.719 | 60.71 | 0.812 | 67.76 | 0.906 | 74.69 |
| 0.365 | 40.48 | 0.500 | 54.75 | 0.594 | 64.43 | 0.719 | 77.03 | 0.844 | 89.29 | 1.000 | 104.13 | 1.125 | 115.65 |
| 0.406 | 53.53 | 0.562 | 73.15 | 0.688 | 88.63 | 0.844 | 107.32 | 1.000 | 125.49 | 1.125 | 139.67 | 1.312 | 160.27 |
| 0.438 | 63.44 | 0.594 | 85.05 | 0.750 | 106.13 | 0.938 | 130.85 | 1.094 | 150.79 | 1.250 | 170.21 | 1.406 | 189.11 |
| 0.500 | 82.77 | 0.656 | 107.50 | 0.844 | 136.61 | 1.031 | 164.82 | 1.219 | 192.43 | 1.438 | 223.64 | 1.594 | 245.25 |
| 0.562 | 104.67 | 0.750 | 138.17 | 0.938 | 170.92 | 1.156 | 207.96 | 1.375 | 244.14 | 1.562 | 274.22 | 1.781 | 308.50 |
| 0.594 | 123.11 | 0.812 | 166.40 | 1.031 | 208.87 | 1.281 | 256.10 | 1.500 | 296.37 | 1.750 | 341.09 | 1.969 | 379.14 |
| | | 0.875 | 197.41 | 1.125 | 250.81 | 1.375 | 302.88 | 1.625 | 353.61 | 1.875 | 403.00 | 2.125 | 451.06 |
| 0.688 | 171.29 | 0.969 | 238.35 | 1.219 | 296.58 | 1.531 | 367.39 | 1.812 | 429.39 | 2.062 | 483.12 | 2.344 | 542.13 |
| | | | | | | | | | | | | | |
| 0.688 | 230.08 | | | | | | | | | | | | |
| 0.688 | 244.77 | | | | | | | | | | | | |
| 0.750 | 282.35 | | | | | | | | | | | | |

# Glossary of piping-related terms

*Alloy steel*—a steel that owes its distinctive properties to elements other than carbon.

*Annealing*—*See* Heat Treatment.

*Arc welding*—a group of welding processes wherein coalescence is produced by heating with an electric arc.†

*Area of a circle*—the measurement of the surface within a circle. To find the area of a circle, multiply the product of the radius times the radius pi (3.142).

*Automatic welding*—a welding operation without manual adjustment of controls.

*Backing ring*—a metal strip in tubular shape, used in butt-weld operations that prevents weld splatter from entering the pipe.

*Bleeder*—a small valve to draw off water or condensation from a run of piping.*

*Blind flange*—a flange used to close the end of a pipe. It produces a dead end (blind end). It may also be called a blank flange (i.e., an undrilled flange).*

*Braze welding or brazing*—A method of joining metals using a nonferrous filler metal, the melting point of which is higher than 800°F but lower than that of the metals to be joined.

*Butt joint*—a pipe joining preparation prior to making a butt weld.

*Butt weld*—A circumferential weld in pipe fusing the abutting pipe walls completely from inside wall to outside wall. The term is also used to describe a pipe manufacturing process wherein the skelp is butted edge to edge.

*Bypass*—a small pipeline around a large valve, used for preheating a pipeline or pressure equalization. Also a connection around a control valve or similar device used during maintenance or emergency repairs.

*Carbon steel*—a steel that owes its distinctive properties chiefly to the various percentages of carbon (as distinguished from the other elements) it contains.*

*Cast iron*—the oldest iron product whose initial use started the Iron Age at around 1,000 B.C. The carbon content is 3.00% to 3.75%.

*From "Piping Handbook" by Sabin Crocker, Copyright 1930, 1931, 1939, 1945, McGraw-Hill Book Company. Used with the permission of McGraw-Hill Book Company.

†Reprinted from the ANSI/ASME Power Piping Code B31.1-1980. Used with permission of the American Society of Mechanical Engineers.

*Circumference of a circle*—the measurement around the perimeter of a circle. To find the circumference, multiply pi (3.142) by the diameter (commonly written as $\pi$d).

*Coefficient of expansion*—a number indicating the degree of expansion or contraction of a substance.

*Cold spring*—this term refers to metallic piping that is fabricated shorter than required and during erection is forcefully pulled into place. This will reduce stresses as the piping systems heat from installation to operating temperature.

*Corrosion*—the gradual destruction or alteration of a metal or alloy caused by direct chemical attack or by electro-chemical reaction.

*Coupling*—a sleeve used to connect two pipes. Couplings have internal threads to fit external threads on pipe or are bored to accommodate a pipe that is to be joined by socketwelding.*

*Creep*—the plastic flow of pipe within a system; the permanent set in metal caused by stresses at high temperatures, generally associated with a time rate of deformation.

*Diameter of a circle*—a straight line drawn through the center of a circle from one extreme edge to the other, equal to twice the radius.

*Double extra strong*—refers to certain standards of carbon steel and alloy pipe.

*Ductile iron*—combines the physical strength of mild steel with the long life of gray cast iron. It also has a limited ability to ben under stress.

*Ductility*—the property of elongation above the elastic limit but under the tensile strength. A measure of ductility is the percentage of elongation of the fractured piece over its original length.

*Elastic limit*—the greatest stress a material can withstand without a permanent deformation after release of the stress.

*Engineering design*—the detailed design developed to suit process requirements, including all necessary drawings and specifications governing a piping installation.†

*Equipment connection*—an integral part of such equipment as pressure vessels, heat exchangers, pumps, etc., designed for attachment of pipe or piping components.†

*Erection*—the complete installation of a piping system, including any field assembly, fabrication, testing, and inspection of the system.†

*Erosion*—the gradual destruction of metal or other material by the abrasive action of liquids, gases, solids, or mixtures thereof.

*Extra heavy*—formerly used to designate cast-iron flanges and fittings suitable for a maximum working steam pressure of 250-lb gauge. The term is now obsolete and using it instead of extra strong is incorrect.*

*Extra strong*—refers to certain standards of carbon steel and alloy pipe. In sizes 8 in. and smaller, extra-strong pipe is identical with Schedule 80 pipe.*

*Filler metal*—metal to be added in welding, soldering, or brazing.†

*Fillet weld*—a weld of approximately triangular cross section joining two surfaces approximately at right angles to each other.†

*Furnace weld*—a term applied to the process of making butt-welded or lap-welded pipe in which the skelp is heated in a furnace.*

*Fusion*—the melting together of filler material and base material or of base material only that results in coalescence.†

*Fusion weld*—the union of metals by fusion using an oxyacetylene torch, the electric arc, or other means. With the two described methods, the edges to be joined usually are beveled to provide an angle of repose, which is filled with metal from a welding rod.*

*Galvanizing*—a process by which metal piping is covered internally and externally with a layer of zinc.

*Gas welding*—a welding process that uses heating with any of various gas flames, with or without the use of filler metal.

*Heat treatments*†

*annealing, full*—heating a metal or alloy to a temperature above the critical temperature range and holding above the range for a proper period of time, followed by cooling to below that range.

*stress relieving*—uniform heating of a structure or portion thereof to a sufficient temperature to relieve the major portion of the residual stresses, followed by uniform cooling.

*Malleable iron*—cast iron that has been heat treated in an oven to relieve its brittlemess.*

*Mill length*—also known as random length. The usual run-of-the-mill pipe is 16 to 22 feet in length. Double random-length pipe is made in lengths of 30 to 40 feet.*

*Miter*—two or more straight sections of pipe matched and joined on a line bisecting the angle of junction so as to produce a change in direction.†

*Nipple*—(or pipe nipple) is a threaded piece of pipe less than 12 in. long. A close nipple is about twice the length of a standard pipe thread and without any shoulder. A shoulder nipple can be of any length up to 12 in. and has a shoulder between the pipe threads.

*Nozzle*—if applied to piping, refers to a short neck or connecting pipe on a vessel, tank, or manifold. The end preparation is suitable for joining with other piping and may be flanged, threaded, bevelled, or otherwise prepared for appropriate assembly.

*Oxygen cutting*—a metal cutting process wherein the severing is effected by means of the chemical reaction of oxygen with the base metal at elevated temperatures.

*Peening*—the mechanical working of metals by means of hammer blows.†

*Preheating*—the application of heat to the base metal prior to welding or cutting.

*Radius of a circle*—a straight line drawn from the center to the extreme edge of a circle.

*Resistance weld*—wherein coalescence is produced by the heat obtained from resistance to the flow of electric current in a circuit of which the material to be welded is a part.

*Schedule numbers*—indicate approximate values of the expression 1,000 × P/S, where P is the service pressure and S is the allowable stress, both expressed in pounds per square inch.*

*Seamless*—pipe formed by piercing and rolling a solid billet or cupping from a plate.*

*Semisteel*—a high grade of cast iron made by the addition of steel scrap to pig iron in the melting process. It is used to some extent for valve bodies and fittings.*

*Shielded-metal arc welding*—an arc-welding process wherein coalescence is produced by heating with an electric arc between a covered metal electrode and the work.†

*Skelp*—slit metal having the thickness and width required for forming and manufacturing pipe.

*Socket-welding fitting*—a fitting used to join pipe in which the pipe is inserted into the fitting. A fillet weld is then made around the edge of the fitting and the outside wall of the pipe at the junction of the pipe and fitting.

*Soldering*—a method of joining metals using fusable alloys, usually tin and lead, having melting points under 700°F.

*Spiral welded*—a method of manufacturing pipe by coiling a plate into a helix and fusion welding the overlapped or abutted edges.*

*Stainless steel*—an alloy steel having unusual corrosion-resisting properties, usually imparted by nickel and chromium.*

*Standard*—formerly used to designate cast-iron valves, flanges, and fittings suitable for a maximum working steam pressure of 125-lb gauge. The use of this word is obsolete today or at best confusing.*

*Standard weight*—a schedule of carbon steel and alloy pipe weights in common use. Standard-weight pipe in sizes 10 in. and smaller is identical with Schedule 40 pipe.*

*Strain*—change of shape or size of a body produced by the action of a stress.

*Stress*—the intensity of the internal distributed forces that resist a change in the form of a body. When external forces act on a body, they are resisted by reactions within the body that are termed stresses. A *tensile stress* is one that resists a force tending to pull a body apart. A *compressive stress* is one that resists a force tending to crush a body. A *shearing stress* is one that resists a force tending to make one layer of a body slide across another layer. A *torsional stress* is one that resists forces tending to twist a body.

*Stress relieving*—*see* Heat Treatments.†

*Submerged-arc welding*—a process wherein coalescence is produced by heating with an electric arc or arcs between a bare metal electrode or electrodes and the work.†

*Tack weld*—a weld made to hold parts of a weldment in proper alignment until the final welds are made.†

*Tensile strength*—the maximum tensile stress a material will develop. The tensile strength is usually considered to be the load in pounds per square inch at which a test specimen ruptures.

*Torque*—a moment of forces that produces or tends to produce rotation or torsion.

*Turbulence*—any deviation from parallel flow in a pipe due to rough inner walls, obstructions, or directional changes.

*Van Stone*—also called lap joint, when the end of a pipe is formed to provide a shoulder and gasketing surface for a lap-joint flange.

*Velocity*—time rate of motion in a given direction and sense, usually expressed in feet per second.

*Volume of a pipe*—the measurement of the space within the walls of the pipe. To find the volume of a pipe, multiply the length (or height) of the pipe by the product of the inside radius times the inside radius by pi (3.142).

*Welding*—a process of joining metals by heating until they are fused together or by heating and applying pressure until there is a plastic joining action. Filler metal may or may not be used.

*Welding fittings*—metallic fittings beveled for buttwelding connections; *See also* Socket Welding Fittings.

*Wrought iron*—iron refined in a plastic state in a puddling furnace. It is characterized by the presence of about three percent of slag irregularly mixed with pure iron and about 0.5 percent carbon and other elements in solution.*

*Wrought pipe*—term that refers to both wrought steel and wrought iron. Wrought in this sense means worked, as in the process of forming pipe from skelp or seamless pipe from billets. The expression wrought pipe is thus used as a distinction from cast pipe. Wrought pipe in this sense should not be confused with wrought-iron pipe, which is one now nearly obsolete variety of wrought pipe.*

*Yield strength*—the stress at which a material exhibits a specified limiting permanent set.

# Glossary of common valve terms

*Actuator*—the mechanical, hydraulic, electric, or pneumatic device or mechanism used to open, position, or close a valve.

*Angle valve*—a variant of the globe valve in which the body ends are at right angles to each other.

*Automatic control*—the combination of a valve, an actuator, and sensing and operating circuits programmed to regulate fluid flow by varying the valve mechanism position without additional intervention.

*Ball valve*—a valve that uses a rotatable spherical closure member.

*Blowoff valve*—a valve used to vent pressure or fluids at the surface or bottom of a drum-type boiler.

*Body*—the principal part of the valve that has ends adapted for connection to piping or tubing lines.

*Body-bonnet connection*—the connection of the body to the bonnet. This may be threaded, union, breech lock, bolted, welded, or pressure-seal type or a combination thereof with a capability for seal welding.

*Bonnet*—that upper part of the valve body assembly that guides the stem and contains the stem packing assembly.

*Butterfly valve*—a valve that uses a rotatable disc or vane as a closure member and obstructs flow when the closure is at a right angle to the flow.

*By-pass*—a piping loop attached to a valve body to provide a means to divert pressure or flow around the seat of the valve.

*Check valve*—a one-directional valve opened by the fluid flow in one direction and that closes automatically when the flow stops or reverses direction.

*Closure member*—that part of the valve that is positioned in the flow stream to permit flow or to obstruct flow, depending on closure position. In specific designs, it may also be called a disc, wedge, plug, ball, gate, or other functionally similar expression.

*Control valve*—a power-actuated valve forming a control element in a system, consisting of a body subassembly linked to one or more actuators and containing internal means for changing the rate of flow in response to a signal transmitted from a controlling instrument.

*Dead-tight*—a term meaning no leaks visible by ordinary observation without instrumentation.

*Diaphragm valve*—a bidirectional valve containing a captive elastomeric or plastic sheet or metal sheet to which mechanical, pneumatic, or fluid forces may be applied to deform the sheet and thus interrupt the fluid flow.

*Disc*—a circular, conical segment or cylindrically shaped element of a globe, check, gate, or butterfly valve that provides closure or control of the line fluid.

*Gate valve*—a valve whose closure element is a wedge or gate which may be lifted out of the flow stream.

*Globe valve*—a valve whose closure element is a disc or plug which seals line fluid on a seat which is generally parallel to line flow.

*Packing*—a sealing system consisting of deformable material or one or more mating deformable elements contained in a chamber that may have a manually adjustable compression means to obtain or maintain an effective leak-proof seal.

*Pinch valve*—a valve that has a flexible center tube or hose that is pinched to effect closure.

*Plug*—that part of the valve that closes the flow passageway.

*Plug valve*—a valve whose closure element is a cylindrical or truncated cone and which is positioned from open to closed in a 90-degree turn.

*Pressure seal joint*—a removable mechanical connection or valve bonnet closure that uses the fluid pressure to effect a seal. Internal pressure acting on the underside of the bonnet forces it against a gasket or seal and increases tightness as pressure increases.

*Quarter-turn valve*—a valve whose closure member can be moved from full open to full closed with a 90-degree rotation of the stem.

*Relief valve*—a self-operated, quick-opening valve used primarily in liquid fluid systems, arranged to open and relieve internal static pressure when the pressure exceeds the force holding the valve closed.

*Safety valve*—an automatic pressure-relieving device actuated by the static pressure upstream of the valve characterized by rapid full opening; used for gas or vapor services.

*Seat*—that portion of a valve against which the closure presses to effect a shutoff.

*Seating surfaces*—the contacting surfaces of the closure member and seat that effect valve closure.

*Shutoff valve*—a valve designed to function either fully open or fully closed but not at intermediate flow-throttling positions.

*Sleeved plug valve*—a plug valve with a second element between the plug and body that provides a sealing and bearing surface and facilitates rotation of the plug.

*Stem*—the rod, shaft, or spindle to which motion is imparted outside the valve assembly to move the closure member inside the valve.

*Stop check or nonreturn valve*—a valve which automatically closes when flow reverses and that can also be mechanically closed into a stop or closed position.

*Through conduit*—any of several types of valves that, when fully open, constitute a smooth extension of the internal surface of the piping system.

*Valve*—a device that isolates or controls fluid direction or flow rate.

*Wedge*—a valve closure member whose seating surfaces are inclined to the direction of thrust so that they can be mechanically forced into an intimate sealing contract.

*Wheel*—a circular or circular-segment device connected to the valve stem to permit manual operation.

*Trim*—describes the material being used for disc and body seatrings in any given valve.

# Piping terminology and abbreviations

## MATERIALS

| | |
|---|---|
| CS | Carbon steel, cast steel |
| WI | Wrought iron |
| CI | Cast iron |
| DI | Ductile iron |
| MI | Malleable iron |
| SS | Stainless steel |

## FABRICATION OR CONNECTING METHODS

| | |
|---|---|
| BW | Butt weld |
| LW | Lap weld |
| ERW | Electric-resistance welded |
| EFW | Electric-fusion welded |
| AW | Automatic welded (without filler material) |
| TIG Welding | Gas tungsten–arc welding |
| MIG Welding | Gas metal–arc welding |
| FF | Full finish |
| SF | Semifinish |
| SMLS | Seamless |
| SW | Socket weld |
| Scrd | Screwed |
| Flgd | Flanged |

## DETAILS OF MANUFACTURE

| | |
|---|---|
| WN | Welding neck (flange) |
| SO | Slip on (flange) |
| LJ | Lap Joint (flange) |
| F&D | Faced & drilled (flange) |
| RF | Raised face (flange) |
| FF | Flat face (flange) |
| B&S | Bell & spigot |
| PE | Plain end (pipe) |
| BE | Bevelled end (pipe) |
| WE | Weld end (pipe or valve) |
| IBBM | Iron-body bronze mounted (valve) |
| OS&Y | Outside stem and yoke (valve) |

## PIPE WALL THICKNESS

| | |
|---|---|
| Sched | Schedule |
| Std | Standard weight |

| | |
|---|---|
| XH | Extra strong |
| XXH | Double extra strong |
| NPS | Nominal pipe size |
| IPS | Iron pipe size (obsolete) |

## MEASUREMENTS

| | |
|---|---|
| gal | Gallon |
| bbl | Barrel |
| C | Centigrade |
| F | Fahrenheit |
| lbs | Pounds |
| cu ft. | Cubic feet |
| cfm | Cubic feet per minute |
| gpm | Gallons per minute |
| psi | Pounds per square inch |
| SSP | Steam service pressure |
| WWP | Working water pressure |
| WOG (OWG) | Water, oil, gas |
| diam | Diameter |
| ID | Inside diameter |
| OD | Outside diameter |
| C to F | Center to face |
| F to F | Face to face |

These abbreviations are reprinted by courtesy of the American Society of Mechanical Engineers (ASME). A complete listing of abbreviations is provided as "Abbreviations for use on drawings and in text," ANSI Specification Y1.1 -1972.

# REFERENCES

1. *Piping Handbook* by Sabin Crocker
   McGraw-Hill Book Co.
2. *Industrial Piping* by Charles E. Littleton
   McGraw-Hill Book Co.
3. *Non-Destructive Testing Handbook*
   Edited for the Society of Non-Destructive Testing
   by Robert Charles McMaster
4. *American Cast Iron Handbook*
   by the Cast Iron Research Institute
5. *ISA Handbook of Control Valves*
   Edited by J.W. Hutchinson
   Instrument Society of America
6. *Valves for Industry*
   Valve Manufacturers Association
7. *Expansion Joint Standards*
   Expansion Joint Manufacturing Association
8. *Piping Pointers*
   Crane Company
9. *Choosing the Right Valve*
   Crane Company
10. *Steam Trapping and Air Venting*
    Sarco Co. Inc.
11. *Hook-Up Design for Steam and Fluid Systems*
    Sarco Co. Inc.
12. *An Introduction to Rupture Disk Technology*
    B, S & B Safety Systems, Inc.
13. *Installation and Performance of Safety and Relief Valves*
    by C. G. Weber Teledyne Farris Engineering
14. *Manual on Installation of Refinery Instruments and Control Systems*
    American Petroleum Institute
15. Piping Report''
    *Chemical Engineering* 6/17/1968
16. ''Piping Report''
    *Chemical Engineering* 2/26/1973
17. ''Selecting and Specifying Valves for New Plants''
    *Chemical Engineering* 9/13/1976
18. ''Materials of Construction Report''
    *Chemical Engineering* 11/20/1978

19. "Mechanical Couplings"
*Plant Engineering* 9/14/1978
20. "Valve Standards"
*Heating/Piping/Airconditioning* October 1978
21. "Valve Manual"
*Power Engineering Magazine* Reprint
22. "Piping Expansion Devices"
*Power Engineering Magazine* Reprint

# Index